VOLUME ONE HUNDRED AND SIXTY THREE

ADVANCES IN
CANCER RESEARCH

Cutting Edge Artificial Intelligence,
Spatial Transcriptomics and Proteomics
Approaches to Analyze Cancer

VOLUME ONE HUNDRED AND SIXTY THREE

ADVANCES IN
CANCER RESEARCH

Cutting Edge Artificial Intelligence,
Spatial Transcriptomics and Proteomics
Approaches to Analyze Cancer

Edited by

ESHA MADAN

Department of Surgery, Virginia Commonwealth University, School of
Medicine; Massey Comprehensive Cancer Center, Virginia Commonwealth
University; VCU Institute of Molecular Medicine, Department of Human and
Molecular Genetics, Virginia Commonwealth University, School of Medicine,
Richmond, Virginia, United States

PAUL B. FISHER

Department of Human and Molecular Genetics, VCU Institute of Molecular
Medicine, VCU Massey Comprehensive Cancer Center, Virginia Commonwealth
University, School of Medicine, Richmond, Virginia, United States

RAJAN GOGNA

Department of Human and Molecular Genetics, Department of Surgery,
VCU Institute of Molecular Medicine, VCU Massey Comprehensive Cancer
Center, Virginia Commonwealth University, School of Medicine, Richmond,
Virginia, United States

ACADEMIC PRESS
An imprint of Elsevier

Academic Press is an imprint of Elsevier
50 Hampshire Street, 5th Floor, Cambridge, MA 02139, United States
525 B Street, Suite 1650, San Diego, CA 92101, United States
125 London Wall, London, EC2Y 5AS, United Kingdom

First edition 2024

Copyright © 2024 Elsevier Inc. All rights are reserved, including those for text and data mining, AI training, and similar technologies.

Publisher's note: Elsevier takes a neutral position with respect to territorial disputes or jurisdictional claims in its published content, including in maps and institutional affiliations.

No part of this publication may be reproduced or transmitted in any form or by any means, electronic or mechanical, including photocopying, recording, or any information storage and retrieval system, without permission in writing from the publisher. Details on how to seek permission, further information about the Publisher's permissions policies and our arrangements with organizations such as the Copyright Clearance Center and the Copyright Licensing Agency, can be found at our website: www.elsevier.com/permissions.

This book and the individual contributions contained in it are protected under copyright by the Publisher (other than as may be noted herein).

Notices
Knowledge and best practice in this field are constantly changing. As new research and experience broaden our understanding, changes in research methods, professional practices, or medical treatment may become necessary.

Practitioners and researchers must always rely on their own experience and knowledge in evaluating and using any information, methods, compounds, or experiments described herein. In using such information or methods they should be mindful of their own safety and the safety of others, including parties for whom they have a professional responsibility.

To the fullest extent of the law, neither the Publisher nor the authors, contributors, or editors, assume any liability for any injury and/or damage to persons or property as a matter of products liability, negligence or otherwise, or from any use or operation of any methods, products, instructions, or ideas contained in the material herein.

ISBN: 978-0-443-29650-5
ISSN: 0065-230X

For information on all Academic Press publications
visit our website at https://www.elsevier.com/books-and-journals

Publisher: Zoe Kruze
Editorial Project Manager: Naiza Ermin Mendoza
Production Project Manager: James Selvam
Cover Designer: Gopalakrishnan Venkatraman
Typeset by MPS Limited, India

Contents

Contributors	*xi*
Preface	*xv*

1. Deep learning-based multimodal spatial transcriptomics analysis for cancer   **1**

Pankaj Rajdeo, Bruce Aronow, and V.B. Surya Prasath

1. Introduction	2
1.1 Deep learning and its significance in oncology	2
1.2 Multimodal data analysis in cancer research	4
1.3 Synergistic potential of deep learning and multimodal approaches	5
2. Core concepts of multimodal spatial transcriptomics	8
2.1 Basics of spatial transcriptomics	8
2.2 Multimodal spatial transcriptomics: integrating varied data types	10
2.3 Unraveling cancer complexity with multimodal spatial transcriptomics	14
3. Deep learning approaches in multimodal spatial transcriptomics	15
3.1 Deep learning models for spatial transcriptomics data	16
3.2 Successful applications of deep learning in multimodal spatial transcriptomics	20
4. Impact of multimodal spatial transcriptomics on cancer diagnostics and therapeutics	26
4.1 Enhancing cancer diagnosis	26
4.2 Personalized treatment strategies	27
4.3 Deep learning's contribution to ST-based precision oncology	28
5. Challenges, future outlook, and ethical considerations	29
5.1 Challenges and future directions	29
5.2 Ethical dimensions and responsible AI use	31
6. Conclusion	32
References	33

2. Data enhancement in the age of spatial biology   **39**

Linbu Liao, Patrick C.N. Martin, Hyobin Kim, Sanaz Panahandeh, and Kyoung Jae Won

1. Introduction	40
2. Data enhancement by integrating other sequencing modalities for ST data	43
2.1 Deconvolution sequencing-based ST data using reference scRNAseq	43

v

2.2 Data imputation for missing data	47
2.3 Mapping single cells to spatial data for Visium data enhancement	48
2.4 Spatial alignment and integrated analysis	50
2.5 Multi-modality of spatial data	53
3. Data enhancement by integrating tissue image data for ST data	55
3.1 Cell segmentation boosted by integrating image data	55
3.2 Data enhancement by integrating image data	56
3.3 Gene prediction from tissue image data	57
4. Data enhancement by synthetic data	58
5. Perspectives	58
Acknowledgments	62
Disclosures	62
References	62

3. Current computational methods for spatial transcriptomics in cancer biology   71

Jaewoo Mo, Junseong Bae, Jahanzeb Saqib, Dohyun Hwang,
Yunjung Jin, Beomsu Park, Jeongbin Park, and Junil Kim

1. Introduction	72
2. Overview of spatial transcriptomics techniques	73
2.1 Imaging-based methods	75
2.2 NGS-based methods	77
2.3 Commercial methods	79
2.4 Cell segmentation	82
3. Application of spatial transcriptomics for cancer study	84
3.1 Deconvolution of NGS-based spatial transcriptomics data	84
3.2 Tissue domain analysis	93
3.3 Niche-gene analysis	96
4. Discussion	97
Acknowledgments	99
References	99

4. Applications of spatial transcriptomics and artificial intelligence to develop integrated management of pancreatic cancer   107

Rishabh Maurya, Isha Chug, Vignesh Vudatha, and
António M. Palma

1. Introduction	108
2. Overview of cancer	109

Contents vii

3. Spatial transcriptomics 111
4. Artificial intelligence 118
5. Spatial transcriptomics contribution to understanding the 121
heterogeneity of pancreatic cancer tumors
6. The use of Artificial Intelligence to diagnose and define pancreatic 126
cancer therapy
7. Conclusions 130
Acknowledgments 131
References 131

5. Advancing cancer therapeutics: Integrating scalable 3D cancer models, extracellular vesicles, and omics for enhanced therapy efficacy — 137

Pedro P. Gonçalves, Cláudia L. da Silva, and Nuno Bernardes

1. Introduction 138
2. Beyond 2D: exploring the potential of 3D models in cancer research 140
   2.1 Leveraging spheroids as reliable preclinical cancer models 141
   2.2 Scalable strategies for tumor spheroid production: bioreactor 143
   systems for robust preclinical modeling
   2.3 Advances and challenges in cancer organoid versus spheroid 153
   culture
3. Leveraging omics technologies for robust validation and application 159
   of 3D cancer models
4. Beyond cells: extracellular vesicles as drug delivery systems for 162
   targeted cancer therapy
   4.1 From production to therapeutic application: omics in EV research 166
5. Conclusions and future perspectives 169
Acknowledgments 169
References 169
Further reading 185

6. Crosstalk between tumor and microenvironment: Insights from spatial transcriptomics — 187

Malvika Sudhakar, Harie Vignesh, and Kedar Nath Natarajan

1. Introduction 188
2. Cancer heterogeneity and landscape 189
   2.1 Unique cancer genomes 189

2.2	Intertumor heterogeneity	191
2.3	Intratumor heterogeneity	192
3.	Spatial transcriptomics and insights	192
3.1	Spatial technologies	192
3.2	Analysis of spatial data	196
4.	Evolution of tumor	201
4.1	Clonal architecture	201
4.2	Tumor evolution types	205
4.3	Insights from spatial transcriptomics	206
4.4	Metastasis	207
5.	Tumor microenvironment (TME)	208
5.1	Dynamics of cancer associated fibroblasts in cancer progression	209
5.2	Insights about the tumor microenvironment from single cell and ST studies	210
6.	Integrating ST and scRNA-seq	212
7.	Challenges and future directions	213
	Acknowledgments	214
	Conflict of interest statement	214
	References	214

7. Modular formation of in vitro tumor models for oncological research/therapeutic drug screening   **223**

Weiwei Wang and Hongjun Wang

1.	Introduction	224
2.	Why do we need 3D tumor models?	225
2.1	Spheroid	226
2.2	Hydrogels	227
2.3	Microfluidics	227
2.4	Polymeric scaffolds	228
3.	Challenges with traditional tissue engineering strategies to form tumorous tissues	229
3.1	Comparison of top-down and bottom-up strategy	230
3.2	Modular tissue engineering	231
4.	Application of modular strategy for tumor model formation	232
5.	Future perspective and concluding remarks	244
	Acknowledgment	245
	References	246

Contents ix

8. Unraveling the complexity: Advanced methods in analyzing DNA, RNA, and protein interactions
251

Maria Leonor Peixoto and Esha Madan

1. Introduction	252
2. Intramolecular DNA-DNA and RNA-RNA interactions	254
2.1 DNA	256
2.2 RNA	257
2.3 X-Ray crystallography	258
2.4 Nuclear magnetic resonance spectroscopy	259
2.5 Cryo-electron microscopy	260
2.6 Circular dichroism	261
2.7 Single-molecule fluorescence resonance energy transfer (smFRET)	261
2.8 Further techniques	262
2.9 In silico predictions and combined approaches	263
3. Intermolecular DNA-RNA interactions	263
3.1 DNA-RNA interactions	263
3.2 Protein-DNA interactions	271
3.3 Protein-RNA interactions	278
4. Conclusion	284
Acknowledgments	286
References	286

9. Multi-omics based artificial intelligence for cancer research
303

Lusheng Li, Mengtao Sun, Jieqiong Wang, and Shibiao Wan

1. Introduction	304
2. Artificial intelligence and machine learning	305
2.1 Supervised learning	306
2.2 Unsupervised learning	309
2.3 Semi-supervised learning	312
2.4 Deep learning	313
3. Omics techniques	314
3.1 Genomics	316
3.2 Epigenomics	317
3.3 Transcriptomics	318
3.4 Proteomics	318
3.5 Metabolomics	319
4. Application of AI and multi-omics in cancer research	320
4.1 Early cancer detection	320
4.2 Diagnosis	323

4.3	Prognosis	325
4.4	Treatment response prediction	328
4.5	Biomarker identification	329
4.6	Pathology identification	332
5.	Multi-omics integration	334
5.1	Early integration	335
5.2	Late integration	340
5.3	Intermediate integration	341
6.	Challenges	343
6.1	Data harmonization	343
6.2	Interpretability	343
6.3	Ethical considerations	344
Acknowledgments		345
Funding		345
Authors' contributions		346
Competing interests		346
References		346

Contributors

Bruce Aronow
Division of Biomedical Informatics, Cincinnati Children's Hospital Medical Center; Department of Pediatrics, College of Medicine, University of Cincinnati, Cincinnati, OH, United States

Junseong Bae
Interdisciplinary Program of Genomic Data Science; Graduate School of Medical AI, Pusan National University, Yangsan, Republic of Korea

Nuno Bernardes
Department of Bioengineering and iBB - Institute for Bioengineering and Biosciences at Instituto Superior Técnico; Associate Laboratory i4HB - Institute for Health and Bioeconomy at Instituto Superior Tecnico, Universidade de Lisboa, Lisboa, Portugal

Isha Chug
Massey Comprehensive Cancer Center, Virginia Commonwealth University, Richmond, VA, United States

Pedro P. Gonçalves
Department of Bioengineering and iBB - Institute for Bioengineering and Biosciences at Instituto Superior Técnico; Associate Laboratory i4HB - Institute for Health and Bioeconomy at Instituto Superior Tecnico, Universidade de Lisboa, Lisboa, Portugal

Dohyun Hwang
Department of Information Convergence Engineering, Pusan National University, Yangsan, Republic of Korea

Yunjung Jin
School of Systems Biomedical Science, Soongsil University, Dongjak-Gu, Seoul, Republic of Korea

Hyobin Kim
Department of Computational Biomedicine, Cedars-Sinai Medical Center, Los Angeles, CA, United States

Junil Kim
School of Systems Biomedical Science, Soongsil University, Dongjak-Gu, Seoul, Republic of Korea

Lusheng Li
Department of Genetics, Cell Biology and Anatomy, University of Nebraska Medical Center, Omaha, NE, United States

Linbu Liao
Biotech Research and Innovation Centre (BRIC), University of Copenhagen, Denmark; Samuel Oschin Cancer Center, Cedars-Sinai Medical Center, Los Angeles, CA, United States

Esha Madan
Department of Surgery, Virginia Commonwealth University, School of Medicine; Massey Comprehensive Cancer Center, Virginia Commonwealth University; VCU Institute of Molecular Medicine, Department of Human and Molecular Genetics, Virginia Commonwealth University, School of Medicine, Richmond, VA, United States

Pankaj Rajdeo
Division of Biomedical Informatics, Cincinnati Children's Hospital Medical Center, Cincinnati, OH, United States

Patrick C.N. Martin
Department of Computational Biomedicine, Cedars-Sinai Medical Center, Los Angeles, CA, United States

Rishabh Maurya
Department of Surgery, Virginia Commonwealth University, School of Medicine; Massey Comprehensive Cancer Center, Virginia Commonwealth University, Richmond, VA, United States

Jaewoo Mo
School of Systems Biomedical Science, Soongsil University, Dongjak-Gu, Seoul, Republic of Korea

Kedar Nath Natarajan
DTU Bioengineering, Technical University of Denmark, Kongens Lyngby, Denmark

António M. Palma
Massey Comprehensive Cancer Center; VCU Institute of Molecular Medicine, Department of Human and Molecular Genetics; Department of Human and Molecular Genetics, Virginia Commonwealth University, School of Medicine, Richmond, VA, United States

Sanaz Panahandeh
Department of Computational Biomedicine, Cedars-Sinai Medical Center, Los Angeles, CA, United States

Beomsu Park
School of Systems Biomedical Science, Soongsil University, Dongjak-Gu, Seoul, Republic of Korea

Jeongbin Park
Interdisciplinary Program of Genomic Data Science; Department of Information Convergence Engineering; School of Biomedical Convergence Engineering, Pusan National University, Yangsan, Republic of Korea

V.B. Surya Prasath
Division of Biomedical Informatics, Cincinnati Children's Hospital Medical Center; Department of Pediatrics, College of Medicine; Department of Biomedical Informatics, College of Medicine; Department of Computer Science, University of Cincinnati, Cincinnati, OH, United States

Maria Leonor Peixoto
Champalimaud Center for the Unknown; Instituto Superior Técnico, Universidade de Lisboa, Lisbon, Portugal

Contributors

Jahanzeb Saqib
School of Systems Biomedical Science, Soongsil University, Dongjak-Gu, Seoul, Republic of Korea

Malvika Sudhakar
DTU Bioengineering, Technical University of Denmark, Kongens Lyngby, Denmark

Mengtao Sun
Department of Genetics, Cell Biology and Anatomy, University of Nebraska Medical Center, Omaha, NE, United States

Harie Vignesh
DTU Bioengineering, Technical University of Denmark, Kongens Lyngby, Denmark

Vignesh Vudatha
Department of Surgery, Virginia Commonwealth University, School of Medicine; Massey Comprehensive Cancer Center, Virginia Commonwealth University, Richmond, VA, United States

Shibiao Wan
Department of Genetics, Cell Biology and Anatomy, University of Nebraska Medical Center, Omaha, NE, United States

Hongjun Wang
Department of Biomedical Engineering; Semcer Center for Healthcare Innovation, Stevens Institute of Technology, Hoboken, NJ, United States

Jieqiong Wang
Department of Neurological Sciences, University of Nebraska Medical Center, Omaha, NE, United States

Weiwei Wang
Department of Biomedical Engineering, Stevens Institute of Technology, Hoboken, NJ, United States; School of Life Sciences, Yantai University, Yantai, Shandong, P.R. China

Kyoung Jae Won
Department of Computational Biomedicine, Cedars-Sinai Medical Center, Los Angeles, CA, United States

Cláudia L. da Silva
Department of Bioengineering and iBB - Institute for Bioengineering and Biosciences at Instituto Superior Técnico; Associate Laboratory i4HB - Institute for Health and Bioeconomy at Instituto Superior Tecnico, Universidade de Lisboa, Lisboa, Portugal

Preface

In recent years, the revolutionary technology of Spatial Transcriptomics (ST) has transformed modern biology, enabling researchers with an unprecedented ability to unravel intricate gene expression patterns within tissues. Although traditional bulk RNA sequencing provides valuable insights into the transcriptome landscape of tissues yet its lack of spatial resolution makes it challenging to interpret the spatial variability observed in complex biological systems like cancers. This limitation is overcome by ST, which allows researchers to map the patterns of gene expression inside tissue sections while preserving the spatial information of RNA molecules. Stahl et al.'s (2016) spatially resolved transcriptomics, or Seurat, is one of the ground-breaking methods in the field of spatial transcriptomics. This technique enables the high-resolution mapping of gene expression patterns by capturing and sequencing RNA molecules from tissue using spatially barcoded mRNA capture beads. Since its introduction, Seurat, has been widely employed and has greatly enhanced our knowledge of spatial gene expression dynamics in a range of biological contexts, including cancer.

Deep learning technologies, a subset of artificial intelligence (AI), are emerging as a promising option for early cancer detection and treatment modalities. When combined with multimodal ST, deep learning algorithms offer unprecedented insights into the biology of malignancies that were previously inaccessible. Through the amalgamation of genetic, transcriptomic, proteomic, imaging, and clinical data, this integration offers thorough data analysis that reveals minute details regarding tumor heterogeneity, interactions with the tumor microenvironment (TME), and treatment outcomes. Such methodologies enhance diagnostic, and prognostic capabilities and allow for personalized treatment planning and prognosis modeling. As with many innovative technologies, constant integration of newer approaches enhances applications of genetic and epigenetic information to refine our understanding of complex physiological processes, including cancer.

Cancer is a complex and evolving condition; hence it is essential to advance our understanding of cellular behavior within the self-organizing systems of multicellular organisms. Computational techniques for ST have played a crucial role in clarifying these complexities in cancer biology. By employing mathematical and statistical methods, scientists seek to interpret the complex interactions that exist within the TME, providing insight into the mechanisms that underlie the progression of cancer and the response to treatment. In particular, for complex cancers such as pancreatic, ST is especially useful in combination with AI. Precise localization of cells based on gene expression allow discovery of novel biomarkers, therapeutic targets, and improve existing drug treatment plans that lead to better patient outcomes. Transformative approaches including scalable 3D cancer models, omics integration, and extracellular vesicles (EVs) also present promising opportunities for successful therapeutic interventions and drug screening procedures in the quest for more potent and less toxic cancer treatments.

Transcriptomics and proteomics are two fundamental disciplines in cancer research, unraveling complex biological mechanisms underlying cancer initiation and progression. Transcriptome analysis, analogous to RNA sequencing and microarray techniques, elucidates patterns of gene expression in different cancer types and stages. Alternatively, proteomic techniques identify aberrant proteins within cancer cells, offering important clues for selection of potential targets for both diagnosis and treatment. Additionally, important methodologies like protein microarray analysis and mass spectrometry are indispensable in cancer research. In recent years, integrative omics approaches merging transcriptomics, proteomics, and other omics data have become essential tools. They offer a holistic understanding of cancer's molecular landscapes, opening new possibilities for early detection and treatment of cancer. Additionally, single-cell omics technologies such as single-cell proteomics and single-cell RNA sequencing (scRNA-seq), provide advanced insights into tumor progression, therapeutic resistance, and heterogeneity of cancer cells.

Overall, combining artificial intelligence, machine learning, spatial transcriptomics and proteomics holds significant potential for enhancing our knowledge of cancer biology, improving cancer detection and prognosis, and developing personalized treatment strategies.

Each chapter in this thematic issue has been systematically designed to offer comprehensive insights into how these state-of-the-art approaches and methodologies work together effectively and synergistically to solve the mysteries of cancer.

Chapter 1, Rajdeo, Aronow and Prasath ("Deep Learning-based multimodal spatial transcriptomics analysis for cancer") discuss how deep learning methods combined with multimodal spatial transcriptomics analysis can transform cancer treatment and diagnosis by interpreting intricate patterns of gene expression within tumors. Next, Chapter 2, Liao, Martin, Kim, Panahandeh and Won ("Integrating single cells to spatial transcriptomics to understand cell communication") explores spatial transcriptomics and highlights how it uncovers new perspectives for cancer biology understanding, thereby contributing to personalized therapy. In Chapter 3, Mo, Bae, Saqib, Hwang, Jin, B. Park, J. Park and Kim ("Current computational methods for spatial transcriptomics in cancer biology") the computational techniques that underpin spatial transcriptomics analysis in cancer biology are explained, with a focus on deciphering the intricacies of cancer. In Chapter 4, Maurya, Chug, Cudatha and Palma ("Applications of spatial transcriptomics and artificial intelligence to develop integrated management of pancreatic cancer") the emphasis shifts to uses of artificial intelligence and spatial transcriptomics in pancreatic cancer treatment, improving early detection and personalized treatment plans. Further Chapter 5, Goncalves, de Silva and Bernardes ("Advancing cancer therapeutics: integrating scalable 3D cancer models, extracellular vesicles, and omics for enhanced therapy efficacy") discusses the revolutionary potential of extracellular vesicles, omics profiling, and scalable 3D cancer models in developing cancer treatments, which will enhance our knowledge of tumor biology and improve the efficacy of treatment. In addition, Chapter 6 Sudhakar, Vignesh and Natarajan ("Crosstalk between tumor and microenvironment: insights from spatial transcriptomics") sheds light on tumor heterogeneity and interactions with the milieu by highlighting the dynamic interplay between tumors and the microenvironment. In Chapter 7 W. Wang and H. Wang ("Modular formation of in vitro tumor models for oncological research/therapeutic drug screening") describe creation of modular *in vitro* tumor models is examined as a means of enabling personalized therapy and drug screening. Chapter 8 Peixoto and Madan ("Unraveling the complexity: advanced methods in analyzing DNA, RNA, and protein interactions") delves into sophisticated techniques for examining molecular interactions, advancing our knowledge of cancer biology. In conclusion, Chapter 9 Li, Sun, Wang and Wan ("Multi-omics based artificial intelligence for cancer research") explores integration of multi-omics data and artificial intelligence as a potential avenue to improve cancer research and clinical treatment.

These chapters together emphasize the revolutionary impact of cutting-edge approaches and technologies on cancer research, advancing our ability to detect and treat cancer with greater effectiveness.

ESHA MADAN, PhD
Department of Surgery
Faculty Advisor for Center Engagement
VCU Institute of Molecular Medicine (VIMM)
VCU Massey Comprehensive Cancer Center
Virginia Commonwealth University
School of Medicine
Richmond, Virginia, United States

PAUL B. FISHER, MPh, PhD, FNAI
Director, VCU Institute of Molecular Medicine (VIMM)
Thelma Newmeyer Corman Endowed Chair in Cancer Research
VCU Massey Comprehensive Cancer Center
Professor, Department of Human and Molecular Genetics
Virginia Commonwealth University
School of Medicine
Molecular Medicine Research Building (MMRB)
Richmond, Virginia, United States

RAJAN GOGNA, PhD
Department of Human & Molecular Genetics
and Department of Surgery
Member, VCU Institute of Molecular Medicine (VIMM)
VCU Massey Comprehensive Cancer Center
Virginia Commonwealth University
School of Medicine
Molecular Medicine Research Building (MMRB)
Richmond, Virginia, United States

CHAPTER ONE

Deep learning-based multimodal spatial transcriptomics analysis for cancer

Pankaj Rajdeo[a], Bruce Aronow[a,b], and V.B. Surya Prasath[a,b,c,d,*]
[a]Division of Biomedical Informatics, Cincinnati Children's Hospital Medical Center, Cincinnati, OH, United States
[b]Department of Pediatrics, College of Medicine, University of Cincinnati, Cincinnati, OH, United States
[c]Department of Biomedical Informatics, College of Medicine, University of Cincinnati, Cincinnati, OH, United States
[d]Department of Computer Science, University of Cincinnati, Cincinnati, OH, United States
*Corresponding author. e-mail address: surya.prasath@cchmc.org

Contents

1. Introduction	2
1.1 Deep learning and its significance in oncology	2
1.2 Multimodal data analysis in cancer research	4
1.3 Synergistic potential of deep learning and multimodal approaches	5
2. Core concepts of multimodal spatial transcriptomics	8
2.1 Basics of spatial transcriptomics	8
2.2 Multimodal spatial transcriptomics: integrating varied data types	10
2.3 Unraveling cancer complexity with multimodal spatial transcriptomics	14
3. Deep learning approaches in multimodal spatial transcriptomics	15
3.1 Deep learning models for spatial transcriptomics data	16
3.2 Successful applications of deep learning in multimodal spatial transcriptomics	20
4. Impact of multimodal spatial transcriptomics on cancer diagnostics and therapeutics	26
4.1 Enhancing cancer diagnosis	26
4.2 Personalized treatment strategies	27
4.3 Deep learning's contribution to ST-based precision oncology	28
5. Challenges, future outlook, and ethical considerations	29
5.1 Challenges and future directions	29
5.2 Ethical dimensions and responsible AI use	31
6. Conclusion	32
References	33

Abstract

The advent of deep learning (DL) and multimodal spatial transcriptomics (ST) has revolutionized cancer research, offering unprecedented insights into tumor biology.

Advances in Cancer Research, Volume 163
ISSN 0065-230X, https://doi.org/10.1016/bs.acr.2024.08.001
Copyright © 2024 Elsevier Inc. All rights are reserved, including those for text and data mining, AI training, and similar technologies.

This book chapter explores the integration of DL with ST to advance cancer diagnostics, treatment planning, and precision medicine. DL, a subset of artificial intelligence, employs neural networks to model complex patterns in vast datasets, significantly enhancing diagnostic and treatment applications. In oncology, convolutional neural networks excel in image classification, segmentation, and tumor volume analysis, essential for identifying tumors and optimizing radiotherapy.

The chapter also delves into multimodal data analysis, which integrates genomic, proteomic, imaging, and clinical data to offer a holistic understanding of cancer biology. Leveraging diverse data sources, researchers can uncover intricate details of tumor heterogeneity, microenvironment interactions, and treatment responses. Examples include integrating MRI data with genomic profiles for accurate glioma grading and combining proteomic and clinical data to uncover drug resistance mechanisms.

DL's integration with multimodal data enables comprehensive and actionable insights for cancer diagnosis and treatment. The synergy between DL models and multimodal data analysis enhances diagnostic accuracy, personalized treatment planning, and prognostic modeling. Notable applications include ST, which maps gene expression patterns within tissue contexts, providing critical insights into tumor heterogeneity and potential therapeutic targets.

In summary, the integration of DL and multimodal ST represents a paradigm shift towards more precise and personalized oncology. This chapter elucidates the methodologies and applications of these advanced technologies, highlighting their transformative potential in cancer research and clinical practice.

1. Introduction
1.1 Deep learning and its significance in oncology

Deep learning (DL), a subset of machine learning, uses artificial neural networks to model complex patterns in data. In oncology, this technology has revolutionized the analysis of vast datasets, including medical images, genomic data, and clinical records, enhancing applications from diagnosis to treatment planning. These capabilities significantly improve cancer diagnosis, treatment planning, and prognostication (Li, Jiang, Zhang, & Zhu, 2023; Lipkova et al., 2022; Steyaert et al., 2023b).

For instance, convolutional neural networks (CNNs) excel in tasks such as image classification, segmentation, and tumor volume segmentation. These capabilities are crucial for accurately identifying tumors and optimizing radiotherapy planning. By enabling precise mapping of tumors, CNNs facilitate a shift towards precision medicine, where treatment strategies are meticulously tailored to the individual characteristics of each patient's tumor, thereby significantly enhancing treatment outcomes (Sharma, Nayak, Balabantaray, Tanveer, & Nayak, 2024).

Additionally, advancements in neural network architectures, particularly through the use of transfer learning and data augmentation, have tailored DL tools to specific oncology needs, enhancing both research and clinical applications.

To illustrate these advancements, Table 1 summarizes some studies that demonstrate DL's transformative role in oncology:

DL's continued integration has significantly enhanced the accuracy and efficiency of cancer care. This is particularly evident in the use of CNNs for

Table 1 Studies on deep learning in oncology.

Title	Key insights
Cancer Detection Using Convolutional Neural Network (Dabral, Singh, & Kumar, 2021)	Developed a CNN-based method for classifying cancer cells into benign and malignant, enhancing early cancer detection accuracy
Deep convolutional neural network with transfer learning for rectum toxicity prediction in cervical cancer radiotherapy: a feasibility study (Zhen et al., 2017)	Used a pre-trained convolutional neural network (VGG-16 CNN) and transfer learning to predict rectum toxicity in cervical cancer radiotherapy, showing significant opportunities for deep learning in radiation oncology
Exascale deep learning to accelerate cancer research (Patton et al., 2019)	Demonstrated that neural network architectures could be automatically generated and tailored for specific applications, significantly accelerating cancer pathology research
A comparative analysis of two deep learning architectures for the automatic segmentation of vestibular schwannoma (Kattau, Glocker, & Darambara, 2021)	Investigated the performance of two existing deep learning frameworks for tumor segmentation, highlighting advancements in automatic segmentation techniques for treatment planning
Prediction of five-year survival rate for rectal cancer using Markov models of convolutional features of RhoB expression on tissue microarray (Pham, 2023)	Combined deep learning, data coding, and probabilistic modeling to predict five-year survival rates in rectal cancer patients, achieving higher accuracy compared to traditional methods

precise tumor segmentation and the application of transfer learning. Transfer learning involves adapting pre-trained models—originally developed for one task—to new, related tasks. This approach is highly effective in oncology, where models trained on large, diverse datasets are fine-tuned to recognize specific cancer types. By using transfer learning, clinicians can leverage the knowledge gained from vast amounts of existing data, thereby reducing the need for extensive labeled datasets specific to each cancer type. This accelerates the development of diagnostic tools and enables more rapid and accurate customization of treatment plans, which helps further the goals of the precision medicine (Aneja, Aneja, Abas, & Naim, 2021; Hanczar, Bourgeais, & Zehraoui, 2022; Luo & Bocklitz, 2023).

Ultimately, the significance of multimodal data analysis cannot be overstated. It represents a substantial shift towards more comprehensive and integrated methods in cancer research and treatment, which is the primary focus of this chapter.

1.2 Multimodal data analysis in cancer research

Multimodal data analysis involves integrating diverse data types—such as genomic, proteomic, imaging, and clinical data—to gain a holistic understanding of complex biological systems (Acosta, Falcone, Rajpurkar, & Topol, 2022). This foundational approach enhances the depth and accuracy of insights into cancer biology by leveraging the strengths of varied data sources.

- **Genomic data** provide insights into genetic mutations and variations.
- **Proteomic data** reveal details about protein expression and modifications.
- **Imaging data** offer a detailed view of tumor morphology and microenvironment.
- **Clinical data** provide context about patient history and treatment outcomes.

For a more detailed look at the practical applications of these integrations in cancer research, particularly in the context of spatial transcriptomics (ST), see Section 2.2.

1.2.1 Key examples and benefits of integrating various data types

- **Genomic and imaging data integration:** This integration enhances tumor subtype identification and treatment response prediction. For example, integrating MRI data with genomic profiles can distinguish between different glioma grades, leading to more accurate treatment decisions (Cebula et al., 2020).

- **Proteomic and clinical data integration:** This combination provides insights into drug resistance mechanisms and potential therapeutic targets. It also helps monitor disease progression and evaluate treatment effectiveness in real-time (Price et al., 2022).
- **Genomic, proteomic, and imaging data integration:** This comprehensive approach uncovers complex interactions between genes, proteins, and the tumor microenvironment (TME), facilitating the development of multi-targeted therapies and improving patient outcomes (Boehm, Khosravi, Vanguri, Gao, & Shah, 2022).

Refer to Table 2 for additional examples of multimodal data integration in cancer research.

1.3 Synergistic potential of deep learning and multimodal approaches

The synergy between DL and multimodal data analysis holds immense potential for advancing precision medicine and enhancing patient outcomes in oncology. CNNs, in particular, excel in processing and interpreting complex medical data, such as imaging, genomics, and clinical records.

One notable application is the use of ST, which allows the mapping of gene expression patterns within the spatial context of tissues. This provides critical insights into tumor heterogeneity, microenvironment interactions, and treatment response. Combining this with multimodal approaches, which incorporate various data types like genomic, proteomic, and imaging data, enables DL models to uncover more comprehensive and actionable information for cancer diagnosis and treatment (Halawani, Buchert, & Chen, 2023; Hu, Sajid, Lv, Liu, & Sun, 2022; Li, Li, You, Wei, & Xu, 2023).

Fig. 1 illustrates the comprehensive workflow for analyzing cancer using DL-based multimodal ST:

This figure illustrates the comprehensive workflow for analyzing cancer using multimodal ST and DL.

A. Tissue sample preparation, including sectioning, staining, and ST profiling.

B. Data acquisition phase, capturing ST data, imaging data (e.g., hematoxylin and eosin [H&E] staining), and optional genomic/proteomic data.

C. Preprocessing and integration of multimodal data, ensuring consistency and compatibility.

D. Application of DL models, such as CNNs, autoencoders, transformers, and GANs, for cell type identification, spatial domain

Table 2 Examples of multimodal data integration in cancer research.

Title	Key insights
Building and sustaining a comprehensive pediatric oncology care team: The roles and integration of psychosocial and rehabilitative team members (Price et al., 2022)	Highlighted the importance of integrating psychosocial and rehabilitative care services in pediatric oncology to improve patient outcomes and quality of life
The cryo-immunologic effect: A therapeutic advance in the treatment of glioblastomas? (Cebula et al., 2020)	Discussed the potential of combining cryotherapy with immunotherapy to enhance the immune response against glioblastomas, demonstrating the benefits of a multimodal therapeutic approach
Predicting response to chemotherapy in patients with newly diagnosed high-risk neuroblastoma (Mayampurath et al., 2021)	Utilized convolutional neural networks and clinical models to predict chemotherapy response in high-risk neuroblastoma patients, highlighting the effectiveness of integrating imaging and clinical data for treatment planning
Development of a cancer rehabilitation dashboard to collect data on physical function in cancer patients and survivors (Cristian et al., 2024)	Described the creation of a dashboard that integrates self-reported and objective physical function data, facilitating personalized rehabilitation plans and improving patient management in oncology
RadioPathomics: Multimodal learning in non-small cell lung cancer for adaptive radiotherapy (Tortora et al., 2023)	Developed a multimodal late fusion approach combining radiomics, pathomics, and clinical data to predict radiotherapy outcomes in non-small-cell lung cancer patients, demonstrating improved precision in treatment planning through data integration

prediction, gene expression pattern analysis, and integration of multi-omics features.

E. Biological interpretation of results, revealing insights into tumor heterogeneity, tumor-microenvironment interactions, and potential therapeutic targets, advancing precision oncology.

Deep learning-based multimodal spatial transcriptomics analysis for cancer

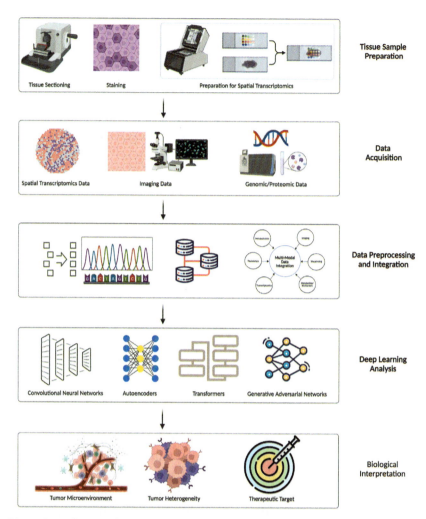

Fig. 1 Deep learning-based multimodal spatial transcriptomics workflow for cancer analysis.

1.3.1 Benefits of this synergy

Enhanced diagnostic accuracy is achieved through DL models that analyze vast imaging data volumes to identify subtle cancer patterns, which might be missed by human observers. By integrating genomic and proteomic data, these models correlate molecular signatures with imaging findings, leading to more accurate diagnoses (Jiang, Hu, Wang, & Zhang, 2023). Personalized treatment planning benefits from multimodal data providing a holistic view of a patient's condition, enabling DL models to predict treatment responses and outcomes more effectively. For example, combining imaging data with genomic profiles can tailor chemotherapy or radiotherapy plans to the individual patient's tumor characteristics, increasing treatment efficacy and reducing adverse effects (Joo et al., 2021). Improved prognostic models emerge from DL models that integrate longitudinal data from various sources to predict disease progression and patient survival more accurately. This capability is crucial for developing personalized follow-up strategies and improving long-term patient management (Cascarano et al., 2023). Real-time decision support is facilitated by the synergy between DL and multimodal approaches, allowing real-time clinical decision support during surgery or radiotherapy. Real-time data from imaging modalities processed by DL algorithms guide procedures, ensuring precise targeting of cancerous tissues while sparing healthy ones (Steyaert et al., 2023b). Accelerated research and development are achieved by harnessing DL to analyze multimodal datasets, helping researchers uncover new biomarkers and therapeutic targets more efficiently. This acceleration leads to faster development of innovative treatments and diagnostics, ultimately benefiting patients (Johnson et al., 2021).

The synergy between DL and multimodal data analysis, especially through the integration of ST, signifies a paradigm shift in oncology. This transition is paving the way for a new era of precision medicine characterized by enhanced diagnostic accuracy, more personalized treatment strategies, and overall improved patient outcomes.

Having explored the synergistic potential of DL and multimodal approaches, we now have established a solid foundation for understanding how these advanced methodologies are revolutionizing the field of oncology. Next, we will delve deeper into the core concepts of multimodal ST. This cutting-edge area further exemplifies the transformative power of integrating diverse data types to unravel the complexities of cancer, which is offering new avenues for research and treatment that promise to enhance the precision and effectiveness of oncology therapies.

2. Core concepts of multimodal spatial transcriptomics
2.1 Basics of spatial transcriptomics
ST is a transformative technique in molecular biology that allows for the visualization and quantification of gene expression within tissue sections, preserving the spatial context of the cells. This innovative approach

provides essential insights into the variability of gene expression across different regions of tissues, enhancing our understanding of cellular functions and tissue organization in both health and disease (Ståhl et al., 2016).

Key principles of ST include spatial resolution, which maintains spatial information about where gene expression occurs within the tissue, providing crucial context for understanding cell interactions within their microenvironments and how these interactions influence overall tissue function. High-throughput sequencing is utilized by ST to analyze the transcriptome across thousands of spatially resolved locations within a tissue sample, enabling a comprehensive analysis of gene expression patterns. Additionally, the integration with imaging combines transcriptomic data with high-resolution tissue imaging, creating a detailed map of gene expression patterns corresponding to specific histological features, which helps to correlate morphological characteristics with molecular data, enhancing the precision of biological insights.

2.1.1 Methods

- **In situ hybridization:** Techniques like MERFISH and seqFISH use fluorescent probes to detect RNA molecules within tissues, allowing for high spatial resolution localization of gene expression (Du et al., 2023).
- **Spatially resolved transcriptomics platforms:** Technologies like 10x Genomics' Visium and Slide-seq use spatial barcoding to map gene expression. These platforms facilitate precise mapping of RNA molecules to their spatial coordinates within the tissue (Williams, Lee, Asatsuma, Vento-Tormo, & Haque, 2022).
- **Laser capture microdissection (LCM):** LCM involves using a laser to precisely cut out specific regions of tissue for RNA extraction and sequencing. This method provides high spatial resolution and is particularly useful for analyzing small or rare cell populations within a tissue (Emmert-Buck et al., 1996).

2.1.2 Importance in mapping the transcriptome to tissue locations

Mapping the transcriptome to specific tissue locations is transformative for several reasons (Williams et al., 2022). Understanding tissue heterogeneity is enhanced as ST allows researchers to identify distinct gene expression profiles across different tissue regions, shedding light on how various cell types contribute to tissue function and pathology. In terms of disease mechanisms, ST reveals interactions between cancer cells and surrounding stromal and immune cells, providing insights into tumor growth and metastasis mechanisms. In developmental biology, ST is invaluable for studying gene expression during tissue development, aiding in understanding the processes of tissue formation, differentiation, and organization. Case studies illustrate the application of ST in uncovering complex cellular interactions and identifying therapeutic targets in diseases, particularly in cancer research. In colorectal cancer, a recent study by Peng et al. (2023b) utilized ST to explore the interactions between fibroblasts and myeloid cells. By integrating single-cell RNA sequencing, ST, and bulk RNA sequencing data, the researchers identified a pro-tumorigenic interaction between MFAP5+ fibroblasts and C1QC+ macrophages. These interactions were mapped spatially within the tumor, highlighting specific signaling pathways that contribute to the malignant behavior of colorectal cancer. This study underscores the importance of ST in elucidating the complex cellular interactions within the TME and identifying potential therapeutic targets. In glioblastoma, another example is the work by Liu et al. (2023), who integrated single-cell RNA sequencing with ST to analyze the cellular heterogeneity. They identified distinct clusters of malignant cells with unique transcriptional and functional properties. ST allowed them to map these clusters within the tumor, revealing their spatial colocalization and interactions with the TME. This integrated approach provided novel insights into the mechanisms of tumor progression and resistance to therapy in glioblastoma.

These examples underscore how ST, by preserving the spatial context of gene expression, helps deepen our understanding of tissue architecture and cellular interactions, which is crucial for developing targeted therapies and improving clinical outcomes.

The following case studies illustrate the application of ST in uncovering complex cellular interactions and identifying therapeutic targets in diseases, particularly in cancer research.

2.2 Multimodal spatial transcriptomics: integrating varied data types

As discussed in Section 1.3, multimodal ST marks a significant advancement in studying complex biological systems, especially in the context of cancer research. By integrating diverse data types such as genomics, proteomics, imaging, and clinical data, researchers can achieve a holistic view of disease mechanisms. This section focuses on how these integrations are uniquely beneficial in the context of ST.

2.2.1 Integrating genomics and proteomics with spatial transcriptomics

Genomic data integration: Combining ST with genomic data offers a comprehensive view of the genetic landscape within tissues. This integration is crucial for revealing mutations and structural variations that drive cancer progression. Tools like MethCNA integrate DNA methylation and copy number variation data, aiding in the identification of oncogenic drivers within their spatial context in tissues (Deng, Yang, Zhang, Xiao, & Cai, 2018). For example, integrating ST with genomic perturbation data enhances gene regulatory network inference, allowing researchers to predict how specific genomic alterations impact gene expression across different tissue regions (Liang, Young, Hung, Raftery, & Yeung, 2019). This helps in understanding how mutations within the same tumor can lead to varied phenotypic outcomes.

Proteomic data integration: Proteomics complements ST by providing insights into protein expression and modifications, the functional executors of genetic information. Proteogenomic approaches integrate patient-specific genomic and proteomic data to characterize the mutational landscape and protein signaling pathways in tumors, identifying patient-specific drug targets and resistance mechanisms (Schmitt et al., 2021).

2.2.2 Complementing ST with imaging and clinical data

Imaging data integration: Integrating imaging data with ST allows for correlating morphological features with molecular changes, enhancing the understanding of tumor phenotypes and treatment responses. For instance, the FDTrans model developed by Cai et al. (2023) integrates histopathological images with genomic data to classify lung cancer subtypes, demonstrating the potential of this approach.

Clinical data integration: Combining ST with clinical data helps identify biomarkers predictive of therapy response and disease prognosis. Integrating clinical information with molecular profiles, as done by Wilson, Li, Yu, Kuan, and Wang (2019) identifies gene sets relevant to prognostic predictions across cancer types.

Fig. 2 illustrates the application of ST techniques in a cancer tissue section, highlighting the detailed visualization methods used to study the TME.

a. **Multiplexed immunofluorescence image:** This image shows the spatial distribution of three different proteins within the tissue, each protein is denoted by a specific color.

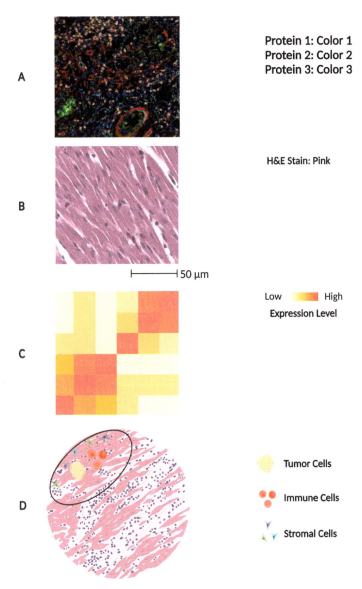

Fig. 2 Visualization of multimodal spatial transcriptomics data in a cancer tissue section.

b. **H&E stained tissue section:** Displays the overall tissue morphology and cellular structure, providing a visual context for the underlying tissue architecture.

c. ST heatmap: This heatmap demonstrates the relative expression levels of a gene of interest across different regions of the tissue, indicating areas of high and low expression.

d. Integrated view: This circle contains a zoomed-in view of the H&E-stained tissue, overlaid with color-coded indicators for tumor cells (yellow), immune cells (red), and stromal cells (green and purple), illustrating the cellular composition and interaction within the microenvironment.

2.2.3 Successful multimodal ST studies in cancer research

The following case studies demonstrate the successful application of multimodal ST in cancer research, providing insights that guide the development of more effective and personalized treatment strategies. Harikumar, Quinn, Rana, Gupta, and Venkatesh (2021) proposed a framework to predict gene responses to drugs using patient-specific single-cell expression data and population-level drug-response data. This dual-channel approach integrates single-cell RNA-seq with drug response data, providing personalized predictions of drug efficacy. This framework has been applied to glioblastoma, illustrating how multimodal integration can enhance personalized treatment strategies.

Another notable example is the FDTrans model by Cai et al. (2023), which exemplifies successful multimodal integration by combining histopathological images with genomic data to classify lung cancer subtypes. This model achieved high accuracy and AUC scores, demonstrating the potential of integrating imaging and genomic data to improve diagnostic precision and treatment planning. Similarly, Schmitt et al. (2021) used proteogenomics to study the mutational landscape of melanoma patients. By integrating genomic and proteomic data, they identified pathways associated with melanoma development and immunotherapy response, highlighting the importance of combining multiple data types to uncover molecular mechanisms and identify therapeutic targets.

In a comprehensive study by Wei, Zhang, Weng, Chen, and Cai (2021), various computational methods for integrating multi-omics data across pan-cancer datasets were assessed. This study demonstrated the ability of integrative methods to identify distinct tumor compositions and molecular patterns, providing valuable insights for pan-cancer analysis and highlighting the importance of multimodal data integration in cancer research. In summary, multimodal ST provides a comprehensive framework for understanding the complex molecular landscape of cancers, enhancing our knowledge of tumor biology and paving the way for personalized medicine.

2.3 Unraveling cancer complexity with multimodal spatial transcriptomics

Multimodal ST empowers researchers to dissect tumor heterogeneity, analyze the TME, and evaluate therapy responses with unprecedented detail and precision. These capabilities lead to several key insights.

2.3.1 Key insights from multimodal ST

Tumor heterogeneity: One of the primary insights gained from multimodal ST is the detailed characterization of tumor heterogeneity. Tumors are composed of a mosaic of genetically distinct cell populations, each contributing differently to disease progression and therapy resistance. ST, combined with single-cell sequencing, allows for the mapping of these heterogeneous cell populations within their spatial context. For instance, Halawani et al. (2023) highlighted how DL frameworks applied to single-cell and ST data can unravel the complex gene transcription profiles and mutation spectra within tumors. This approach has been crucial in identifying subclonal populations that drive disease progression and metastasis. In esophageal cancer, the integration of multi-omics data has provided insights into the cellular and genetic heterogeneity within tumors. Li et al. (2023) discussed how artificial intelligence (AI) can analyze and interpret multi-omics data, which enables a comprehensive understanding of tumor heterogeneity and its implications for disease progression and treatment strategies. This high-dimensional characterization helps in identifying novel cell types and understanding their roles in tumor biology.

TME: The TME plays a pivotal role in cancer development and therapy response. Multimodal ST allows for the spatial profiling of immune and stromal cells within the TME, providing insights into their interactions with tumor cells. Hu et al. (2022) emphasized the importance of spatial profiling technologies in evaluating the transcriptional activity of immune and stromal cells within the TME. This multidimensional classification facilitates the study of tumor-immune interactions, the evolutionary trajectory of tumors, and the identification of immune evasion mechanisms. A study by Zhang et al. (2024a) on hepatocellular carcinoma revealed how single-cell sequencing and ST can uncover the spatial distribution and functional states of tumor-infiltrating lymphocytes. The identification of stimulatory dendritic cells and macrophages as potential biomarkers underscores the importance of understanding the spatial organization of immune cells in predicting therapy responses.

Therapy response: Understanding the spatial and molecular determinants of therapy response is crucial for developing effective cancer treatments. Multimodal ST provides a platform for identifying biomarkers and therapeutic targets by linking molecular profiles to spatial contexts. For example, Lyubetskaya et al. (2022) demonstrated how ST can capture key tumor features such as hypoxia, necrosis, and vasculature, which are critical for understanding therapy responses. The ability to map these features within the tumor landscape helps in identifying regions of therapeutic resistance and potential targets for intervention. In breast cancer, Bottosso et al. (2024) highlighted the use of precision medicine to predict drug sensitivity based on a deeper molecular understanding of the disease. The integration of genomic, transcriptomic, and imaging data allows for the identification of biomarkers that predict response to specific therapies and can aid in the personalization of treatment strategies.

Having established the foundational concepts and diverse applications of multimodal ST, we now shift our focus to the innovative role of DL in this domain. This transition marks a pivotal move towards harnessing sophisticated computational models to further unravel the complexities of cancer at a molecular and spatial level.

3. Deep learning approaches in multimodal spatial transcriptomics

Building upon our exploration of multimodal ST, Section 3 delves into the transformative role of DL in refining and advancing this field. As depicted in Fig. 3, these models integrate diverse data types—including genomic, proteomic, imaging, clinical, and ST—to enable comprehensive analysis and interpretation of cancer data. Here, we focus on cutting-edge deep learning models that bring unprecedented precision to analyzing ST data. These models not only interpret complex biological relationships within tissues but also push the boundaries of diagnostic and prognostic capabilities, marking a significant evolution in the tools available to oncologists and researchers.

This figure illustrates the process of multimodal data fusion and the application of DL to achieve improved diagnostic accuracy, personalized treatments, and a better understanding of tumor heterogeneity and the TME.

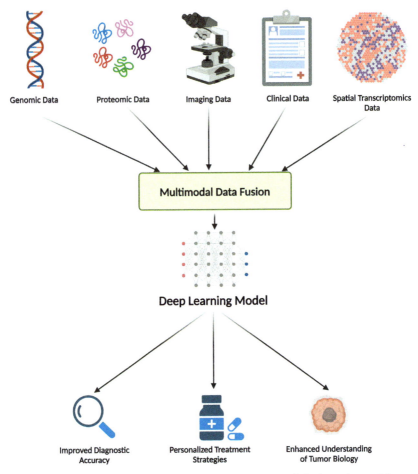

Fig. 3 Integration of genomic, proteomic, imaging, clinical, and spatial transcriptomics data using deep learning models in cancer research.

3.1 Deep learning models for spatial transcriptomics data

ST technologies provide a detailed map of gene expression patterns within the spatial architecture of tissues. The complexity of this data necessitates the use of advanced DL models that can identify and interpret intricate spatial patterns and relationships.

This section introduces various DL models suitable for analyzing ST data, including graph neural networks (GNNs), CNNs, transformer models, and hybrid approaches (see Fig. 4).

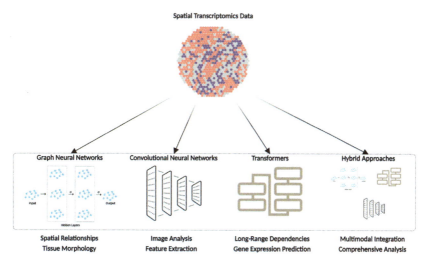

Fig. 4 Different deep learning models for analyzing spatial transcriptomics data.

This figure highlights the key DL models used in ST, including GNNs, CNNs, transformer models, and hybrid approaches, showcasing their unique processing techniques and applications in cancer research.

3.1.1 Graph-based deep learning models

Graph-based models excel at contextualizing the complex morphology and structure within whole slide images by representing spatial relationships through graph structures. These models efficiently encode the intricate patterns of tissue architecture, enabling a detailed analysis of ST data. This approach facilitates the exploration of cellular interactions and molecular pathways across diverse tissue regions, enhancing our understanding of disease mechanisms and progression (Wu et al., 2022; Yuan & Bar-Joseph, 2020). Azher et al. (2023) demonstrated the potential of leveraging ST data with a contrastive crossmodal pretraining mechanism to improve graph-based learning tasks. This approach enhanced the extraction of molecular and histological information, which improved outcomes in cancer staging, lymph node metastasis prediction, survival prediction, and tissue clustering analyses. The integration of spatial omics data significantly improved the performance of graph-based models in pathology workflows, thus highlighting their potential to enhance cancer diagnostics and prognostics.

Peng, He, Peng, Li, and Zhang (2023a) developed STGNNks, a method that combines GNNs, denoising auto-encoders, and k-sums clustering to process spatially resolved transcriptomics data. This model constructs a hybrid adjacency matrix and integrates gene expressions with spatial context, mapping learned features to a low-dimensional space for clustering. STGNNks significantly outperformed existing ST analysis algorithms and provided valuable insights into tumor progression through spatial trajectory inference and differentially expressed gene detection. Additionally, Wu et al. (2022) applied GNNs to model TMEs as local subgraphs using spatial protein profiles. This strategy captures distinctive cellular interactions associated with differential clinical outcomes. The model demonstrated substantial improvements in predicting patient outcomes compared to traditional DL approaches, underscoring its potential to enhance the understanding of TME dynamics.

3.1.2 Convolutional neural networks (CNNs)

CNNs are adept at analyzing histopathological images and predicting gene expression patterns due to their ability to extract high-level features from complex visual data. These networks use multiple layers of filters to process spatial hierarchies in images and capture details from cell morphology to tissue architecture (Dabeer, Khan, & Islam, 2019). This enables CNNs to identify subtle patterns and structures that are crucial for accurate gene expression prediction, which enhances our understanding of the molecular underpinnings of cancer and other diseases. Zheng, Carrillo-Perez, Pizurica, Heiland, and Gevaert (2023) developed a DL model to predict transcriptional subtypes of glioblastoma cells from histology images. This model phenotypically analyzed millions of tissue spots and identified consistent associations between tumor architecture and prognosis across independent cohorts. The approach underscored the critical connection between spatial cellular architecture and clinical outcomes, thus offering a scalable method to predict gene expression and understand the spatial organization of tumor cells.

In a related advancement, Wang, Zhou, Kong, and Lu (2023) proposed a CNN-based method for enhancing spot resolution in ST, termed superresolved ST. Their approach, utilizing a shift-predict operation, achieved 9× superresolution and outperformed traditional superresolution techniques. This method provides a deeper understanding of gene expression patterns and their underlying biological significance, highlighting the potential of CNNs to revolutionize ST by improving spatial resolution and analytical precision.

3.1.3 Transformer models

Transformers excel in capturing long-range dependencies, which enhances gene expression predictions from histology images by leveraging the self-attention mechanism introduced in the seminal paper *Attention Is All You Need* (Vaswani et al., 2023). This approach allows transformers to simultaneously process distant parts of an image, identifying crucial patterns and correlations for accurate prediction. This mechanism enables the comprehensive analysis of the whole tissue section, improving the mapping of gene expressions and aiding in the study of complex interactions within the TME. Liu, Zhang, and Luo (2024) proposed a contrastive learning-based framework to infer molecular subtypes and clinical outcomes of breast cancer from unannotated whole slide images. By leveraging patch-level features and a gated attention mechanism, the model produced slide-level predictions. This method effectively established high-order genotype-phenotype associations, enhancing the application of digital pathology in clinical practice. Xiao, Kong, Li, Wang, and Lu (2024) introduced TCGN, a model combining convolutional layers, transformer encoders, and GNNs to estimate gene expressions from histopathological images. TCGN operates with a single spot image input, maintaining interpretability while enhancing accuracy. The model provided a powerful tool for inferring gene expressions in precision health applications.

The versatility of transformers to capture long-range dependencies complements the localized feature extraction of CNNs. Combining these approaches can further enhance model performance, leading us to hybrid approaches.

3.1.4 Hybrid approaches

Hybrid models combine multiple DL architectures to enhance the analysis of ST data. By integrating the unique advantages of each model, they achieve a more comprehensive and accurate interpretation of complex biological data. This synergy allows for a deeper understanding of the spatial and molecular dynamics within tissues, aiding in the identification of key biological insights and potential therapeutic targets (Song et al., 2024; Yan et al., 2019).

Steyaert et al. (2023a) emphasized the importance of developing multimodal fusion approaches to integrate complementary data types such as molecular, histopathology, radiology, and clinical records. By leveraging DL models that can handle multiple modalities, researchers can capture the heterogeneity of complex diseases more accurately, advancing precision medicine. Chen, Zhang, Tang, Liu, and Huang (2024) proposed the Edge-relational Window-attentional GNN (ErwaNet) for predicting gene expression from standard tissue images. This model constructs hetero-geneous graphs to model local window interactions and incorporates an attention mechanism for global information analysis. ErwaNet stands out as a cost-effective and accurate method for gene expression prediction, offering a significant advantage in cancer research by providing more efficient and accessible analytical paradigms.

By integrating the strengths of different models, hybrid approaches offer a robust framework for ST analysis, combining the localized feature extraction of CNNs, the contextual understanding of transformers, and the relational mapping of GNNs.

3.2 Successful applications of deep learning in multimodal spatial transcriptomics

This subsection showcases how DL significantly enhances the analysis of ST by enabling precise cell-type identification, biomarker discovery, and treatment response prediction. Each case study directly correlates with the models discussed, illustrating the practical impact of these technological advancements.

Table 3 lists recent DL applications in ST, emphasizing diverse meth-odologies and their impacts on cancer research. Each entry will be suc-cinctly described, highlighting its relevance and connection to the DL models introduced earlier.

These case studies illustrate the diverse applications of DL in enhancing ST analysis. By integrating multimodal data, such as histological images and ST, DL models have significantly improved our understanding of cellular and molecular mechanisms in various diseases.

Reflecting on this comprehensive review, some key areas highlight the significant strides where DL optimizes multimodal ST:

DL methods, such as STGNNks and RESEPT, have revolutionized cell-type identification and spatial domain mapping in tissue samples. By precisely characterizing cellular heterogeneity and tissue organization, these advancements open new avenues for understanding disease mechanisms and developing targeted therapies tailored to specific cell populations (Chang et al., 2022; Peng et al., 2023a). Additionally, DL approaches, exemplified by the work of Halawani et al. (2023) and Liu et al. (2024), have accelerated biomarker discovery by analyzing spatial gene expression patterns and tumor heterogeneity. The identification of novel molecular

Table 3 Recent applications of deep learning in multimodal spatial transcriptomics.

Title	Key applications	Methodology	Key findings	Impact
Spatial omics driven crossmodal pretraining applied to graph-based deep learning for cancer pathology analysis (Azher et al., 2023)	Cancer Pathology Analysis	Graph Neural Network, Crossmodal Pretraining	Enhances extraction of molecular and histological information, improving cancer staging and survival prediction	Improved pathology workflow accuracy
STGNNks: Identifying cell types in spatial transcriptomics data based on graph neural network, denoising auto-encoder, and k-sums clustering (Peng et al., 2023a)	Cell Type Identification	Graph Neural Network, Auto-encoder, K-sums Clustering	Outperforms other clustering methods, providing insights into tumor progression and spatial gene expression	Superior spatial transcriptomics data analysis and disease diagnosis
Graph deep learning for the characterization of tumor microenvironments from spatial protein profiles in tissue specimens (Wu et al., 2022)	Tumor Microenvironment Analysis	Graph Neural Network	Identifies spatial motifs associated with cancer recurrence and patient survival, outperforming traditional models	Enhanced understanding of tumor microenvironments

(continued)

Table 3 Recent applications of deep learning in multimodal spatial transcriptomics. (*cont'd*)

Title	Key applications	Methodology	Key findings	Impact
Single-cell omics traces the heterogeneity of prostate cancer cells and the tumor microenvironment (Yu, Liu, Gao, Wang, & Zhang, 2023)	Tumor Heterogeneity Analysis	Single-cell Omics, Deep Learning	Analyzes heterogeneity and spatial distribution of prostate cancer cells, revealing new insights into tumor biology	Advanced prostate cancer research and personalized treatment strategies
Spatial cellular architecture predicts prognosis in glioblastoma (Zheng et al., 2023)	Prognosis Prediction	Deep Learning, Histology Analysis	Links tumor architecture with prognosis, identifying gene expression patterns associated with survival.	Improved prognosis prediction for glioblastoma patients
Contrastive learning-based histopathological features infer molecular subtypes and clinical outcomes of breast cancer from unannotated whole slide images (Liu et al., 2024)	Molecular Subtype Prediction	Contrastive Learning, Attention Mechanism	Predicts molecular subtypes and clinical outcomes, establishing genotype–phenotype associations	Enhanced molecular subtype prediction for breast cancer

Transformer with convolution and graph-node co-embedding: An accurate and interpretable vision backbone for predicting gene expressions from local histopathological image (Xiao et al., 2024)	Gene Expression Prediction	Transformer, Convolutional Layers, Graph Neural Networks	Operates with single spot image input, maintaining interpretability while enhancing accuracy	Improved gene expression prediction from histopathological images
Detecting spatially co-expressed gene clusters with functional coherence by graph-regularized convolutional neural network (Song et al., 2022)	Gene Cluster Detection	Convolutional Neural Network, Graph Regularization	Effectively detects spatially co-expressed gene clusters, providing biological insights into tissue morphology	Enhanced understanding of tissue morphology and gene functionality
Deep learning exploration of single-cell and spatially resolved cancer transcriptomics to unravel tumour heterogeneity (Halawani et al., 2023)	Tumor Heterogeneity Analysis	Deep Learning Frameworks	Enhances precision oncology applications by analyzing single-cell and spatial transcriptomics data to investigate tumor heterogeneity	Advanced tumor characterization and personalized oncology

(continued)

Table 3 Recent applications of deep learning in multimodal spatial transcriptomics. (*cont'd*)

Title	Key applications	Methodology	Key findings	Impact
Superresolved spatial transcriptomics transferred from a histological context (Wang et al., 2023)	Spatial Resolution Enhancement	Convolutional Neural Network	Achieves superresolution in spatial transcriptomics data, improving gene expression pattern analysis	Enhanced spatial resolution and gene expression analysis
THItoGene: a deep learning method for predicting spatial transcriptomics from histological images (Jia, Liu, Chen, Zhao, & Wang, 2024)	Gene Expression Prediction	Dynamic Convolutional and Capsule Networks	Demonstrates superior performance in spatial gene expression prediction, linking pathology image phenotypes to gene expression regulation	Improved link between pathology images and gene expression
Define and visualize pathological architectures of human tissues from spatially resolved transcriptomics using deep learning (Chang et al., 2022)	Tissue Architecture Visualization	Deep Learning Framework (RESEPT)	Accurately infers and visualizes tissue architecture, aiding in the identification of pathological features in diseases like Alzheimer's and cancer	Better visualization and understanding of tissue architecture

ScribbleDom: Using scribble-annotated histology images to identify domains in spatial transcriptomics data (Rahman et al., 2023)	Domain Identification	Semi-Supervised Convolutional Neural Network	Improves spatial domain identification by integrating human annotations with computational power	Enhanced spatial domain identification and integration of human expertise
Cross-linking breast tumor transcriptomic states and tissue histology (Dawood et al., 2023)	Gene Expression Prediction	Graph Neural Network	Captures gene expression states from whole slide images, linking histological phenotypes with gene expression patterns	Better linking of histological and gene expression data
DeepST: identifying spatial domains in spatial transcriptomics by deep learning (Xu et al., 2022)	Spatial Domain Identification	Deep Learning Framework	Provides accurate and scalable spatial domain identification, outperforming existing methods	More accurate spatial domain identification

markers through these techniques offers the potential for earlier disease detection, more accurate prognosis, and the development of targeted therapies that address the unique molecular signatures of individual tumors.

Furthermore, DL models, as demonstrated by Wu et al. (2022) and Zheng et al. (2023), have shown remarkable promise in predicting patient outcomes and treatment responses. By analyzing spatial relationships and cellular interactions within tumors, these models offer a powerful tool for tailoring treatment strategies to individual patients, ultimately leading to improved clinical outcomes and more effective cancer care.

These advancements underscore the transformative potential of DL in ST, which is paving the way for more precise and personalized medical interventions. The integration of multimodal data, supported by robust DL frameworks is continuing to drive significant progress in cancer research and precision medicine. By leveraging these advanced techniques, researchers can significantly improve the accuracy, robustness, and interpretability of ST analysis, ultimately accelerating the development of novel diagnostics and therapeutics in precision medicine.

4. Impact of multimodal spatial transcriptomics on cancer diagnostics and therapeutics

In the preceding sections, we explored how multimodal ST integrates various data types to revolutionize cancer research. This section will focus on how these integrations enhance cancer diagnostics and therapeutics, thus highlighting specific instances where ST has uniquely contributed to advancements in these areas. Here, we emphasize novel applications and methodologies that demonstrate the transformative potential of ST in clinical settings.

4.1 Enhancing cancer diagnosis

Recent advancements have demonstrated the power of pathogenomics in integrating advanced molecular diagnostics from genomic data with morphological information from histopathological imaging and codified clinical data. Feng et al. (2024) describe how this integration leads to the discovery of new multimodal cancer biomarkers, and thus enhances the precision of oncology diagnoses. Their study emphasizes the importance of synthesizing complementary modalities with emerging multimodal AI methods in pathogenomics, including the correlation and fusion of histology and genomics profiles of cancer. This approach overcomes the limitations of unimodal data and propels precision oncology into the next decade.

Similarly, Xi et al. (2022) presented a modality-correlation embedding model for breast tumor diagnosis using mammography and ultrasound images. By optimizing the correlation between these modalities, their model improved tumor classification accuracy significantly, with a sensitivity of 91.67% and a specificity of 95.83%. This study demonstrates the importance of maintaining pairwise closeness of multimodal data in a common label space with consistent diagnostic results from multimodal images of the same patient.

Abdollahyan et al. (2023) highlight the role of dynamic biobanking in advancing breast cancer research. The Breast Cancer Now Tissue Bank at the Barts Cancer Institute serves as a dynamic biobanking ecosystem and integrates longitudinal biospecimens with multimodal data, including EHRs, genomic, and imaging data. This ecosystem supports precision medicine efforts in breast cancer research by offering high-quality annotated biospecimens and advanced analytics tools, demonstrating how integrated data can inform diagnostic accuracy.

4.2 Personalized treatment strategies

Chen et al. (2023) introduced the Multimodal Data Fusion Diagnosis Network (MDFNet), a framework that effectively fuses clinical skin images with patient clinical data for skin cancer classification. MDFNet establishes a mapping among heterogeneous data features which addresses issues of feature paucity and richness that arise from using single-mode data. Their model showed an improvement in diagnostic accuracy by about 9% and illustrates the fusion advantages exhibited by MDFNet and its potential as an effective auxiliary diagnostic tool for skin cancer, enhancing clinical decision-making.

Yan et al. (2019) developed a hybrid DL method for breast cancer classification by integrating pathological images and structured data from clinical EMRs. This approach significantly outperforms state-of-the-art methods and demonstrates the importance of combining multimodal data for precise patient stratification and personalized therapy. Their method provides a practical tool for breast cancer diagnosis in real clinical practice.

Rani, Ahmad, Masood, and Saxena (2023) highlighted the importance of using machine learning models on unimodal and multimodal datasets for diagnosing breast cancer molecular subtypes. By operating on multimodal data samples for each patient, their custom DL-based model pipeline achieved 94% accuracy and thus showed the superiority of multimodal data over unimodal in breast cancer subtype classification. These findings underscore the value of multimodal data in enhancing the precision of treatment strategies.

Li et al. (2022) utilized multi-omics data for lung cancer stage prediction, showing that microbial combinatorial transcriptome fusion analysis had the highest accuracy, reaching 0.809. This study reveals the role of multimodal data and fusion algorithms in accurately diagnosing lung cancer stages, which aids doctors in clinics and emphasizes the potential of multimodal integration in personalized treatment planning.

4.3 Deep learning's contribution to ST-based precision oncology

Steyaert et al. (2023a) discussed the challenges and opportunities in multimodal data fusion for cancer biomarker discovery with DL. They highlighted the need for developing effective multimodal fusion approaches as a single modality might not be sufficient to capture the heterogeneity of complex diseases like cancer. By tackling data sparsity and scarcity, enhancing multimodal interpretability, and standardizing datasets, DL can significantly advance the analysis of biomedical data to the discovery of key cancer biomarkers.

Lobato-Delgado, Priego-Torres, and Sanchez-Morillo (2022) combined molecular, imaging, and clinical data analysis for predicting cancer prognosis and demonstrated how integration of these data types using machine learning and DL techniques improves patient stratification and clinical management. This review emphasizes that multimodal models can better stratify patients and can contribute to personalized medicine and provide valuable insights into cancer biology and progression.

Zhang et al. (2024b) explored the use of AI in liver imaging, summarizing AI methodologies and their applications in managing liver diseases. They stressed the importance of feature interpretability, multimodal data integration, and multicenter study in enhancing the clinical applications of AI, supporting precise disease detection, diagnosis, and treatment planning.

Shao, Ma, Zhang, Li, and Wang (2023) discussed predicting gene mutation status via AI technologies based on multimodal integration to advance precision oncology. Their review synthesized the general framework of MMI for molecular intelligent diagnostics and summarized emerging applications of AI in predicting mutational profiles of common cancers. This work highlights the potential of AI as a decision-support tool to aid oncologists in future cancer treatment management and the challenges and prospects of AI in medical fields.

As we conclude our exploration of the impact that multimodal ST and DL have on enhancing cancer diagnostics and therapeutics, we recognize the transformative advancements these technologies bring to precision oncology. The integration of diverse data types, coupled with sophisticated analytical models, has not only revolutionized our approach to understanding and treating cancer but also set the stage for addressing the complexities of implementation in clinical settings.

Looking ahead, Section 5 will delve into the challenges, future outlook, and ethical considerations of these technologies. We will explore the barriers that must be overcome to fully realize their potential, the ethical frameworks that need to be established for their responsible use, and the directions future research might take to continue advancing the field of cancer research and treatment.

5. Challenges, future outlook, and ethical considerations

As we delve into the transformative world of AI-driven multimodal ST in cancer research and treatment, it's critical to navigate the complex challenges and ethical considerations while also looking toward future advancements. By addressing these elements comprehensively, we can fully harness the transformative potential of these technologies.

5.1 Challenges and future directions

5.1.1 Challenges in multimodal spatial transcriptomics

While traditional ST has revolutionized our understanding of tissue architecture and cellular function, it faces several issues that hinder its broader application. These include data sparsity and noise, potential batch effects, and missing data, all of which can obscure meaningful biological insights. Despite significant advancements, the field of multimodal ST continues to encounter numerous challenges that must be addressed to fully unlock its transformative potential. Integrating and interpreting high-dimensional data from diverse sources requires sophisticated computational tools and algorithms. Moreover, the heterogeneity and sparsity of single-cell and ST data add layers of complexity to the analysis.

A major barrier in the field is the lack of standardization and interoperability among different imaging and omics modalities. Standardized protocols and interdisciplinary collaboration are essential to overcome these challenges (Jaiswal, Agarwal, & Jaiswal, 2023). Robust imputation techniques and model-based approaches are critical for addressing missing and sparse data, which are pivotal for maintaining the integrity of ST studies (Rathore, Abdulkadir, & Davatzikos, 2020). Technical variability and batch effects can significantly skew analyses, necessitating robust preprocessing pipelines and advanced batch correction methods to mitigate these effects (Feng et al., 2024; Khalighi et al., 2024). The scalability of data analysis and the ethical, legal, and social implications must be addressed for the responsible implementation of AI and ST in healthcare (Khalighi et al., 2024; Liu et al., 2024).

Standardization and robust imputation techniques are crucial for combating issues of missing data and ensuring interoperability across platforms (Capobianco, 2022; Rathore et al., 2020). These measures will help maintain data integrity and enhance the comparability of results across different studies. Addressing these challenges will pave the way for the more effective and widespread use of multimodal ST in research and clinical settings, ultimately advancing our understanding of complex biological systems and improving patient care.

5.1.2 Future directions

To overcome the existing challenges and pave the way for new discoveries, research in multimodal ST is focusing on several key areas:

Increasing sequencing depth and using robust preprocessing pipelines are necessary strategies to enhance the signal-to-noise ratio and address data sparsity issues (Akhoundova & Rubin, 2022; Feng et al., 2024). Implementing advanced batch correction methods and developing multimodal fusion techniques are essential to mitigate batch effects and improve data integration (Khalighi et al., 2024; Steyaert et al., 2023a). Ensuring the reliability and generalizability of models through practices like cross-validation and external validation is crucial. Additionally, incorporating interpretability techniques, such as attention mechanisms and SHAP values, helps translate complex ST data into actionable insights (Steyaert et al., 2023a; Yu et al., 2023). Rigorous biological validation of model predictions is essential to ensure that identified patterns reflect true biological phenomena. Integrating ST data with experimental data, such as functional assays, is crucial for confirming the biological significance of the findings.

Promoting collaborative efforts and establishing standardized workflows for data sharing are key to advancing ST research and developing standardized analytical tools (Tortora et al., 2023). Continued advancements in computational models, including DL and Bayesian statistics, will enhance data analysis capabilities and offer new insights into complex biological systems (Li, 2023). Building comprehensive databases and repositories for multimodal data will facilitate data sharing and collaborative research, which are crucial for accelerating discoveries in cancer research. Integrating clinical data with multimodal ST is vital for translating research into clinical practice and developing more effective and personalized therapies.

By addressing these challenges and leveraging future directions, the field of multimodal ST is poised for significant advancements, promising to enhance our understanding of complex biological systems and improve therapeutic strategies in precision medicine.

5.2 Ethical dimensions and responsible AI use

The use of AI and ST in healthcare introduces significant ethical considerations that must be managed to ensure responsible application:

5.2.1 Ethical considerations in data collection and analysis

Ensuring that patient data is collected and used in a manner that respects consent and confidentiality is paramount to balancing the benefits of AI-driven diagnostics against potential risks to patient autonomy and privacy (Love, 2023). Building trust among patients and healthcare providers requires transparent AI algorithms that provide clear explanations of their decision-making processes, ensuring that decisions are made in the best interests of patients. Ethical guidelines for data anonymization and sharing should be established to protect patient identities and outline the purposes for which data can be used, fostering responsible research practices.

5.2.2 Addressing biases and ensuring fairness

Training AI algorithms on diverse datasets is essential to prevent biases that could lead to healthcare disparities (Wang et al., 2024). Ensuring that AI models provide equitable care and identifying any biases that could disadvantage certain patient groups are crucial for fostering fair healthcare practices (Fusar-Poli et al., 2022). Developing comprehensive ethical frameworks to address the challenges and risks associated with AI in healthcare is necessary to guide ethical implementation and ensure model accuracy and fairness (Sundaramurthy & Vaithiyalingam, 2023).

In summary, addressing the challenges in ST data analysis demands a multifaceted approach that includes improving data quality, developing robust validation strategies, enhancing model interpretability, and performing rigorous biological validation. By tackling these challenges head-on, researchers can fully leverage the potential of ST technologies to revolutionize precision medicine and deepen our understanding of complex biological systems. By addressing these challenges and ethical considerations, we can enhance the robustness, interpretability, and application of ST in cancer research. Continued development and integration of these advanced tools promise to transform our understanding of cancer, improve diagnostic and therapeutic strategies, and ultimately enhance patient outcomes.

6. Conclusion

The integration of multimodal ST and DL has already begun to transform cancer research, offering deep insights into tumor biology and enabling personalized treatment strategies. As we reflect on the journey through this chapter, the diverse applications and potential of multimodal ST have been thoroughly explored, emphasizing its critical role in enhancing diagnostic precision, guiding personalized therapies, and advancing our understanding of cancer at the molecular level.

Key takeaways revisited.

- **Enhanced diagnostic accuracy:** The integration of various data types through multimodal ST has proven to improve diagnostic accuracy and precision, highlighting the importance of comprehensive views of tumor biology.
- **Personalized treatment strategies:** By utilizing the detailed molecular and spatial information provided by ST, researchers have developed targeted therapeutic strategies, enhancing personalized medicine approaches.
- **DL and AI integration:** DL models, such as CNNs and GNNs, have been essential in analyzing complex ST data, supporting the advancement of precision oncology.
- **Ethical considerations and responsible AI use:** Ensuring ethical practices in AI and ST integration remains crucial for maintaining transparency, fairness, and patient data protection in clinical settings.

Transformative potential of multimodal ST and DL.

The advancements in multimodal ST and DL are not merely incremental; they are transformative, enabling a holistic understanding of cancer. These technologies facilitate the development of more effective and less toxic treatments by providing insights into the TME and the interactions of different cell types and molecular pathways within the spatial context of tissues.

As we look to the future, the promise of these technologies to revolutionize cancer research, improve diagnostic and therapeutic strategies, and ultimately enhance patient outcomes is immense. Addressing the remaining ethical and technical challenges will be crucial to fully realize the potential of these cutting-edge tools in cancer research and clinical practice.

In conclusion, the journey through multimodal ST and DL in cancer research has been enlightening, demonstrating the profound impact these technologies will continue to have on advancing our understanding of cancer and improving patient care. As we continue to develop and integrate these tools, the future of cancer research looks brighter than ever.

References

Abdollahyan, M., Gadaleta, E., Asif, M., Oscanoa, J., Barrow-McGee, R., Jones, S., et al. (2023). Dynamic biobanking for advancing breast cancer research. *Journal of Personalized Medicine, 13*(2), 360. https://doi.org/10.3390/jpm13020360.

Acosta, J. N., Falcone, G. J., Rajpurkar, P., & Topol, E. J. (2022). Multimodal biomedical AI. *Nature Medicine, 28*(9), 1773–1784. https://doi.org/10.1038/s41591-022-01981-2.

Akhoundova, D., & Rubin, M. A. (2022). Clinical application of advanced multi-omics tumor profiling: Shaping precision oncology of the future. *Cancer Cell, 40*(9), 920–938. https://doi.org/10.1016/j.ccell.2022.08.011.

Aneja, S., Aneja, N., Abas, P. E., & Naim, A. G. (2021). Transfer learning for cancer diagnosis in histopathological images. *arXiv preprint*, arXiv:2112.15523. https://doi.org/10.48550/arXiv.2112.15523.

Azher, Z., Fatemi, M., Lu, Y., Srinivasan, G., Diallo, A., Christensen, B., et al. (2023). Spatial omics driven crossmodal pretraining applied to graph-based deep learning for cancer pathology analysis. *bioRxiv: The Preprint Server for Biology*, 2023.07.30.551187. https://doi.org/10.1101/2023.07.30.551187.

Boehm, K. M., Khosravi, P., Vanguri, R., Gao, J., & Shah, S. P. (2022). Harnessing multimodal data integration to advance precision oncology. *Nature Reviews. Cancer, 22*(2), 114–126. https://doi.org/10.1038/s41568-021-00408-3.

Bottosso, M., Mosele, F., Michiels, S., Cournède, P.-H., Dogan, S., Labaki, C., et al. (2024). Moving toward precision medicine to predict drug sensitivity in patients with metastatic breast cancer. *ESMO Open, 9*(3), 102247. https://doi.org/10.1016/j.esmoop.2024.102247.

Cai, M., Zhao, L., Hou, G., Zhang, Y., Wu, W., Jia, L., et al. (2023). FDTrans: Frequency domain transformer model for predicting subtypes of lung cancer using multimodal data. *Computers in Biology and Medicine, 158*, 106812. https://doi.org/10.1016/j.compbiomed.2023.106812.

Capobianco, E. (2022). High-dimensional role of AI and machine learning in cancer research. *British Journal of Cancer, 126*(4), 523–532. https://doi.org/10.1038/s41416-021-01689-z.

Cascarano, A., Mur-Petit, J., Hernández-González, J., Camacho, M., de Toro Eadie, N., Gkontra, P., et al. (2023). Machine and deep learning for longitudinal biomedical data: A review of methods and applications. *Artificial Intelligence Review, 56*(2), 1711–1771. https://doi.org/10.1007/s10462-023-10561-w.

Cebula, H., Noel, G., Garnon, J., Todeschi, J., Burckel, H., de Mathelin, M., et al. (2020). The cryo-immunologic effect: A therapeutic advance in the treatment of glioblastomas? *Neurochirurgie, 66*(6), 455–460. https://doi.org/10.1016/j.neuchi.2020.06.135.

Chang, Y., He, F., Wang, J., Chen, S., Li, J., Liu, J., et al. (2022). Define and visualize pathological architectures of human tissues from spatially resolved transcriptomics using deep learning. *Computational and Structural Biotechnology Journal, 20*, 4600–4617. https://doi.org/10.1016/j.csbj.2022.08.029.

Chen, C., Zhang, Z., Tang, P., Liu, X., & Huang, B. (2024). Edge-relational window-attentional graph neural network for gene expression prediction in spatial transcriptomics analysis. *Computers in Biology and Medicine, 174*, 108449. https://doi.org/10.1016/j.compbiomed.2024.108449.

Chen, Q., Li, M., Chen, C., Zhou, P., Lv, X., & Chen, C. (2023). MDFNet: Application of multimodal fusion method based on skin image and clinical data to skin cancer classification. *Journal of Cancer Research and Clinical Oncology, 149*(7), 3287–3299. https://doi.org/10.1007/s00432-022-04180-1.

Cristian, A., Rubens, M., Orada, R., DeVries, K., Syrkin, G., DePiero, M. T., et al. (2024). Development of a cancer rehabilitation dashboard to collect data on physical function in cancer patients and survivors. *American Journal of Physical Medicine & Rehabilitation, 103*(3S Suppl 1), S36–S40. https://doi.org/10.1097/PHM.0000000000002424.

Dabeer, S., Khan, M. M., & Islam, S. (2019). Cancer diagnosis in histopathological image: CNN based approach. *Informatics in Medicine Unlocked, 16*, 100231. https://doi.org/10.1016/j.imu.2019.100231.

Dabral, I., Singh, M., & Kumar, K. (2021). Cancer detection using convolutional neural network. In M. Tripathi, & S. Upadhyaya (Eds.). *Conference Proceedings of ICDLAIR2019* (pp. 290–298)Springer International Publishing. https://doi.org/10.1007/978-3-030-67187-7_30.

Dawood, M., Eastwood, M., Jahanifar, M., Young, L., Ben-Hur, A., Branson, K., et al. (2023). Cross-linking breast tumor transcriptomic states and tissue histology. *Cell Reports. Medicine, 4*(12), 101313. https://doi.org/10.1016/j.xcrm.2023.101313.

Deng, G., Yang, J., Zhang, Q., Xiao, Z.-X., & Cai, H. (2018). MethCNA: A database for integrating genomic and epigenomic data in human cancer. *BMC Genomics, 19*(1), 138. https://doi.org/10.1186/s12864-018-4525-0.

Du, J., Yang, Y.-C., An, Z.-J., Zhang, M.-H., Fu, X.-H., Huang, Z.-F., et al. (2023). Advances in spatial transcriptomics and related data analysis strategies. *Journal of Translational Medicine, 21*, 330. https://doi.org/10.1186/s12967-023-04150-2.

Emmert-Buck, M. R., Bonner, R. F., Smith, P. D., Chuaqui, R. F., Zhuang, Z., Goldstein, S. R., et al. (1996). Laser capture microdissection. *Science (New York, N.Y.), 274*(5289), 998–1001. https://doi.org/10.1126/science.274.5289.998.

Feng, X., Shu, W., Li, M., Li, J., Xu, J., & He, M. (2024). Pathogenomics for accurate diagnosis, treatment, prognosis of oncology: A cutting edge overview. *Journal of Translational Medicine, 22*(1), 131. https://doi.org/10.1186/s12967-024-04915-3.

Fusar-Poli, P., Manchia, M., Koutsouleris, N., Leslie, D., Woopen, C., Calkins, M. E., et al. (2022). Ethical considerations for precision psychiatry: A roadmap for research and clinical practice. *European Neuropsychopharmacology, 63*, 17–34. https://doi.org/10.1016/j.euroneuro.2022.08.001.

Halawani, R., Buchert, M., & Chen, Y.-P. P. (2023). Deep learning exploration of single-cell and spatially resolved cancer transcriptomics to unravel tumour heterogeneity. *Computers in Biology and Medicine, 164*, 107274. https://doi.org/10.1016/j.compbiomed.2023.107274.

Hanczar, B., Bourgeais, V., & Zehraoui, F. (2022). Assessment of deep learning and transfer learning for cancer prediction based on gene expression data. *BMC Bioinformatics, 23*(1), 262. https://doi.org/10.1186/s12859-022-04807-7.

Harikumar, H., Quinn, T. P., Rana, S., Gupta, S., & Venkatesh, S. (2021). Personalized single-cell networks: A framework to predict the response of any gene to any drug for any patient. *BioData Mining, 14*(1), 37. https://doi.org/10.1186/s13040-021-00263-w.

Hu, B., Sajid, M., Lv, R., Liu, L., & Sun, C. (2022). A review of spatial profiling technologies for characterizing the tumor microenvironment in immuno-oncology. *Frontiers in Immunology, 13*, 996721. https://doi.org/10.3389/fimmu.2022.996721.

Jaiswal, H., Agarwal, A., & Jaiswal, V. (2023). Expanding the scope of oncological understanding via multimodal imaging. *Onkologia i Radioterapia, 17*(10), 652–661.

Jia, Y., Liu, J., Chen, L., Zhao, T., & Wang, Y. (2024). THItoGene: A deep learning method for predicting spatial transcriptomics from histological images. *Briefings in Bioinformatics, 25*(1), bbad464. https://doi.org/10.1093/bib/bbad464.

Jiang, X., Hu, Z., Wang, S., & Zhang, Y. (2023). Deep learning for medical image-based cancer diagnosis. *Cancers, 15*(14), 3608. https://doi.org/10.3390/cancers15143608.

Johnson, K. B., Wei, W., Weeraratne, D., Frisse, M. E., Misulis, K., Rhee, K., et al. (2021). Precision medicine, AI, and the future of personalized health care. *Clinical and Translational Science, 14*(1), 86–93. https://doi.org/10.1111/cts.12884.

Joo, S., Ko, E. S., Kwon, S., Jeon, E., Jung, H., Kim, J.-Y., et al. (2021). Multimodal deep learning models for the prediction of pathologic response to neoadjuvant chemotherapy in breast cancer. *Scientific Reports, 11*(1), 18800. https://doi.org/10.1038/s41598-021-98408-8.

Kattau, M., Glocker, B., & Darambara, D. (2021). A comparative analysis of two deep learning architectures for the automatic segmentation of vestibular schwannoma. *IEEE Nuclear Science Symposium and Medical Imaging Conference (NSS/MIC), 2021*, 1–3. https://doi.org/10.1109/NSS/MIC44867.2021.9875733.

Khalighi, S., Reddy, K., Midya, A., Pandav, K. B., Madabhushi, A., & Abedalthagafi, M. (2024). Artificial intelligence in neuro-oncology: Advances and challenges in brain tumor diagnosis, prognosis, and precision treatment. *NPJ Precision Oncology, 8*(1), 1–12. https://doi.org/10.1038/s41698-024-00575-0.

Li, J., Li, L., You, P., Wei, Y., & Xu, B. (2023). Towards artificial intelligence to multi-omics characterization of tumor heterogeneity in esophageal cancer. *Seminars in Cancer Biology, 91*, 35–49. https://doi.org/10.1016/j.semcancer.2023.02.009.

Li, M., Jiang, Y., Zhang, Y., & Zhu, H. (2023). Medical image analysis using deep learning algorithms. *Frontiers in Public Health, 11*, 1273253. https://doi.org/10.3389/fpubh.2023.1273253.

Li, Q. (2023). AI-powered Bayesian statistics in biomedicine. *Statistics in Biosciences, 15*(3), 737–749. https://doi.org/10.1007/s12561-023-09400-x.

Li, W., Liu, B., Wang, W., Sun, C., Che, J., Yuan, X., et al. (2022). Lung cancer stage prediction using multi-omics data. *Computational and Mathematical Methods in Medicine, 2022*, 2279044. https://doi.org/10.1155/2022/2279044.

Liang, X., Young, W. C., Hung, L.-H., Raftery, A. E., & Yeung, K. Y. (2019). Integration of multiple data sources for gene network inference using genetic perturbation data. *Journal of Computational Biology, 26*(10), 1113–1129. https://doi.org/10.1089/cmb.2019.0036.

Lipkova, J., Chen, R. J., Chen, B., Lu, M. Y., Barbieri, M., Shao, D., et al. (2022). Artificial intelligence for multimodal data integration in oncology. *Cancer Cell, 40*(10), 1095–1110. https://doi.org/10.1016/j.ccell.2022.09.012.

Liu, H., Zhang, Y., & Luo, J. (2024). Contrastive learning-based histopathological features infer molecular subtypes and clinical outcomes of breast cancer from unannotated whole slide images. *Computers in Biology and Medicine, 170*, 107997. https://doi.org/10.1016/j. compbiomed.2024.107997.

Liu, Y., Wu, Z., Feng, Y., Gao, J., Wang, B., Lian, C., et al. (2023). Integration analysis of single-cell and spatial transcriptomics reveal the cellular heterogeneity landscape in glioblastoma and establish a polygenic risk model. *Frontiers in Oncology, 13*, 1109037. https://doi.org/10.3389/fonc.2023.1109037.

Lobato-Delgado, B., Priego-Torres, B., & Sanchez-Morillo, D. (2022). Combining molecular, imaging, and clinical data analysis for predicting cancer prognosis. *Cancers, 14*(13), 3215. https://doi.org/10.3390/cancers14133215.

Love, C. S. (2023). "Just the facts ma'am": Moral and ethical considerations for artificial intelligence in medicine and its potential to impact patient autonomy and hope. *The Linacre Quarterly, 90*(4), 375–394. https://doi.org/10.1177/00243639231162431.

Luo, R., & Bocklitz, T. (2023). A systematic study of transfer learning for colorectal cancer detection. *Informatics in Medicine Unlocked, 40*, 101292. https://doi.org/10.1016/j.imu. 2023.101292.

Lyubetskaya, A., Rabe, B., Fisher, A., Lewin, A., Neuhaus, I., Brett, C., et al. (2022). Assessment of spatial transcriptomics for oncology discovery. *Cell Reports Methods, 2*(11), 100340. https://doi.org/10.1016/j.crmeth.2022.100340.

Mayampurath, A., Ramesh, S., Michael, D., Liu, L., Feinberg, N., Granger, M., et al. (2021). Predicting response to chemotherapy in patients with newly diagnosed high-risk neuroblastoma: A report from the International Neuroblastoma Risk Group (Scopus) *JCO Clinical Cancer Informatics, 5*, 1181–1188. https://doi.org/10.1200/CCI.21.00103.

Patton, R. M., Johnston, J. T., Young, S. R., Schuman, C. D., Potok, T. E., Rose, D. C., . & Saltz, J. (2019). Exascale deep learning to accelerate cancer research. In 2019 IEEE International Conference on Big Data (Big Data) (pp. 1488-1496). IEEE. https://doi. org/10.1109/BigData47090.2019.9006467.

Peng, L., He, X., Peng, X., Li, Z., & Zhang, L. (2023a). STGNNks: Identifying cell types in spatial transcriptomics data based on graph neural network, denoising auto-encoder, and k-sums clustering. *Computers in Biology and Medicine, 166*, 107440. https://doi.org/ 10.1016/j.compbiomed.2023.107440.

Peng, Z., Ren, Z., Tong, Z., Zhu, Y., Zhu, Y., & Hu, K. (2023b). Interactions between MFAP5 + fibroblasts and tumor-infiltrating myeloid cells shape the malignant micro-environment of colorectal cancer. *Journal of Translational Medicine, 21*(1), 405. https:// doi.org/10.1186/s12967-023-04281-6.

Pham, T. D. (2023). Prediction of five-year survival rate for rectal cancer using Markov models of convolutional features of RhoB expression on tissue microarray. *IEEE/ACM Transactions on Computational Biology and Bioinformatics, 20*(5), 3195–3204. https://doi. org/10.1109/TCBB.2023.3274211.

Price, K., Westfall, A. D., Beemer, C., Casper, L. H., Dierkes, D., Hickey, E. M., et al. (2022). Building and sustaining a comprehensive pediatric oncology care team: The roles and integration of psychosocial and rehabilitative team members. *Horizons in Cancer Research, 84*, 1–63.

Rahman, M. N., Noman, A. A., Turza, A. M., Abrar, M. A., Samee, M. A. H., & Rahman, M. S. (2023). ScribbleDom: Using scribble-annotated histology images to identify domains in spatial transcriptomics data. *Bioinformatics (Oxford, England), 39*(10), btad594. https://doi.org/10.1093/bioinformatics/btad594.

Rani, S., Ahmad, T., Masood, S., & Saxena, C. (2023). Diagnosis of breast cancer molecular subtypes using machine learning models on unimodal and multimodal datasets. *Neural Computing and Applications, 35*(34), 24109–24121. https://doi.org/10.1007/s00521-023-09005-x.

Rathore, S., Abdulkadir, A., & Davatzikos, C. (2020). Analysis of MRI data in diagnostic neuroradiology. *Annual Review of Biomedical Data Science, 3*(1), 365–390. https://doi.org/10.1146/annurev-biodatasci-022620-015538.

Schmitt, M., Sinnberg, T., Niessner, H., Forschner, A., Garbe, C., Macek, B., et al. (2021). Individualized proteogenomics reveals the mutational landscape of melanoma patients in response to immunotherapy. *Cancers, 13*(21), 5411. https://doi.org/10.3390/cancers13215411.

Shao, J., Ma, J., Zhang, Q., Li, W., & Wang, C. (2023). Predicting gene mutation status via artificial intelligence technologies based on multimodal integration (MMI) to advance precision oncology. *Seminars in Cancer Biology, 91*, 1–15. https://doi.org/10.1016/j.semcancer.2023.02.006.

Sharma, P., Nayak, D. R., Balabantaray, B. K., Tanveer, M., & Nayak, R. (2024). A survey on cancer detection via convolutional neural networks: Current challenges and future directions. *Neural Networks, 169*, 637–659. https://doi.org/10.1016/j.neunet.2023.11.006.

Song, J., Lamstein, J., Ramaswamy, V. G., Webb, M., Zada, G., Finkbeiner, S., et al. (2024). Enhancing spatial transcriptomics analysis by integrating image-aware deep learning methods. *Pacific Symposium on Biocomputing, 29*, 450–463.

Song, T., Markham, K. K., Li, Z., Muller, K. E., Greenham, K., & Kuang, R. (2022). Detecting spatially co-expressed gene clusters with functional coherence by graph-regularized convolutional neural network. *Bioinformatics (Oxford, England), 38*(5), 1344–1352. https://doi.org/10.1093/bioinformatics/btab812.

Ståhl, P. L., Salmén, F., Vickovic, S., Lundmark, A., Navarro, J. F., Magnusson, J., et al. (2016). Visualization and analysis of gene expression in tissue sections by spatial transcriptomics. *Science (New York, N. Y.), 353*(6294), 78–82. https://doi.org/10.1126/science.aaf2403.

Steyaert, S., Pizurica, M., Nagaraj, D., Khandelwal, P., Hernandez-Boussard, T., Gentles, A. J., et al. (2023a). Multimodal data fusion for cancer biomarker discovery with deep learning. *Nature Machine Intelligence, 5*(4), 351–362. https://doi.org/10.1038/s42256-023-00633-5.

Steyaert, S., Qiu, Y. L., Zheng, Y., Mukherjee, P., Vogel, H., & Gevaert, O. (2023b). Multimodal deep learning to predict prognosis in adult and pediatric brain tumors. *Communications Medicine, 3*(1), 1–15. https://doi.org/10.1038/s43856-023-00276-y.

Sundaramurthy, A., & Vaithiyalingam, C. (2023). Ethical dimensions and future prospects of artificial intelligence in decision making systems for oncology: A comprehensive analysis and reference scheme. In: *2023 International conference on intelligent technologies for sustainable electric and communications systems (iTech SECOM)* (pp. 59–63). https://doi.org/10.1109/iTechSECOM59882.2023.10435323.

Tortora, M., Cordelli, E., Sicilia, R., Nibid, L., Ippolito, E., Perrone, G., et al. (2023). RadioPathomics: Multimodal learning in non-small cell lung cancer for adaptive radiotherapy. *IEEE Access, 11*, 47563–47578. https://doi.org/10.1109/ACCESS.2023.3275126.

Vaswani, A., Shazeer, N., Parmar, N., Uszkoreit, J., Jones, L., Gomez, A.N., & Polosukhin, I. (2023). Attention is all you need. *arXiv preprint*, arXiv:1706.03762. https://doi.org/10.48550/arXiv.1706.03762.

Wang, J., Li, B., Luo, M., Huang, J., Zhang, K., Zheng, S., et al. (2024). Progression from ductal carcinoma in situ to invasive breast cancer: Molecular features and clinical significance. *Signal Transduction and Targeted Therapy, 9*(1), 1–28. https://doi.org/10.1038/s41392-024-01779-3.

Wang, S., Zhou, X., Kong, Y., & Lu, H. (2023). Superresolved spatial transcriptomics transferred from a histological context. *Applied Intelligence, 53*(24), 31033–31045. https://doi.org/10.1007/s10489-023-05190-3.

Wei, Z., Zhang, Y., Weng, W., Chen, J., & Cai, H. (2021). Survey and comparative assessments of computational multi-omics integrative methods with multiple regulatory networks identifying distinct tumor compositions across pan-cancer data sets. *Briefings in Bioinformatics, 22*(3), bbaa102. https://doi.org/10.1093/bib/bbaa102.

Williams, C. G., Lee, H. J., Asatsuma, T., Vento-Tormo, R., & Haque, A. (2022). An introduction to spatial transcriptomics for biomedical research. *Genome Medicine, 14*(1), 68. https://doi.org/10.1186/s13073-022-01075-1.

Wilson, C. M., Li, K., Yu, X., Kuan, P.-F., & Wang, X. (2019). Multiple-kernel learning for genomic data mining and prediction. *BMC Bioinformatics, 20*(1), 426. https://doi.org/10.1186/s12859-019-2992-1.

Wu, Z., Trevino, A. E., Wu, E., Swanson, K., Kim, H. J., D'Angio, H. B., et al. (2022). Graph deep learning for the characterization of tumour microenvironments from spatial protein profiles in tissue specimens. *Nature Biomedical Engineering, 6*(12), 1435–1448. https://doi.org/10.1038/s41551-022-00951-w.

Xi, X., Li, W., Li, B., Li, D., Tian, C., & Zhang, G. (2022). Modality-correlation embedding model for breast tumor diagnosis with mammography and ultrasound images. *Computers in Biology and Medicine, 150*, 106130. https://doi.org/10.1016/j.compbiomed.2022.106130.

Xiao, X., Kong, Y., Li, R., Wang, Z., & Lu, H. (2024). Transformer with convolution and graph-node co-embedding: An accurate and interpretable vision backbone for predicting gene expressions from local histopathological image. *Medical Image Analysis, 91*, 103040. https://doi.org/10.1016/j.media.2023.103040.

Xu, C., Jin, X., Wei, S., Wang, P., Luo, M., Xu, Z., et al. (2022). DeepST: Identifying spatial domains in spatial transcriptomics by deep learning. *Nucleic Acids Research, 50*(22), e131. https://doi.org/10.1093/nar/gkac901.

Yan, R., Ren, F., Rao, X., Shi, B., Xiang, M., ... Zhang, L., et al. (2019). *Integration of multimodal data for breast cancer classification using a hybrid deep learning method. Intelligent computing theories and application. ICIC 2019. Lecture notes in computer science.* Cham: Springer460–469. https://doi.org/10.1007/978-3-030-26763-6_44.

Yu, X., Liu, R., Gao, W., Wang, X., & Zhang, Y. (2023). Single-cell omics traces the heterogeneity of prostate cancer cells and the tumor microenvironment. *Cellular & Molecular Biology Letters, 28*(1), 38. https://doi.org/10.1186/s11658-023-00450-z.

Yuan, Y., & Bar-Joseph, Z. (2020). GCNG: Graph convolutional networks for inferring gene interaction from spatial transcriptomics data. *Genome Biology, 21*(1), 300. https://doi.org/10.1186/s13059-020-02214-w.

Zhang, N., Yang, X., Piao, M., Xun, Z., Wang, Y., Ning, C., et al. (2024a). Biomarkers and prognostic factors of PD-1/PD-L1 inhibitor-based therapy in patients with advanced hepatocellular carcinoma. *Biomarker Research, 12*(1), 26. https://doi.org/10.1186/s40364-023-00535-z.

Zhang, P., Gao, C., Huang, Y., Chen, X., Pan, Z., Wang, L., et al. (2024b). Artificial intelligence in liver imaging: Methods and applications. *Hepatology International, 18*(2), 422–434. https://doi.org/10.1007/s12072-023-10630-w.

Zhen, X., Chen, J., Zhong, Z., Hrycushko, B., Zhou, L., Jiang, S., et al. (2017). Deep convolutional neural network with transfer learning for rectum toxicity prediction in cervical cancer radiotherapy: A feasibility study. *Physics in Medicine and Biology, 62*(21), 8246–8263. https://doi.org/10.1088/1361-6560/aa8d09.

Zheng, Y., Carrillo-Perez, F., Pizurica, M., Heiland, D. H., & Gevaert, O. (2023). Spatial cellular architecture predicts prognosis in glioblastoma. *Nature Communications, 14*(1), 4122. https://doi.org/10.1038/s41467-023-39933-0.

CHAPTER TWO

Data enhancement in the age of spatial biology

Linbu Liao[a,b,1], Patrick C.N. Martin[c,1], Hyobin Kim[c], Sanaz Panahandeh[c], and Kyoung Jae Won[c,*]

[a]Biotech Research and Innovation Centre (BRIC), University of Copenhagen, Denmark
[b]Samuel Oschin Cancer Center, Cedars-Sinai Medical Center, Los Angeles, CA, United States
[c]Department of Computational Biomedicine, Cedars-Sinai Medical Center, Los Angeles, CA, United States
*Corresponding author. e-mail address: kyoungjae.won@cshs.org

Contents

1. Introduction	40
2. Data enhancement by integrating other sequencing modalities for ST data	43
2.1 Deconvolution sequencing-based ST data using reference scRNAseq	43
2.2 Data imputation for missing data	47
2.3 Mapping single cells to spatial data for Visium data enhancement	48
2.4 Spatial alignment and integrated analysis	50
2.5 Multi-modality of spatial data	53
3. Data enhancement by integrating tissue image data for ST data	55
3.1 Cell segmentation boosted by integrating image data	55
3.2 Data enhancement by integrating image data	56
3.3 Gene prediction from tissue image data	57
4. Data enhancement by synthetic data	58
5. Perspectives	58
Acknowledgments	62
Disclosures	62
References	62

Abstract

Unveiling the intricate interplay of cells in their native environment lies at the heart of understanding fundamental biological processes and unraveling disease mechanisms, particularly in complex diseases like cancer. Spatial transcriptomics (ST) offers a revolutionary lens into the spatial organization of gene expression within tissues, empowering researchers to study both cell heterogeneity and microenvironments in health and disease. However, current ST technologies often face limitations in either resolution or the number of genes profiled simultaneously. Integrating ST data with complementary sources, such as single-cell transcriptomics and detailed tissue staining images, presents a powerful solution to overcome these limitations.

[1] These authors contributed equally.

Advances in Cancer Research, Volume 163
ISSN 0065-230X, https://doi.org/10.1016/bs.acr.2024.06.008
Copyright © 2024 Elsevier Inc. All rights are reserved, including those for text and data mining, AI training, and similar technologies.

39

This review delves into the computational approaches driving the integration of spatial transcriptomics with other data types. By illuminating the key challenges and outlining the current algorithmic solutions, we aim to highlight the immense potential of these methods to revolutionize our understanding of cancer biology.

1. Introduction

The development of refined optical instruments in the 17th century and the curiosity of a British scientist—Robert Hooke—led to the discovery of highly organized structures in a thin slice of cork. He named these structures cells from the Latin *cellula* signifying a *hollow space*. A decade later, with further optical improvements, Anton van Leeuwenhoek discovered unicellular organisms in a drop of rainwater. Despite these early discoveries, it took nearly two centuries for a *cellular theory* to emerge from the minds of Theodor Schwann, Matthias Jakob Schleiden, and Rudolf Virchow in the mid-19th century. Through careful observation of plants and animal matter, they posited that all living organisms are composed of one or more cells and that cells are the basic unit of life (Mukherjee, 2022).

Since then, the scientific community has studied the intricacies of cellular organization within multicellular organisms. A key concept taken from decades of research is that cells are not placed at random within a tissue; their identity and function are affected by their location (Kim, Lövkvist, et al., 2023; Martin, Kim, Lövkvist, Hong, & Won, 2022). Cells continuously communicate with each other and their environment. While the matter is still subject to intense research, it is generally accepted that molecular signals permeate developing tissues. The signals induce changes in cells, driving them to differentiate, resulting in the formation of complex tissues (Arnosti, Barolo, Levine, & Small, 1996; Joshi, Sipani, & Bakshi, 2022; Small, Blair, & Levine, 1992). Prompted by changes in gene expression, the precise spatial distribution of these signals is essential to coordinate proper development. During early embryonic development in *Drosophila,* the expression of a small set of genes is responsible for the dorso-lateral patterning of fly embryos (Joshi et al., 2022; Petkova, Tkačik, Bialek, Wieschaus, & Gregor, 2019). In fact, disruption in these expression patterns will lead to abnormal and even Lovecraftian morphologies with eyes growing in place of wings or antennae (Brand, Manoukian, & Perrimon, 1994; Halder, Callaerts, & Gehring, 1995;

Phelps & Brand, 1998). Interestingly, spatial geometry and mechanical constraints can also induce cellular differentiation (Gjorevski et al., 2022; Karzbrun, Kshirsagar, Cohen, Hanna, & Reiner, 2018).

The importance of spatial patterning and cellular distribution is also observed in cases where disease disrupts homeostasis. Earlier investigations have established that tumor microenvironment is closely linked with the growth and invasion of tumor cells (Anderson & Simon, 2020; Clark & Vignjevic, 2015; Kiemen et al., 2020; Ravi et al., 2021; Thrane, Eriksson, Maaskola, Hansson, & Lundeberg, 2018). Knowledge of the spatial distribution of cells is essential to comprehensively investigate the cellular microenvironment and gain insights into the mechanisms underlying cancer progression (Janesick et al., 2023; Li et al., 2022).

The emergence of multicellular tissue structures and how cells communicate with each other at varying spatial scales remains mysterious, especially when it comes to the genetic mechanism at play. In the early days of molecular biology, the product of single genes was marked using fluorescence protein tags such as green fluorescent protein (GFP) (Tsien, 1998). Alongside fluorescence microscopy, these tools enabled the investigation of cell-cell interactions (Bitgood & McMahon, 1995) and their spatial distribution in tissue sections (Richardson et al., 2004). However, observing a single gene at a time is insufficient for providing a holistic view of a cell in a tissue.

Recent advances in sequencing protocols provide highly sensitive methods to capture and measure the expression of genes at a single-cell resolution (Cao et al., 2017; Fan, Fu, & Fodor, 2015; F. Tang et al., 2009). Studies powered by single-cell RNA sequencing (scRNAseq) found that cells are very heterogeneous and many cell types were found not to have monolithic expression profiles but could be discerned into sub-cell types characterized by subtle changes in the expression of key genes (Fischer & Gillis, 2021; Franzén, Gan, & Björkegren, 2019; Kim, Tam, & Yang, 2021; Morris, 2019; Regev et al., 2017). However, scRNAseq cannot effectively explain the reason for heterogeneity and how cells change their expression according to their microenvironment. The advent of spatial transcriptomic (ST) technologies attempts to remedy this by measuring the transcriptome of a cell while retaining it's spatial location within a tissue. This powerful approach has highlighted how the transcriptome of cells is indeed modulated by spatial location and local cellular neighborhood (Kim, Lövkvist, et al., 2023; Martin et al., 2022).

The technology for spatial biology is still evolving in terms of coverage and resolution to provide a more comprehensive view of cells in a tissue.

Spatial technology can be separated into two main flavors: image-based technologies, that rely on pre-selected gene panels, and sequencing-based technologies that capture mRNA transcripts in an unbiased manner (Moses & Pachter, 2022; Rao, Barkley, França, & Yanai, 2021; Tian, Chen, & Macosko, 2023). Image-based technologies, including SeqFISH (Shah, Lubeck, Zhou, & Cai, 2017) or MERFISH (Xia, Fan, Emanuel, Hao, & Zhuang, 2019), excel at capturing transcriptome at a cellular—even sub-cellular—resolution with great sensitivity (Chen, Boettiger, Moffitt, Wang, & Zhuang, 2015; Eng et al., 2019; Raj, Van Den Bogaard, Rifkin, Van Oudenaarden, & Tyagi, 2008; Shah, Lubeck, Zhou, & Cai, 2016). For instance, MERFISH enhances gene detection capability by integrating a sequence of fluorescence signals, increasing the number of detectable genes in a single run, and improving error tolerance. Initially, MERFISH had the capability to detect approximately 1000 genes (Biancalani et al., 2020; Raj et al., 2008), which was subsequently expanded to 10,000 genes by adding more encoding options (Lubeck & Cai, 2012). Sequencing-based technologies allow for an unbiased exploration of transcriptional landscapes as these methods can capture over 20,000 genes simultaneously (Chen et al., 2021; Cho et al., 2021; Rodriques et al., 2019; Ståhl et al., 2016; Stickels et al., 2020; Stickels et al., 2021; Vickovic et al., 2019). Visium, as the first and most widely commercialized sequence-based ST technique, offers an affordable price with the compromise of resolution and sensitivity (Ståhl et al., 2016). A Visium spot is 55 μm in diameter and is spaced 110 μm apart from its neighboring spots. The transcriptomes from multiple cells can be captured by a Visium spot. The efficiency of sequencing and the number of detectable molecules can vary depending on the species, tissue type, etc. SlideSeq (Stickels et al., 2021) and DBiT-seq (Liu et al., 2020) profile transcriptome at a near cellular resolution with a diameter of 10 μm. Other sequencing-based approaches such as Stereo-seq (Chen et al., 2021) and Seq-Scope (Cho et al., 2021) offer transcriptomic profiles even at a sub-cellular resolution, providing insights into more intricate cellular structures.

Despite the undeniable insights gained through ST, our understanding of cellular organization is hindered by certain technical limitations. Indeed, image-based technologies are limited by the breadth of genes captured simultaneously. The use of fluorescence probes limits the number of unique genes that can be measured simultaneously due to concerns with overlapping fluorescent signals (Zhuang, 2021). The task of assigning measured transcripts to their respective cells relies on the ability of cell segmentation algorithms to accurately distinguish cell boundaries, a feat that still remains challenging (Haoran Chen & Murphy, 2023). Sequencing-based technologies may be better suited to capture subtle changes in the transcriptional landscape of cells; however, these techniques cannot guarantee cellular resolution measurement of gene transcripts. Popular techniques such as Visium (Ståhl et al., 2016) capture the expression of multiple cells within a single spot, which becomes akin to a spatial mini-bulk rather than a single-cell measurement. High resolution techniques such as Slide-seq (Stickels et al., 2020, 2021) or even ultra-high resolution methods such as Stereo-seq (Chen et al., 2021) cannot promise to capture transcripts from only one cell. Conversely, for large cells, transcripts will need to be combined to better represent the transcriptome of a single cell.

Computational advancements are addressing limitations in ST by integrating other data modalities like scRNAseq and tissue staining images. Certain steps have become standard for analyzing ST data. For example, high-resolution sequencing-based ST often requires cell segmentation to define cell boundaries for accurate RNA quantification. Conversely, Visium data analysis often relies on scRNAseq-assisted spatial deconvolution. This review explores advancements in machine learning and artificial intelligence that address these limitations by enhancing spatial data through various computational strategies.

2. Data enhancement by integrating other sequencing modalities for ST data

ST data integration extends beyond gene expression, offering powerful insights when combined with various sequencing modalities. The abundance of public scRNAseq data provides a rich resource for integration. Additionally, integrating ST data with spatial proteomics, spatial epigenomics, or even other ST datasets enhances its knowledge base. This integration unlocks diverse functionalities, including data deconvolution, imputation, single-cell mapping via location inference, and spatial data alignment. (Fig. 1).

2.1 Deconvolution sequencing-based ST data using reference scRNAseq

Deconvolution is the process of separating a mixed signal or dataset into its components. While often unnecessary for image-based ST data (where measurements are taken directly at cellular resolution), it is crucial for

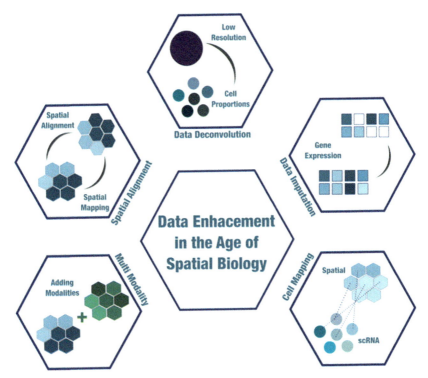

Fig. 1 ST data enhancement through the integration of other sequencing-based modalities. ST data can be enhanced through various techniques. (1) Cell type deconvolution: This approach utilizes single-cell RNA sequencing (scRNAseq) data to identify the specific cell types and their portions present within the ST data. (2) Data imputation: Missing data points within the ST data are estimated and replaced using appropriate strategies, leading to improved spatial resolution. (3) Cell mapping: This technique also aims to enhance spatial resolution by integrating external scRNAseq with the ST data. (4) Spatial alignment: Aligning other spatial transcriptomics data ST data comparison. (5) Adding another modality.

sequencing-based ST assays like Visium (Ståhl et al., 2016). SlideSeq (Stickels et al., 2021), which offers near-cellular resolution, still requires deconvolution to isolate individual cell transcriptomes accurately.

Methods such as RCTD (Cable et al., 2022), Cell2location (Kleshchevnikov et al., 2022), Stereoscope (Andersson et al., 2020), DestVI (Lopez et al., 2022), SpatialDecon (Danaher et al., 2022), and STRIDE (Sun, Liu, Li, Wu, & Wang, 2022) are based on probabilistic models. RCTD obtains the mean expression of marker genes for each cell type from a scRNAseq reference (Cable et al., 2022). Given the reference cell type expression and observed gene counts in ST data, RCTD uses maximum likelihood estimation (MLE) to estimate cell type proportions. Cell2location computes cell type-specific expression profiles from a scRNAseq reference through negative binomial regression (Kleshchevnikov et al., 2022). Based on the reference expression, Cell2location uses variational inference to predict the number of cells for each cell type within individual spots. Stereoscope assumes that both scRNAseq reference and spatial transcriptomics data follow negative binomial distributions (Andersson et al., 2020). To estimate the distribution of each cell type in the single cell reference data, Stereoscope performs MLE. Then, it infers cell type proportions through maximum a posteriori (MAP) estimation. DestVI leverages variational autoencoder (VAE) to obtain two latent variable models (Lopez et al., 2022). A VAE is an artificial neural network composed of encoding and decoding components similar to an autoencoder (Kingma & Welling, 2019). Encoding components map the input variable to a latent space with reduced dimension, and the decoding components have an opposite function. Unlike standard autoencoders, VAEs do not simply output a single point in the latent space. Instead, they generate a probability distribution over the space. This allows the model to capture the natural variability inherent in data, enabling diverse outputs and enriching its understanding. The first model for DestVI is trained on a scRNAseq reference. The same decoder from the first model is maintained to fit the second model. The second one is a spatial latent variable model, which estimates cell type proportions using MAP.

SpatialDecon was designed to correct the skewness and unequal variance of gene expression data by modeling expression variability on the log-scale (Danaher et al., 2022). SpatialDecon performs log-normal regression, and the least squares approach to estimate absolute cell counts in each spot. STRIDE deconvolves cell types in each spot by using Latent Dirichlet Allocation (LDA), which is a generative probabilistic approach for topic modeling (Sun et al., 2022). STRIDE obtains cell type-associated topics from a scRNAseq reference. Then, the pre-trained topic model is applied to predict the topic distributions at each spatial location. Combining the topic-by-location distribution and the cell type-associated topic distribution, STRIDE estimates cell type proportions in each spot.

Methods such as SPOTlight (Elosua-Bayes, Nieto, Mereu, Gut, & Heyn, 2021), SpatialDWLS (Dong & Yuan, 2021), NMFreg (Rodriques et al., 2019), and CARD (Ma & Zhou, 2022) employ non-negative matrix factorization (NMF). SPOTlight integrates spatial transcriptomics data and scRNAseq reference data (Elosua-Bayes et al., 2021). Subsequently,

SPOTlight uses seeded non-negative matrix factorization (NMF) regression and non-negative least squares (NNLS) to infer the cell types present at each spatial location within the tissue. SpatialDWLS leverages cell type signature genes obtained from a scRNAseq reference (Dong & Yuan, 2021). Based on the genes, SpatialDWLS identifies enriched cell types in each spot and applies the dampened weighted least squares (DWLS) approach to estimate cell type composition. NMFreg uses NMF to extract metagenes that capture underlying cell types from a scRNAseq reference (Rodriques et al., 2019). After the gene signatures for individual cell types are acquired, non-negative least squares (NNLS) regression is performed to map scRNAseq cell types onto Slide-seq data. CARD conducts deconvolution of spatial transcriptomics data based on NMF (Ma & Zhou, 2022). CARD actively utilizes spatial correlation in cell type composition across spots. It combines the spatial correlation and cell type-specific expression information from a scRNAseq reference to decompose cell types.

Tools such as DSTG (Song & Su, 2021) and Tangram (Biancalani et al., 2021) are deep learning-based methods. Deep learning is a branch of machine learning that draws inspiration from the structure and function of the human brain (Svozil, Kvasnicka, & Pospichal, 1997). It leverages a specific type of algorithm called an artificial neural network, which mimics the interconnected network of neurons in our brains. These artificial neural networks are comprised of multiple layers, each acting as a building block in the learning process. Just like the human brain progressively extracts and refines information as it travels through different regions, each layer in a deep learning network performs increasingly intricate feature extraction from the previous layer. This layered approach allows deep learning models to learn complex representations of data, capturing intricate patterns and relationships that would be difficult or impossible to identify with simpler methods. Deep learning has been used in various areas, including image recognition, natural language processing, and prediction models.

DSTG uses a semi-supervised graph convolutional network (GCN) (Song & Su, 2021). It generates pseudo-ST spots representing cell mixtures from a scRNAseq reference. Through the canonical correlation analysis (CCA), the pseudo-ST and real-ST data are projected to a common latent space. In the low-dimension space, k-nearest neighbor (KNN) is performed to identify mutual nearest neighbors between the two data and construct a link graph. Using the link graph as input, DSTG learns a latent representation from the already known pseudo-ST spots and predicts unknown cell type proportions in each spot of real ST data. Tangram aligns scRNAseq or snRNA-seq data to different spatial transcriptomics data based on non-convex optimization (Biancalani et al., 2021). The optimization function was designed to maximize cell density and gene expression similarities between the two datasets. Mapping single cells to each spatial spot allows the prediction of cell type composition.

scRNAseq reference-free methods offer an alternative for decomposing cell types when a suitable reference is unavailable. As a probabilistic-based method, STdeconvolve predicts a transcriptional profile for each cell type and their proportions without scRNAseq references (Miller, Huang, Atta, Sahoo, & Fan, 2022). STdeconvolve selects genes with higher-than-expected expression variance across spatial spots or significantly overdispersed genes. For LDA analysis, it guides the estimation of the optimal number of topics indicating cell types. After applying LDA modeling with the appropriate number of cell types, STdeconvole uses a variational expectation-maximization approach to infer cell type-specific transcriptional profiles and cell type proportions for the selected genes in each spot.

2.2 Data imputation for missing data

Data imputation is a technique used to address the issue of missing data. Data imputation is useful for both image-based and sequence-based ST technologies. Image-based technologies cannot capture the entire spectrum of gene expression changes induced by spatial positioning, even though their capturing sensitivity is very high. Conversely, sequencing-based technologies attempt to cover the entire transcriptome; however, their lower transcript capture sensitivity may miss lowly expressed genes. A proposed solution to this issue is to impute missing gene expression.

For image-based ST data, a common approach is to enhance sparse expression profiles by associating them to more complete single cell expression profiles. For instance, SpaGE integrates spatial data with single-cell data to find the single cell that best matches a spatial index (Abdelaal, Mourragui, Mahfouz, & Reinders, 2020). Tools such as gimVI use deep generative models to integrate single cells with spatial data and, by extension, impute gene expression (Lopez et al., 2019). ADEPT learns a latent representation of spatial data sets using a graph autoencoder (Hu et al., 2023). This learned latent space can be used to impute gene expression in similar samples. More recently, TransformerST uses a neural network architecture known as a transformer—commonly used in Large Language Models such as chatGPT—to learn a representation of ST data and generate imputed count matrices (Vaswani et al., 2017; Zhao et al., 2022). A more in-depth overview of mapping single cells to spatial data will be covered in the next section.

Data imputation can also be used to account for the lower spatial resolution of ST technologies such as Visium. BayesSpace utilizes spatial neighborhoods of spots to enhance the resolution to sub-spot resolution (Zhao et al., 2021). To achieve this feat, BayesSpace segments each spot into sub-spots and estimates the likelihood of a sub-spot belonging to the initial spot or to the neighboring spot using a Bayesian model and a Potts model as a spatial prior. TransformerST—the graph transformer model mentioned above – is also able to increase the resolution of spatial data to sub-spot levels using its learned latent space (Zhao et al., 2022). BayesSpace has been used for enhanced resolution clustering for melanoma samples and invasive ductal carcinoma (IDC) to provide spatial expression for tumor cells and other stromal cells (Zhao et al., 2021).

2.3 Mapping single cells to spatial data for Visium data enhancement

Initially, mapping single cells back to their spatial origin was developed to address the lack of spatial context in scRNAseq data. This technique allows researchers to analyze the gene expression profiles of individual cells within the tissue structure, providing valuable insights into cell–cell interactions and local microenvironments.

However, the principle of mapping single cells to their spatial origin has found a second application in enhancing the resolution of Visium data. Unlike scRNAseq, Visium data offers spatial information but at a lower resolution. By leveraging scRNAseq data and its detailed cell identity information, researchers can map these individual cells onto the coarser Visium map, effectively increasing the resolution and enabling the analysis of gene expression at a finer scale within the tissue.

The concept of mapping single cells to ST references was first presented by Satija et al. (Satija, Farrell, Gennert, Schier, & Regev, 2015) and Achim et al. (Achim et al., 2015). Satija et al. initially reconstructed spatial probabilistic expression profiles from a binary in situ hybridization (ISH) dataset, and subsequently aligned genes from scRNAseq data. Achim et al. positions single cells within spatial contexts by aligning their binary gene expression patterns with those in a reference in situ hybridization (ISH) dataset. However, the binary representation of these approaches could limit precise mapping of scRNAseq data.

NovoSpaRc (Moriel et al., 2021; Nitzan, Karaiskos, Friedman, & Rajewsky, 2019) was developed under the hypothesis that cells with similar gene expression profiles are likely to be physically proximate. When no training set is available, NovoSpaRc probabilistically maps single cells onto a target space and infer gene expression patterns across it. When a reference atlas is available, NovoSpaRc calculates a cost matrix that reflects the discrepancy between the gene expression of cells and each location within the atlas.

The algorithm computes three critical cost matrices corresponding to structural correspondence (associating gene expression similarity with physical proximity), reference atlas adherence, and entropy regularization. The first two cost matrices help interpolate between minimizing the deviation from the structural correspondence assumption and any available reference spatial transcriptomics data. Entropy regularization ensures a dispersal mapping of cells across the target space. NovoSpaRc maps cells onto the target space while respecting both gene expression similarities and spatial constraints. After mapping cells to locations, NovoSpaRc predicts the spatial gene expression across the tissue. This involves using the transport matrix to assign a probability distribution of tissue locations to each cell, based on which the overall gene expression profile for the tissue is computed.

SpaOTsc employs three distinct dissimilarity matrices that measure gene expression level differences between single cells, spatial voxels, and their pairwise combinations as the cost of optimal transport (OT) (Cang & Nie, 2020). OT (Peyré & Cuturi, 2019; Thorpe, 2019) is a mathematical framework designed to find the most efficient way to redistribute resources from one configuration to another while minimizing the cost of transport. This concept, originating from the 18th century by Gaspard Monge and later extended by Leonid Kantorovich, has evolved into a powerful tool in various fields, including economics (Galichon, 2016), geography (Métivier, Brossier, Mérigot, Oudet, & Virieux, 2016), and more recently, computational biology (Cang & Nie, 2020; Moriel et al., 2021; Nitzan et al., 2019). By minimizing the dedicated cost, the OT algorithm efficiently aligns single cells with spatial locations to reflect the true cellular organization and spatial expression patterns in a biological tissue. Tangram also aligns scRNAseq or snRNA-seq data to different spatial transcriptomics data based on non-convex optimization (Biancalani et al., 2021).

Deep learning has been widely used for mapping cells. DEEPsc (Maseda, Cang, & Nie, 2021) employs a two-layer fully-connected neural network (FCNN) for mapping scRNAseq data to spatial transcriptomics data. It leverages low-dimensional feature vectors to represent gene expression of single cells and spatial voxels. Based on the vectors of reduced dimensions, it predicts the likelihood of a cell's origin from specific spatial positions. While DEEPsc (Maseda et al., 2021) benefits from its deep learning model, it is similar to older methods in terms of comparing the gene expressions of cells with those in spatial spots to figure out where cells are located. Pairwise comparison between single cells and spatial spots can lead to exponential increase in computational complexity when the input data scale up. An alternative method to predict cell location is to connect gene expression with spatial coordinates.

CeLEry predicts spatial coordinates from gene expression levels to restore the spatial locations of single cells (Zhang et al., 2023). An FCNN of three layers is adopted for this task. Instead of using the Euclidean distance between the predicted and true coordinates as the loss function, CeLEry estimates an uncertain elliptical region with four outputs: two indicating the ellipse's center and two representing its radii. Additionally, CeLEry (Zhang et al., 2023) uses a VAE to simulate ST data. Optionally, the simulated ST data can be served as additional training data to increase the robustness of the location prediction model. However, simulating ST data using a VAE is computationally expensive, and the effectiveness of the augmentation module in increasing the robustness of location prediction is yet to be proven.

SC2Spa addresses this challenge by using a more complex FCNN that includes more layers (Linbu et al., 2023). Initially configured with six layers, SC2Spa's architecture allows for expansion to accommodate even more layers for enhanced learning capabilities. Also, directly mapping gene expression levels to spatial coordinates is efficient in handling large datasets, avoiding the computational complexity that typically increases with expanding scRNAseq and ST data. The two features combined grant SC2Spa the capability of learning from large-scale ST data. The trained SC2Spa model can further extract spatial information from scRNAseq data. SC2Spa uses L1 regularization to penalize an overcomplicated model. L1 regularization can force a deep learning model to focus on important features, preventing the model from disturbances caused by less important genes regarding spatial locations.

2.4 Spatial alignment and integrated analysis

A key step in ST experimental protocol is the preservation of tissue biopsies either via flash-freezing of the sample or by preservation in a parafilm solution. Thin tissue slices (approximately 10 µm) are cut from the original biopsy before being prepared for sequencing. This approach allows for the sectioning and analysis of adjacent tissue slices. Interestingly, a recent study suggests that 3D stacking of spatial data is required to accurately assess tissue composition between healthy and diseased tissue. They demonstrate that histological sections may decay within a span of microns. As such, a single tissue section will not be representative of the neighboring tissue (Forjaz et al., 2023). Furthermore, with the increasing popularity of spatial transcriptomics to study disease, comparing and aligning spatial assays will become extremely relevant. Thankfully, computational tools have been developed to integrate and spatially align these slices into a single cohesive data set.

One of the earliest attempts to align spatial transcriptomic datasets was provided by STutility (Bergenstråhle, Larsson, & Lundeberg, 2020). The alignment procedure relied on the conversion of the tissue area into an image mask that was then used as input to image registration algorithms. This approach is limited to tissue sections that share similar overall shapes as it does not account for inter-sample similarities or differences. A similar approach was implemented in stLearn with the difference that stLearn utilized the H& E images feature to provide improved alignment (Pham et al., 2020). More recently, STAlign leverages diffeomorphic metric mapping to align spatial transcriptomic data sets (Clifton et al., 2023). In this instance, the alignment of data sets to a common coordinate system is achieved by using landmark points and normalized gene expression. The target is then distorted to map to the reference. While these attempts have relied on utilizing methods taken from computer vision and image processing, other approaches have used probabilistic methods to align spatial assays.

GPSA (Gaussian Process Spatial Alignment) attempts to map spatial samples with a two-layer gaussian process (Jones, Townes, Li, & Engelhardt, 2023). The first step maps the layers into a common coordinate system, while the second step matches the expression within the common coordinate system. PASTE leverages optimal transport to align slices by considering both the transcriptional similarity and the spatial position (Zeira, Land, Strzalkowski, & Raphael, 2022). This approach allows for the alignment of adjacent spatial assays into a 3D stack or in the creation of a single consensus slice.

The question of spatial alignment has also been tackled as a data integration and co-clustering problem. For instance, STARGATE creates a spatial neighbor network from spatial coordinates, which can be pruned using cell type labels (Dong & Zhang, 2022). This spatial network is used as an attention layer in a graph attention auto-encoder, which attempts to integrate gene expression profiles with spatial information. This approach was originally designed for spatial domain detection, but the authors extended its use case to 3D alignment by considering multiple 2D spatial neighborhood networks simultaneously. STAligner also leverages a graph attention auto-encoder but uses Mutual Nearest Neighbors to pair spatial indices across samples (Zhou, Dong, & Zhang, 2023). In continuing with the graph auto-encoder approaches, STACI enhances this architecture with the addition of a convolution module that integrates chromatin images to predict and reconstruct gene expression (Zhang, Wang, Shivashankar, & Uhler, 2022).

Another popular approach in spatial domain detection and, by extension adjacent slice integration is the use of graph-based neural networks to create embeddings that account for spatial proximity and gene expression similarities. SPACEL trains a graph neural network to learn a common latent space between adjacent slices and predicts spatial coordinates of spots (Xu et al., 2023). SPACEL then builds a Mutual Nearest Neighbors graph between predicted coordinates and optimizes the overlap between slices. This approach also accounts for potential partial overlaps between slices by explicitly penalizing differences in the proportion of overlapping spots. SpaGCN highlighted the analysis of two consecutive sections of brain tissue using their graph convolution neural network to learn a common embedding between tissue sections; however, these sections are adjacent in the x/y plane rather than in the z/depth plane (Hu et al., 2021). Learning of a common latent space between these slices supposes that vertical alignment of spatial assay would also be feasible.

The co-embedding ethos was also applied in the development of GraphST which employs a graph self-supervised contrastive learning method (Long et al., 2023). GraphST is composed of three graph self-supervised contrastive learning modules each aiming at accomplishing a specific task: spatial clustering, horizontal integration, and vertical integration. GraphST produces a cohesive latent space across samples that can be used for spatial domain detection. It should be noted that the initial alignment of tissue sections in the vertical integration module is performed by the PASTE algorithm. Also leveraging graphs, SPIRAL employs graph-based domain adaptation to remove batch effects and a cluster-aware optimal transport framework for mapping of spots (Guo et al., 2023).

Methods such as PRECAST lean on spatial factor analysis to compute shared distribution between cells or spatial domains (Liu et al., 2023). Concomitantly, the method computes a spatially aware dimensionality reduction step with a conditional autoregressive component used to define cross sample spatial clusters. This method does not explicitly align adjacent slices but assumes that the alignment is not a requirement for integrated analysis.

In many cases, the question of spatial alignment will not apply, at least not directly. Indeed, tissue sections taken from different patients at different time points or under different conditions are unlikely to have matching tissue structures that can be easily aligned. We would face a similar challenge when attempting to align spatial assays from different developmental stages, as is the case with the STOmics data collection in developing mice (MOSTA) (Chen et al., 2022). An alternative approach is mapping cells at a single cell level between data sets while accounting for spatial context. The Spatial Linked Alignment Tool (SLAT) reformulates the slice alignment task as a graph-matching problem (Xia, Cao, Tu, & Gao, 2023). First, the data sets are decomposed into a common lower dimensional space with a Singular Value Decomposition strategy. The latent space embeddings will serve as node features in their multi-layer graph convolutional networks, which will aggregate information across scale to provide a holistic view of the spatial assay from single cells to large scale tissue structures. SLAT then matches graphs between data sets using a minimum cost bipartite matching approach. Moscot attempts to map cells through space and time by formulating the mapping tasks as an optimal transport problem (Klein et al., 2023). Cost estimates of mapping cells are generated using preliminary biological knowledge and spatial information embedded into a common latent space. Next, optimal transport solutions are found using the Sinkhorn Algorithm (Cuturi, 2013). This mapping can then be used for tissue alignment across heterogeneous tissue sections.

2.5 Multi-modality of spatial data

Measuring gene expression in a spatially resolved manner has undeniably yielded many insights into cellular biology and improved our understanding of diseases. Yet, gene expression only provides a limited view into the complex biological mechanisms that regulate cellular organization. Before genes can be expressed, they need to be accessible for regulatory proteins - called transcription factors - to bind (Ali, Renkawitz, & Bartkuhn, 2016; Allen & Taatjes, 2015; Baker, 2011; Bozek et al., 2019; Chen et al., 2020; Pcn & Nr, 2020). With specific exceptions, these proteins bind in regions of open chromatin where tightly coiled DNA is relaxed to leave space for binding. The mechanisms by which chromatin is opened are complex and rely on the conjunct effects of many proteins and chemical alterations of DNA (Baker, 2011; Bannister & Kouzarides, 2011).

Measuring chromatin accessibility to regulatory proteins allows us to gain insights into the expression patterns that are yet to come; the DNA encoding for a gene being opened suggests that this gene is primed to be transcribed. In some cases, chromatin is opened by so-called pioneer transcription factors(Donaghey et al., 2018; Iwafuchi-Doi & Zaret, 2014; Mayran et al., 2019; Zaret & Carroll, 2011). These transcription factors remain bound to the DNA until other transcription factors activate the transcription of the gene. Even once genes have been transcribed, for many genes, they will only play a functional role once translated into a protein. The journey from mRNA transcript to functional protein contains many obstacles and checkpoints. Many proteins undergo post translational modifications that modulate their function (Ramazi & Zahiri, 2021). Proteins will often migrate to different locations within the cell before performing their role. Furthermore, the correlation between mRNA and protein levels tends to be loosely correlated (De Sousa Abreu, Penalva, Marcotte, & Vogel, 2009). This strengthens the need to accurately measure not only the transcriptome but also the epigenome and proteome.

The compaction of DNA can be measured by a variety of techniques such as DNase I hypersensitivity (Song & Crawford, 2010) or ATAC-seq (Buenrostro, Giresi, Zaba, Chang, & Greenleaf, 2013). The scATAC-seq protocol was first successfully adapted to application in single cells (Cusanovich et al., 2015) and later adapted to be used in spatial assays (Deng et al., 2022). More recently, experimental advancements have allowed us to measure gene expression and chromatin accessibility simultaneously (Liu et al., 2022). Other technologies also allow simultaneous measurements of RNA and proteins (Liu et al., 2020).

The measurement of modalities requires the development of specialized algorithms to match cells between modalities. While the context provided in jointly measured spatial assays allows us to match cells directly, the challenge arises when a single modality is measured. There have been several attempts at developing multimodal data integration. Seurat for instance, uses Canonical Correlation Analysis to link scRNA and scATAC under the assumption that scATAC activity is a form of pseudo-gene expression (Stuart et al., 2019). Other tools such as LIGER utilizes integrative Non-Negative Matrix Factorization to find common factors between data sets (Welch et al., 2019). MultiMAP performs multimodal integration using a generalization of Universal Manifold Approximation and Projection (UMAP) by leveraging Riemannian geometry and algebraic topology (Jain et al., 2021). However, there has been little work of including the spatial component during integration of modalities. Moscot maps jointly measured modalities (RNA-seq and ATAC-seq) to a reference spatial assay (RNA-seq) by carrying over the new modality (Klein et al., 2023). Effectively, spatial mapping is only performed using common measurements. The challenges related to data integration were reviewed by (Argelaguet, Cuomo, Stegle, & Marioni, 2021).

3. Data enhancement by integrating tissue image data for ST data

Within the realm of tissue image data, stained data plays a vital role. Specific dyes or chemicals enhance features of interest, such as cell types, organelles, or pathological changes. Hematoxylin and eosin (H&E) staining, a widely used technique, offers a basic overview of tissue morphology. While not always available, H&E data acquisition is often customary for Visium data, especially in cancer research. This integration of data modalities holds significant potential to enhance the analysis of both ST data and H&E data.

3.1 Cell segmentation boosted by integrating image data

Cell segmentation algorithms have been developed to automate and optimize the identification of cells within images. Cell segmentation is essential for image-based ST data analysis to quantify the transcriptomes at the cellular level. Cell segmentation is also required for high-resolution sequencing-based ST data such as Stereo-seq (Chen et al., 2021) and SeqScope (Cho et al., 2021) by accurately merging sequencing spots within defined cell boundaries.

Some of the earliest approaches were based on the watershed transformation that creates a topological map of grayscale gradients (Li et al., 2008). The valleys between cells would be defined as the boundary between cells. However, this approach and its variants struggle if the image of a cell contains more than one color hotspot. In this case, a watershed approach will tend to over-segment. Conversely, if the gradients between cells are not clear cut this approach might under segment cells. Building on the idea of topological maps, Cellpose generates topological maps using a diffusion process within a boundary determined by a human annotator (Stringer, Wang, Michaelos, & Pachitariu, 2021). Their deep neural network based on U-Net architecture, predicts multi-directional gradients that form a vector field. All pixels within this vector field that converge to the same point are grouped together and are considered to belong to the same cellular structure.

Other approaches have been designed specifically for use in ST data. Baysor assumes that neighboring transcripts are likely to belong to the same cell since transcripts tend to cluster together (Petukhov et al., 2022). They leverage Markov Random Field priors on simple tessellation graphs to capture this clustering tendency. Interestingly, this approach was designed as a cell segmentation free approach to transcript assignment, but they demonstrate how this approach correlates well with cell straining images. SCS (Sub-cellular spatial transcriptomics Cell Segmentation) is another cell segmentation method that first identifies nuclei from image using the watershed algorithm and then expands the cell boundaries using a transformer-based model (Hao Chen, Li, & Bar-Joseph, 2023).

For Stereo-seq (Chen et al., 2021) and SeqScope (Cho et al., 2021), image data from adjacent tissue sections are used for cell segmentation. VisiumHD, in contrast, uses the same tissue section for both imaging and the spatial assay, with a spot diameter of 2 µm.

3.2 Data enhancement by integrating image data

While Visium technology offers valuable insights into spatial gene expression within tissue samples, it has limitations. It cannot directly provide information at the level of individual cells, which is often crucial for understanding biological processes. Therefore, enhancing the quality and resolution of Visium data is crucial. Visium data often comes paired with H&E stained images. These images provide high-resolution information about cell morphology (shape) and abundance within the tissue. Researchers are developing computational approaches that combine Visium data with detailed cell information from H&E images. These methods aim to infer gene expression at a higher cellular resolution than what Visium data directly offers.

For instance, Bergenstrahle et al. suggested XFuse, a deep generative model that fuses image-based spatial assays with histology images to impute gene expression landscapes at high resolution (Bergenstråhle et al., 2022). More recently, iStar employs a hierarchical vision transformer (HViT), which associates image patches of varying size (from 16×16 to 256×256 pixels) to gene expression taken from low resolution Visium data sets (Zhang et al., 2024). SpatialScope also leverages a deep generative model to increase ST resolution (Wan et al., 2023). First, SpatialScope decomposes low resolution spots into estimated cell compositions and enhances the resolution by generating cellular resolution spots from spot-level expression profiles. For image-based methods, SpatialScope will associate learned expression patterns from scRNAseq data to the sparse expression profiles in the image assay. The generative model will then fill in the gaps in gene expression. siGra employs a graph representation of the original spatial data and a hybrid graph transformer model to better model gene expression patterns and enhance the spatial resolution of ST data (Tang et al., 2023).

These strategies greatly help in handling Visium data for cancers. For instance, iStar applied to breast cancer data produced finer resolution ST data, providing information about intratumoral heterogeneity and recapitulating the tissue architecture from Visium datasets (Zhang et al., 2024).

3.3 Gene prediction from tissue image data

While ST technologies offer a powerful toolkit, their application in clinical settings with large patient pools can be limited by cost and practical considerations. Conducting spatial technology experiments on numerous patients simultaneously might be prohibitively expensive. Additionally, acquiring high-quality tissue biopsies suitable for ST experiments may not always be feasible. Therefore, the ability to predict gene expression directly from H&E image data represents a valuable alternative. This approach could potentially leverage ST data to integrate gene expression information with readily available H&E images.

Early attempts, such as HE2RNA (Schmauch et al., 2020) and hist2RNA (Mondol et al., 2023), attempted to predict gene expression profiles from images and associated bulk RNA sequencing. However, this approach does conserve spatially resolved gene expression patterns. To remedy this limitation, several tools have been developed to produce ST-like data from H&E images. One of the earliest attempts, ST-Net, uses a deep learning classifier on small image patches to predict the expression of 250 genes (He et al., 2020). ST-Net quantified the spatial co-location of tumor and immune expression activities directly from H&E images.

While this approach pioneered the idea of gene expression prediction from images, their neural network architecture could only predict expression patterns of a limited number of genes and did not account for global expression patterns. To overcome these limitations, HisToGene exploits the strength of vision transformer architecture to learn local and global expression patterns in expression data and link these patterns to H& E images (Pang, Su, & Li, 2021). Attempting to further refine the use of transformer architecture, Hist2ST introduced three modules: a convmixer architecture module, a transformer module, and a graph neural network module (Zeng et al., 2022). The image is broken down into small patches then fed into the 3 modules to learn the local and global spatial distribution of genes from images only. Other methods, such as DeepSpaCE, use convolution neural networks to learn expression patterns from images (Monjo, Koido, Nagasawa, Suzuki, & Kamatani, 2022). A method called BLEEP uses bi-modal encoders and contrastive learning to effectively link an image latent space with an expression latent space (Jiang, Xie, Tan, Ye, & Nguyen, 2023).

4. Data enhancement by synthetic data

In a realm adjacent to data imputation is the creation of synthetic spatial data sets. The generation of synthetic spatial data may serve multiple purposes. Synthetic data sets provide a known ground truth that can be used for benchmarking novel computational strategies in spatial data. Spider proposes a simulator that leverages cell type labels and the probability of transitioning to that label in neighboring spatial positions (Yang et al., 2023). SRTsim performs primarily reference-based simulations where spatially cohesive permutations of the reference data are generated (Zhu, Shang, & Zhou, 2023). Similarly, scDesign3 also uses a reference-based approach to generate synthetic data but focuses on capturing the statistical properties of the data to generate new data sets (Song et al., 2023). The approach proposed in scDesign3 is applicable to both scRNAseq data and spatial data. Interestingly, scDesign3 has been suggested as a way to generate synthetic null data sets to improve differential gene expression analysis. The idea brought forth with scDesign3 is that comparing gene expression between clusters will always provide differentially expressed genes since differential gene expression is used as the basis to define clusters. The generation of synthetic null data sets provides a more neutral background to compare differentially expressed genes. Synthetic data sets that account for the spatial distribution of cells and associated gene expression patterns have been suggested as a means to estimate spatial statistical power (Baker, Schapiro, Dumitrascu, Vickovic, & Regev, 2023).

5. Perspectives

The evolving landscape of spatial technologies promises to shed light on mechanisms of diseases such as cancer. ST has been extremely useful in visualizing tissue structure, heterogeneity, TME, and spatial gene expression distribution from cancer samples (Yu, Jiang, & Wu, 2022). Experimental development has moved at a fast pace providing increasing spatial resolution and high sensitivity to gene expression patterns. However, the development and improvement of experimental protocols is an extremely challenging and laborious task. Even once these methods have been developed, they will need to be refined until they can be reliably used in a clinical setting. Years might pass until spatial transcriptomics will be routinely used to aid clinicians in making patient related decisions. While single-cell methods have become stable enough to be used in a clinical setting, they do not explicitly provide spatial context and thus can lose precious insights related to the spatial context in which cells find themselves. Fortunately, computational methods have advanced rapidly alongside spatial transcriptomic and scRNAseq development. These algorithms can assist in capitalizing on the information contained within spatial assays and, in many cases, can even improve spatial data. Data enhancement techniques can bring the strengths of scRNAseq to compensate for the limitations incurred by ST. Deep learning and generative models can learn complex patterns from high-dimensional data and untangle the spatial component from expression profiles. In turn, they can be used to perform a variety of tasks, from data imputation to scRNAseq mapping and even multi-modal data integration.

Data integration plays a crucial role in cancer research, enabling the study of complex interactions between different cell types and the tumor microenvironment (Yu et al., 2022). For example, NicheSVM used deconvolution techniques and identified genes influenced by tumor-stromal cell interactions by analyzing Visium data (Saqib et al., 2023). Similarly, CellNeighborEx, which leverages decompositional methods on SlideSeq data, allows researchers to identify genes activated upon contact between liver tumors and monocytes (Kim, Kumar, et al., 2023). These examples highlight how algorithmic development, combined with advanced analytical pipelines, offers powerful tools to investigate diverse aspects of cancer biology.

While ST offers valuable insights, generating such data remains expensive. Therefore, leveraging the readily available and cost-effective H &E images, widely used in histopathology for disease diagnosis, to extract additional information holds significant potential. Integrating ST data with H&E data offers the potential to extract additional information and enrich the knowledge gained from H&E images alone. This approach has already

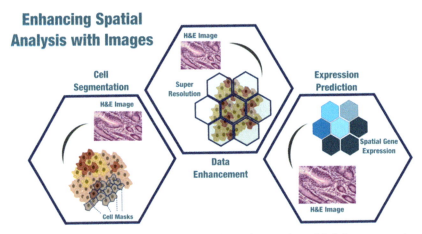

Fig. 2 **Data enhancement by integrating tissue image data.** (1) Cell segmentation, (2) ST data enhancement, and (3) gene expression prediction.

gained traction, with researchers actively exploring methods for such integration (Fig. 2). Research efforts are actively exploring these integration methods. As these technologies advance, the ultimate goal of inferring gene expression at the cellular level from H&E data for cancer studies may become achievable.

Beyond providing deeper insights into gene expression, ST has emerged as a powerful tool for unraveling cell-cell communication. Algorithms like SpaTalk (Shao et al., 2022) and COMMOT (Cang et al., 2023) leverage ST data to analyze these interactions under various conditions, including cancer. Understanding these communication pathways is crucial for deciphering how tumor cells co-opt surrounding healthy cells and drive cancer progression. Furthermore, enhancing data through methods like SC2Spa, which integrates single-cell data with ST analysis using tools like SpaTalk, holds immense potential for even deeper exploration of cell-cell communication. This can lead to the question of whether similar insights can be gleaned from readily available H&E images.

It is of note that certain precautions must be taken when using these techniques in understanding diseases. First, deep learning models have often struggled with providing human interpretable predictions. A model can accurately predict which cell should be mapped to which spot in a spatial dataset, but it remains unclear what factors the model considered to make that prediction. A similar limitation can be said for generative models where the data produced through their generative process might not reflect biological reality in some key aspects. Understanding the biological mechanisms

Data enhancement in the age of spatial biology

governing disease and the causal relationships between gene expression lies at the heart of biomedical research. It is through this understanding that the scientific community can attempt to find promising avenues for the development of therapeutics. There have been many attempts to make deep learning models interpretable, and as this field of research advances, we can hope to make the best use of these powerful models.

Second, we must carefully consider the provenance of the data sets used to train these models. Data sparsity in spatial data may be compensated with the use of scRNAseq data, but the amount of information that is not being accurately captured, especially information related to clonal diversity is unknown. Since different tumors can be characterized by different sets of mutations it is not unreasonable to assume that there will be downstream effects on gene expression. A similar argument could be raised concerning the ethnic background of the patients: to what extent will their transcriptomes match? While these effects might remain minor, the risk of misdiagnosis due to data imbalance and model biases should not be overlooked.

Finally, we must contend with the question of generalization. Many algorithms employ deep learning frameworks that will learn complex patterns found within the dataset, but these patterns might not generalize well to new data sets. A current trend seen in the field of single cells is the use of atlases that contain millions of cells from various tissues and diseases. We discussed how single cells can be used to enhance spatial data, but also highlighted how the transcriptome of a cell will change depending on its local and global neighborhood. It follows that single cell atlases could be enhanced by providing spatially aware cell type labels especially with the advent of single cell foundation models.

Spatial inference proves to be highly beneficial in the integration of single-cell and spatial transcriptomics data. Predicting single-cell coordinates using gene expression levels is a further advancement for extracting spatial information from scRNAseq data. This method addresses the gap between the demand for single-cell whole-transcriptome spatial transcriptomics data and the high costs associated with current commercial technologies.

The addition of inferred spatial coordinates to scRNAseq data increases the power of ligand-receptor interaction analysis by estimating cell-cell spatial proximity. ScRNAseq data were used to infer cell-cell communication based on ligand and receptor expression. Predicting single-cell spatial coordinates enhances our understanding of disease mechanisms through its ability to improve the analysis of ligand-receptor interactions of scRNAseq data.

Acknowledgments

We appreciate the institutional support from Cedars-Sinai Medical Center.

Disclosures

The authors have no conflicts of interest to declare.

References

Abdelaal, T., Mourragui, S., Mahfouz, A., & Reinders, M. J. T. (2020). SpaGE: Spatial gene enhancement using scRNA-seq. e107-e107 *Nucleic Acids Research, 48*(18), https://doi.org/10.1093/nar/gkaa740.

Achim, K., Pettit, J. B., Saraiva, L. R., Gavriouchkina, D., Larsson, T., Arendt, D., & Marioni, J. C. (2015). High-throughput spatial mapping of single-cell RNA-seq data to tissue of origin. *Nature Biotechnology, 33*(5), 503–509. https://doi.org/10.1038/nbt.3209.

Ali, T., Renkawitz, R., & Bartkuhn, M. (2016). Insulators and domains of gene expression. *Current Opinion in Genetics & Development, 37.* https://doi.org/10.1016/j.gde.2015.11.009.

Allen, B. L., & Taatjes, D. J. (2015). The Mediator complex: A central integrator of transcription. *Nature Reviews. Molecular Cell Biology, 16*(3), https://doi.org/10.1038/nrm3951.

Anderson, N. M., & Simon, M. C. (2020). The tumor microenvironment. *Current Biology, 30*(16), R921–R925. https://doi.org/10.1016/j.cub.2020.06.081.

Andersson, A., Bergenstråhle, J., Asp, M., Bergenstråhle, L., Jurek, A., Fernández Navarro, J., & Lundeberg, J. (2020). Single-cell and spatial transcriptomics enables probabilistic inference of cell type topography. *Communications Biology, 3,* 565 (In).

Argelaguet, R., Cuomo, A. S. E., Stegle, O., & Marioni, J. C. (2021). Computational principles and challenges in single-cell data integration. *Nature Biotechnology, 39*(10), 1202–1215. https://doi.org/10.1038/s41587-021-00895-7.

Arnosti, D. N., Barolo, S., Levine, M., & Small, S. (1996). The eve stripe 2 enhancer employs multiple modes of transcriptional synergy. *Development (Cambridge, England), 122*(1), 205–214. https://doi.org/10.1242/dev.122.1.205.

Baker, E. A. G., Schapiro, D., Dumitrascu, B., Vickovic, S., & Regev, A. (2023). In silico tissue generation and power analysis for spatial omics. *Nature Methods, 20*(3), 424–431. https://doi.org/10.1038/s41592-023-01766-6.

Baker, M. (2011). Making sense of chromatin states. *Nature Methods, 8*(9), 717–722. https://doi.org/10.1038/nmeth.1673.

Bannister, A. J., & Kouzarides, T. (2011). Regulation of chromatin by histone modifications. 381-381 *Cell Research, 21*(3), https://doi.org/10.1038/CR.2011.21.

Bergenstråhle, J., Larsson, L., & Lundeberg, J. (2020). Seamless integration of image and molecular analysis for spatial transcriptomics workflows. *BMC Genomics, 21*(1), 1–7. https://doi.org/10.1186/s12864-020-06832-3.

Bergenstråhle, L., He, B., Bergenstråhle, J., Abalo, X., Mirzazadeh, R., Thrane, K., ... Maaskola, J. (2022). Super-resolved spatial transcriptomics by deep data fusion. *Nature Biotechnology, 40*(4), 476–479. https://doi.org/10.1038/s41587-021-01075-3.

Biancalani, T., Scalia, G., Buffoni, L., Avasthi, R., Lu, Z., Sanger, A., ... Zhang, M. (2021). Deep learning and alignment of spatially resolved single-cell transcriptomes with Tangram. *Nature Methods, 18*(11), 1352–1362.

Biancalani, T., Scalia, G., Buffoni, L., Avasthi, R., Lu, Z., Sanger, A., ... Regev, A. (2020). Deep learning and alignment of spatially-resolved whole transcriptomes of single cells in the mouse brain with Tangram. 2020.2008.2029.272831-272020.272808.272829.272831 *bioRxiv.* https://doi.org/10.1101/2020.08.29.272831.

Bitgood, M. J., & McMahon, A. P. (1995). HedgehogandBmpGenes are coexpressed at many diverse sites of cell–cell interaction in the mouse embryo. *Developmental Biology, 172*(1), 126–138. https://doi.org/10.1006/dbio.1995.0010.

Bozek, M., Cortini, R., Storti, A. E., Unnerstall, U., Gaul, U., & Gompel, N. (2019). ATAC-seq reveals regional differences in enhancer accessibility during the establishment of spatial coordinates in the Drosophila blastoderm. *Genome Research.* gr.242362.242118-gr.242362.242118. https://doi.org/10.1101/GR.242362.118.

Brand, A. H., Manoukian, A. S., & Perrimon, N. (1994). *Chapter 33 ectopic expression in drosophila.* Elsevier,635–654.

Buenrostro, J. D., Giresi, P. G., Zaba, L. C., Chang, H. Y., & Greenleaf, W. J. (2013). Transposition of native chromatin for fast and sensitive epigenomic profiling of open chromatin, DNA-binding proteins and nucleosome position. *Nature Methods, 10*(12), 1213–1218.

Cable, D. M., Murray, E., Zou, L. S., Goeva, A., Macosko, E. Z., Chen, F., & Irizarry, R. A. (2022). Robust decomposition of cell type mixtures in spatial transcriptomics. *Nature Biotechnology, 40*(4), 517–526.

Cang, Z., & Nie, Q. (2020). Inferring spatial and signaling relationships between cells from single cell transcriptomic data. *Nature Communications, 11*(1), 2084. https://doi.org/10.1038/s41467-020-15968-5.

Cang, Z., Zhao, Y., Almet, A. A., Stabell, A., Ramos, R., Plikus, M. V., ... Nie, Q. (2023). Screening cell-cell communication in spatial transcriptomics via collective optimal transport. *Nature Methods, 20*(2), 218–228. https://doi.org/10.1038/s41592-022-01728-4.

Cao, J., Packer, J. S., Ramani, V., Cusanovich, D. A., Huynh, C., Daza, R., ... Shendure, J. (2017). Comprehensive single-cell transcriptional profiling of a multicellular organism. *Science (New York, N. Y.), 357*(6352), 661–667. https://doi.org/10.1126/science.aam8940.

Chen, A., Liao, S., Cheng, M., Ma, K., Wu, L., Lai, Y., ... Wang, J. (2022). Spatiotemporal transcriptomic atlas of mouse organogenesis using DNA nanoball-patterned arrays. *Cell, 185*(10), 1777–1792.e1721. https://doi.org/10.1016/j.cell.2022.04.003.

Chen, A., Liao, S., Cheng, M., Ma, K., Wu, L., Lai, Y., ... Xu, X. (2021). Large field of view-spatially resolved transcriptomics at nanoscale resolution Short title: DNA nanoball stereo-sequencing. *bioRxiv.* 2021.2001.2017.427004-422021.427001.427017.427004. https://doi.org/10.1101/2021.01.17.427004.

Chen, C.-H., Zheng, R., Tokheim, C., Dong, X., Fan, J., Wan, C., ... Liu, X. S. (2020). Determinants of transcription factor regulatory range. *Nature Communications, 11*(1), 2472. https://doi.org/10.1038/s41467-020-16106-x.

Chen, H., Li, D., & Bar-Joseph, Z. (2023). SCS: Cell segmentation for high-resolution spatial transcriptomics. *Nature Methods, 20*(8), 1237–1243. https://doi.org/10.1038/s41592-023-01939-3.

Chen, H., & Murphy, R. F. (2023). Evaluation of cell segmentation methods without reference segmentations. *Molecular Biology of the Cell, 34*(6), https://doi.org/10.1091/mbc.e22-08-0364.

Chen, K. H., Boettiger, A. N., Moffitt, J. R., Wang, S., & Zhuang, X. (2015). Spatially resolved, highly multiplexed RNA profiling in single cells. *Science (New York, N. Y.), 348*(6233), aaa6090. https://doi.org/10.1126/science.aaa6090.

Cho, C.-S. S., Xi, J., Si, Y., Park, S.-R. R., Hsu, J.-E. E., Kim, M., ... Lee, J. H. (2021). Microscopic examination of spatial transcriptome using Seq-Scope. *Cell, 184*(13), 3559–3572.e3522. https://doi.org/10.1016/j.cell.2021.05.010.

Cho, C. S., Xi, J., Si, Y., Park, S. R., Hsu, J. E., Kim, M., ... Lee, J. H. (2021). Microscopic examination of spatial transcriptome using Seq-Scope. *Cell, 184*(13), 3559–3572.e3522. https://doi.org/10.1016/j.cell.2021.05.010.

Clark, A. G., & Vignjevic, D. M. (2015). Modes of cancer cell invasion and the role of the microenvironment. *Current Opinion in Cell Biology, 36,* 13–22. https://doi.org/10.1016/j.ceb.2015.06.004.

Clifton, K., Anant, M., Aihara, G., Atta, L., Aimiuwu, O. K., Kebschull, J. M., ... Fan, J. (2023). STalign: Alignment of spatial transcriptomics data using diffeomorphic metric mapping. *Nature Communications, 14*(1), https://doi.org/10.1038/s41467-023-43915-7.

Cusanovich, D. A., Daza, R., Adey, A., Pliner, H. A., Christiansen, L., Gunderson, K. L., ... Shendure, J. (2015). Multiplex single-cell profiling of chromatin accessibility by combinatorial cellular indexing. *Science (New York, N. Y.), 348*(6237), 910–914.

Cuturi, M. (2013). Sinkhorn distances: Lightspeed computation of optimal transportation distances. *arXiv preprint arXiv, 1306,* 0895.

Danaher, P., Kim, Y., Nelson, B., Griswold, M., Yang, Z., Piazza, E., & Beechem, J. M. (2022). Advances in mixed cell deconvolution enable quantification of cell types in spatial transcriptomic data. *Nature communications, 13*(1), 385.

De Sousa Abreu, R., Penalva, L. O., Marcotte, E. M., & Vogel, C. (2009). Global signatures of protein and mRNA expression levels. *Molecular Biosystems, 5*(12), 1512–1526.

Deng, Y., Bartosovic, M., Ma, S., Zhang, D., Kukanja, P., Xiao, Y., ... Fan, R. (2022). Spatial profiling of chromatin accessibility in mouse and human tissues. *Nature, 609*(7926), 375–383. https://doi.org/10.1038/s41586-022-05094-1.

Donaghey, J., Thakurela, S., Charlton, J., Chen, J. S., Smith, Z. D., Gu, H., ... Meissner, A. (2018). Genetic determinants and epigenetic effects of pioneer-factor occupancy. *Nature Genetics, 50*(2), 250–258. https://doi.org/10.1038/s41588-017-0034-3.

Dong, K., & Zhang, S. (2022). Deciphering spatial domains from spatially resolved transcriptomics with an adaptive graph attention auto-encoder. *Nature Communications, 13*(1), 1–12. https://doi.org/10.1038/s41467-022-29439-6.

Dong, R., & Yuan, G.-C. (2021). SpatialDWLS: Accurate deconvolution of spatial transcriptomic data. *Genome Biology, 22*(1), 145.

Elosua-Bayes, M., Nieto, P., Mereu, E., Gut, I., & Heyn, H. (2021). SPOTlight: Seeded NMF regression to deconvolute spatial transcriptomics spots with single-cell transcriptomes. *Nucleic Acids Research, 49*(9), e50.

Eng, C. H. L., Lawson, M., Zhu, Q., Dries, R., Koulena, N., Takei, Y., ... Cai, L. (2019). Transcriptome-scale super-resolved imaging in tissues by RNA seqFISH+. *Nature, 568*(7751), 235–239. https://doi.org/10.1038/s41586-019-1049-y.

Fan, H. C., Fu, G. K., & Fodor, S. P. A. (2015). Combinatorial labeling of single cells for gene expression cytometry. *Science (New York, N. Y.), 347*(6222), 1258367. https://doi.org/10.1126/science.1258367.

Fischer, S., & Gillis, J. (2021). How many markers are needed to robustly determine a cell's type. *iScience, 24*(11), 103292. https://doi.org/10.1016/j.isci.2021.103292.

Forjaz, A., Vaz, E., Romero, V. M., Joshi, S., Braxton, A. M., Jiang, A. C., ... Wirtz, D. (2023). *Three-dimensional assessments are necessary to determine the true spatial tissue composition of diseased tissues.* Cold Spring Harbor Laboratory. Retrieved from: https://doi.org/10.1101/2023.12.04.569986.

Franzén, O., Gan, L. M., & Björkegren, J. L. M. (2019). PanglaoDB: A web server for exploration of mouse and human single-cell RNA sequencing data. *Database, 2019*(1), https://doi.org/10.1093/database/baz046.

Galichon, A. (2016). *Optimal Transport Methods in Economics.* https://doi.org/10.23943/princeton/9780691172767.001.0001.

Gjorevski, N., Nikolaev, M., Brown, T. E., Mitrofanova, O., Brandenberg, N., Delrio, F. W., ... Lutolf, M. P. (2022). Tissue geometry drives deterministic organoid patterning. *Science (New York, N. Y.), 375*(6576), https://doi.org/10.1126/science.aaw9021.

Guo, T., Yuan, Z., Pan, Y., Wang, J., Chen, F., Zhang, M. Q., & Li, X. (2023). SPIRAL: Integrating and aligning spatially resolved transcriptomics data across different experiments, conditions, and technologies. *Genome Biology, 24*(1), https://doi.org/10.1186/s13059-023-03078-6.

Halder, G., Callaerts, P., & Gehring, W. J. (1995). Induction of ectopic eyes by targeted expression of the eyeless gene in Drosophila. *Science (New York, N. Y.), 267*(5205), 1788–1792. Retrieved from. http://www.ncbi.nlm.nih.gov/pubmed/7892602.

He, B., Bergenstråhle, L., Stenbeck, L., Abid, A., Andersson, A., Borg, Å., ... Zou, J. (2020). Integrating spatial gene expression and breast tumour morphology via deep learning. *Nature Biomedical Engineering, 4*(8), 827–834. https://doi.org/10.1038/s41551-020-0578-x.

Hu, J., Li, X., Coleman, K., Schroeder, A., Ma, N., Irwin, D. J., ... Li, M. (2021). SpaGCN: Integrating gene expression, spatial location and histology to identify spatial domains and spatially variable genes by graph convolutional network. *Nature Methods, 18*(11), 1342–1351. https://doi.org/10.1038/s41592-021-01255-8.

Hu, Y., Zhao, Y., Schunk, C. T., Ma, Y., Derr, T., & Zhou, X. M. (2023). ADEPT: Autoencoder with differentially expressed genes and imputation for robust spatial transcriptomics clustering. *iScience, 26*(6), 106792. https://doi.org/10.1016/j.isci.2023.106792.

Iwafuchi-Doi, M., & Zaret, K. S. (2014). Pioneer transcription factors in cell reprogramming. *Genes & Development, 28*(24), 2679–2692. https://doi.org/10.1101/gad.253443.114.

Jain, M. S., Polanski, K., Conde, C. D., Chen, X., Park, J., Mamanova, L., ... Teichmann, S. A. (2021). MultiMAP: Dimensionality reduction and integration of multimodal data. *Genome Biology, 22*(1), https://doi.org/10.1186/s13059-021-02565-y.

Janesick, A., Shelansky, R., Gottscho, A. D., Wagner, F., Williams, S. R., Rouault, M., ... Taylor, S. E. B. (2023). High resolution mapping of the tumor microenvironment using integrated single-cell, spatial and in situ analysis. *Nature Communications, 14*(1), https://doi.org/10.1038/s41467-023-43458-x.

Jiang, Y., Xie, J., Tan, X., Ye, N., & Nguyen, Q. (2023). *Generalization of deep learning models for predicting spatial gene expression profiles using histology images: A breast cancer case study.* Cold Spring Harbor Laboratory. Retrieved from https://doi.org/10.1101/2023.09.20.558624.

Jones, A., Townes, F. W., Li, D., & Engelhardt, B. E. (2023). Alignment of spatial genomics data using deep Gaussian processes. *Nature Methods, 20*(9), 1379–1387. https://doi.org/10.1038/s41592-023-01972-2.

Joshi, R., Sipani, R., & Bakshi, A. (2022). Roles of drosophila hox genes in the assembly of neuromuscular networks and behavior. *Frontiers in Cell and Developmental Biology, 9*. https://doi.org/10.3389/fcell.2021.786993.

Karzbrun, E., Kshirsagar, A., Cohen, S. R., Hanna, J. H., & Reiner, O. (2018). Human brain organoids on a chip reveal the physics of folding. *Nature Physics, 14*(5), 515–522. https://doi.org/10.1038/s41567-018-0046-7.

Kiemen, A., Braxton, A. M., Grahn, M. P., Han, K. S., Mahesh Babu, J., Reichel, R., ... Wirtz, D. (2020). In situ characterization of the 3D microanatomy of the pancreas and pancreatic cancer at single cell resolution. *bioRxiv.* 2020.2012.2008.416909-412020.416912.416908.416909. https://doi.org/10.1101/2020.12.08.416909.

Kim, H., Kumar, A., Lövkvist, C., Palma, A. M., Martin, P., Kim, J., ... Madan, E. (2023). CellNeighborEX: Deciphering neighbor-dependent gene expression from spatial transcriptomics data. *Molecular Systems Biology*e11670.

Kim, H., Lövkvist, C., Palma, A. M., Martin, P., Kim, J., Kumar, A., ... Won, K. J. (2023). CellNeighborEX: Deciphering neighbor-dependent gene expression from spatial transcriptomics data. *bioRxiv.* 2022.2002.2016.480673. https://doi.org/10.1101/2022.02.16.480673.

Kim, H. J., Tam, P. P. L., & Yang, P. (2021). Defining cell identity beyond the premise of differential gene expression. *Cell Regeneration, 10*(1), https://doi.org/10.1186/s13619-021-00083-7.

Kingma, D. P., & Welling, M. (2019). An introduction to variational autoencoders. *Foundations and Trends® in Machine Learning, 12*(4), 307–392.

Klein, D., Palla, G., Lange, M., Klein, M., Piran, Z., Gander, M., ... Theis, F. J. (2023). *Mapping cells through time and space with moscot.* Cold Spring Harbor Laboratory. Retrieved from. https://doi.org/10.1101/2023.05.11.540374.

Kleshchevnikov, V., Shmatko, A., Dann, E., Aivazidis, A., King, H. W., Li, T., ... Gayoso, A. (2022). Cell2location maps fine-grained cell types in spatial transcriptomics. *Nature Biotechnology, 40*(5), 661–671.

Li, G., Liu, T., Nie, J., Guo, L., Chen, J., Zhu, J., ... Wong, S. T. C. (2008). Segmentation of touching cell nuclei using gradient flow tracking. *Journal of Microscopy, 231*(1), 47–58. https://doi.org/10.1111/j.1365-2818.2008.02016.x.

Li, R., Ferdinand, J. R., Loudon, K. W., Bowyer, G. S., Laidlaw, S., Muyas, F., ... Mitchell, T. J. (2022). Mapping single-cell transcriptomes in the intra-tumoral and associated territories of kidney cancer. *Cancer Cell, 40*(12), 1583–1599.e1510. https://doi.org/10.1016/j.ccell.2022.11.001.

Linbu, L., Esha, M., António, M. P., Hyobin, K., Amit, K., Praveen, B., ... Kyoung Jae, W. (2023). SC2Spa: A deep learning based approach to map transcriptome to spatial origins at cellular resolution. *bioRxiv.* 2023.2008.2022.554277. https://doi.org/10.1101/2023.08.22.554277.

Liu, W., Liao, X., Luo, Z., Yang, Y., Lau, M. C., Jiao, Y., ... Liu, J. (2023). Probabilistic embedding, clustering, and alignment for integrating spatial transcriptomics data with PRECAST. *Nature Communications, 14*(1), https://doi.org/10.1038/s41467-023-35947-w.

Liu, Y., DiStasio, M., Su, G., Asashima, H., Enninful, A., Qin, X., ... Fan, R. (2022). Spatial-CITE-seq: Spatially resolved high-plex protein and whole transcriptome co-mapping. *bioRxiv.* 2022.2004.2001.486788-482022.486704.486701.486788. https://doi.org/10.1101/2022.04.01.486788.

Liu, Y., Yang, M., Deng, Y., Su, G., Enninful, A., Guo, C. C., ... Fan, R. (2020). High-spatial-resolution multi-omics sequencing via deterministic barcoding in tissue. *Cell, 183*(6), 1665–1681.e1618. https://doi.org/10.1016/j.cell.2020.10.026.

Long, Y., Ang, K. S., Li, M., Chong, K. L. K., Sethi, R., Zhong, C., ... Chen, J. (2023). Spatially informed clustering, integration, and deconvolution of spatial transcriptomics with GraphST. *Nature Communications, 14*(1), https://doi.org/10.1038/s41467-023-36796-3.

Lopez, R., Li, B., Keren-Shaul, H., Boyeau, P., Kedmi, M., Pilzer, D., ... Wagner, A. (2022). DestVI identifies continuums of cell types in spatial transcriptomics data. *Nature Biotechnology, 40*(9), 1360–1369.

Lopez, R., Nazaret, A., Langevin, M., Samaran, J., Regier, J., Jordan, M. I., & Yosef, N. (2019). A joint model of unpaired data from scRNA-seq and spatial transcriptomics for imputing missing gene expression measurements. *arXiv preprint arXiv, 1905*, 02269.

Lubeck, E., & Cai, L. (2012). Single-cell systems biology by super-resolution imaging and combinatorial labeling. *Nature Methods, 9*(7), 743–748. https://doi.org/10.1038/nmeth.2069.

Ma, Y., & Zhou, X. (2022). Spatially informed cell-type deconvolution for spatial transcriptomics. *Nature Biotechnology, 40*(9), 1349–1359.

Martin, P. C. N., Kim, H., Lövkvist, C., Hong, B. W., & Won, K. J. (2022). Vesalius: High-resolution in silico anatomization of spatial transcriptomic data using image analysis. *Molecular Systems Biology, 18*(9), https://doi.org/10.15252/msb.202211080.

Maseda, F., Cang, Z., & Nie, Q. (2021). DEEPsc: A deep learning-based map connecting single-cell transcriptomics and spatial imaging data. *Front Genet, 12*, 636743. https://doi.org/10.3389/fgene.2021.636743.

Mayran, A., Sochodolsky, K., Khetchoumian, K., Harris, J., Gauthier, Y., Bemmo, A., ... Drouin, J. (2019). Pioneer and nonpioneer factor cooperation drives lineage specific chromatin opening. *Nature Communications, 10*(1), 1–13. https://doi.org/10.1038/s41467-019-11791-9.

Métivier, L., Brossier, R., Mérigot, Q., Oudet, E., & Virieux, J. (2016). An optimal transport approach for seismic tomography: Application to 3D full waveform inversion. *Inverse Problems, 32*(11), 115008. https://doi.org/10.1088/0266-5611/32/11/115008.

Miller, B. F., Huang, F., Atta, L., Sahoo, A., & Fan, J. (2022). Reference-free cell type deconvolution of multi-cellular pixel-resolution spatially resolved transcriptomics data. *Nature Communications, 13*(1), 2339.

Mondol, R. K., Millar, E. K., Graham, P. H., Browne, L., Sowmya, A., & Meijering, E. (2023). hist2RNA: An efficient deep learning architecture to predict gene expression from breast cancer histopathology images. *Cancers, 15*(9), 2569.

Monjo, T., Koido, M., Nagasawa, S., Suzuki, Y., & Kamatani, Y. (2022). Efficient prediction of a spatial transcriptomics profile better characterizes breast cancer tissue sections without costly experimentation. *Scientific Reports, 12*(1), https://doi.org/10.1038/s41598-022-07685-4.

Moriel, N., Senel, E., Friedman, N., Rajewsky, N., Karaiskos, N., & Nitzan, M. (2021). NovoSpaRc: Flexible spatial reconstruction of single-cell gene expression with optimal transport. *Nature Protocols, 16*(9), 4177–4200. https://doi.org/10.1038/s41596-021-00573-7.

Morris, S. A. (2019). The evolving concept of cell identity in the single cell era. *Development, 146*(12), dev169748. https://doi.org/10.1242/dev.169748.

Moses, L., & Pachter, L. (2022). Museum of spatial transcriptomics. *Nature Methods, 19*(5), 534–546. https://doi.org/10.1038/s41592-022-01409-2.

Mukherjee, S. (2022). *The song of the cell: An exploration of medicine and the new human.* Scribner.

Nitzan, M., Karaiskos, N., Friedman, N., & Rajewsky, N. (2019). Gene expression cartography. *Nature, 576*(7785), 132–137. https://doi.org/10.1038/s41586-019-1773-3.

Pang, M., Su, K., & Li, M. (2021). *Leveraging information in spatial transcriptomics to predict super-resolution gene expression from histology images in tumors.* Cold Spring Harbor Laboratory. Retrieved from. https://doi.org/10.1101/2021.11.28.470212.

Pcn, M., & Nr, Z. (2020). Dissecting the binding mechanisms of transcription factors to DNA using a statistical thermodynamics framework. *Computational and Structural Biotechnology Journal, 18*, 3590–3605. https://doi.org/10.1016/J.CSBJ.2020.11.006.

Petkova, M. D., Tkačik, G., Bialek, W., Wieschaus, E. F., & Gregor, T. (2019). Optimal decoding of cellular identities in a genetic network. *Cell, 176*(4), 844–855.e815. https://doi.org/10.1016/j.cell.2019.01.007.

Petukhov, V., Xu, R. J., Soldatov, R. A., Cadinu, P., Khodosevich, K., Moffitt, J. R., & Kharchenko, P. V. (2022). Cell segmentation in imaging-based spatial transcriptomics. *Nature Biotechnology, 40*(3), 345–354. https://doi.org/10.1038/s41587-021-01044-w.

Peyré, G., & Cuturi, M. (2019). Computational optimal transport: With applications to data science. *Foundations and Trends® in Machine Learning, 11*(5-6), 355–607. https://doi.org/10.1561/2200000073.

Pham, D., Tan, X., Xu, J., Grice, L. F., Lam, P. Y., Raghubar, A., ... Nguyen, Q. (2020). stLearn: Integrating spatial location, tissue morphology and gene expression to find cell types, cell-cell interactions and spatial trajectories within undissociated tissues. *bioRxiv.* 2020.2005.2031.125658-122020.125605.125631.125658. https://doi.org/10.1101/2020.05.31.125658.

Phelps, C. B., & Brand, A. H. (1998). Ectopic gene expression in Drosophila using GAL4 system. *Methods (San Diego, Calif.), 14*(4), 367–379. https://doi.org/10.1006/meth.1998.0592.

Raj, A., Van Den Bogaard, P., Rifkin, S. A., Van Oudenaarden, A., & Tyagi, S. (2008). Imaging individual mRNA molecules using multiple singly labeled probes. *Nature Methods, 5*(10), 877–879. https://doi.org/10.1038/nmeth.1253.

Ramazi, S., & Zahiri, J. (2021). Post-translational modifications in proteins: Resources, tools and prediction methods. *Database, 2021.* https://doi.org/10.1093/database/baab012.

Rao, A., Barkley, D., França, G. S., & Yanai, I. (2021). Exploring tissue architecture using spatial transcriptomics. *Nature, 596*(7871), 211–220. https://doi.org/10.1038/s41586-021-03634-9.

Ravi, V. M., Will, P., Kueckelhaus, J., Sun, N., Joseph, K., Salié, H., ... Henrik Heiland, D. (2021). Spatiotemporal heterogeneity of glioblastoma is dictated by microenvironmental interference. *bioRxiv.* 2021.2002.2016.431475-432021.431402.431416.431475. https://doi.org/10.1101/2021.02.16.431475.

Regev, A., Teichmann, S. A., Lander, E. S., Amit, I., Benoist, C., Birney, E., ... Yosef, N. (2017). The human cell atlas. *Elife, 6.* https://doi.org/10.7554/elife.27041.

Richardson, G. D., Robson, C. N., Lang, S. H., Neal, D. E., Maitland, N. J., & Collins, A. T. (2004). CD133, a novel marker for human prostatic epithelial stem cells. *Journal of Cell Science, 117*(16), 3539–3545. https://doi.org/10.1242/jcs.01222.

Rodriques, S. G., Stickels, R. R., Goeva, A., Martin, C. A., Murray, E., Vanderburg, C. R., ... Macosko, E. Z. (2019). Slide-seq: A scalable technology for measuring genome-wide expression at high spatial resolution. *Science, 363*(6434), 1463–1467.

Saqib, J., Park, B., Jin, Y., Seo, J., Mo, J., & Kim, J. (2023). Identification of Niche-specific gene signatures between malignant tumor microenvironments by integrating single cell and spatial transcriptomics data. *Genes (Basel), 14*(11), https://doi.org/10.3390/genes14112033.

Satija, R., Farrell, J. A., Gennert, D., Schier, A. F., & Regev, A. (2015). Spatial reconstruction of single-cell gene expression data. *Nature Biotechnology, 33*(5), 495–U206. https://doi.org/10.1038/nbt.3192.

Schmauch, B., Romagnoni, A., Pronier, E., Saillard, C., Maillé, P., Calderaro, J., ... Wainrib, G. (2020). A deep learning model to predict RNA-Seq expression of tumours from whole slide images. *Nature Communications, 11*(1), https://doi.org/10.1038/s41467-020-17678-4.

Shah, S., Lubeck, E., Zhou, W., & Cai, L. (2016). In situ transcription profiling of single cells reveals spatial organization of cells in the mouse hippocampus. *Neuron, 92*(2), 342–357. https://doi.org/10.1016/j.neuron.2016.10.001.

Shah, S., Lubeck, E., Zhou, W., & Cai, L. (2017). seqFISH accurately detects transcripts in single cells and reveals robust spatial organization in the hippocampus. e751 *Neuron, 94*(4), 752–758. https://doi.org/10.1016/j.neuron.2017.05.008.

Shao, X., Li, C., Yang, H., Lu, X., Liao, J., Qian, J., ... Fan, X. (2022). Knowledge-graph-based cell-cell communication inference for spatially resolved transcriptomic data with SpaTalk. *Nature Communications, 13*(1), 4429. https://doi.org/10.1038/s41467-022-32111-8.

Small, S., Blair, A., & Levine, M. (1992). Regulation of even-skipped stripe 2 in the Drosophila embryo. *EMBO Journal, 11*(11), 4047–4057. Retrieved from ⟨http://www.ncbi.nlm.nih.gov/pubmed/1327756⟩.

Song, D., Wang, Q., Yan, G., Liu, T., Sun, T., & Li, J. J. (2023). scDesign3 generates realistic in silico data for multimodal single-cell and spatial omics. *Nature Biotechnology.* https://doi.org/10.1038/s41587-023-01772-1.

Song, L., & Crawford, G. E. (2010). DNase-seq: A high-resolution technique for mapping active gene regulatory elements across the genome from mammalian cells. *Cold Spring Harbor Protocols, 5*(2), https://doi.org/10.1101/pdb.prot5384.

Song, Q., & Su, J. (2021). DSTG: Deconvoluting spatial transcriptomics data through graph-based artificial intelligence. *Briefings in Bioinformatics, 22*(5) bbaa414.

Ståhl, P. L., Salmén, F., Vickovic, S., Lundmark, A., Navarro, J. F., Magnusson, J., ... Frisén, J. (2016). *Visualization and analysis of gene expression in tissue sections by spatial transcriptomics. Science, 353,* American Association for the Advancement of Science, 78–82.

Stickels, R. R., Murray, E., Kumar, P., Li, J., Marshall, J. L., Bella, D. D., ... Chen, F. (2020). Sensitive spatial genome wide expression profiling at cellular resolution. *bioRxiv.* 2020.2003.2012.989806-982020.989803.989812.989806. https://doi.org/10.1101/2020.03.12.989806.

Data enhancement in the age of spatial biology

Stickels, R. R., Murray, E., Kumar, P., Li, J., Marshall, J. L., Di Bella, D. J., ... Chen, F. (2021). Highly sensitive spatial transcriptomics at near-cellular resolution with Slide-seqV2. *Nature Biotechnology, 39*(3), 313–319. https://doi.org/10.1038/s41587-020-0739-1.

Stringer, C., Wang, T., Michaelos, M., & Pachitariu, M. (2021). Cellpose: A generalist algorithm for cellular segmentation. *Nature Methods, 18*(1), 100–106. https://doi.org/10.1038/s41592-020-01018-x.

Stuart, T., Butler, A., Hoffman, P., Stoeckius, M., Smibert, P., Satija, R., ... Hao, Y. (2019). Comprehensive integration of single-cell data resource comprehensive integration of single-cell data. *Cell, 177*, 1888–1902.e1821. https://doi.org/10.1016/j.cell.2019.05.031.

Sun, D., Liu, Z., Li, T., Wu, Q., & Wang, C. (2022). STRIDE: Accurately decomposing and integrating spatial transcriptomics using single-cell RNA sequencing. *Nucleic Acids Research, 50*(7) e42-e42.

Svozil, D., Kvasnicka, V., & Pospichal, J. I. (1997). Introduction to multi-layer feed-forward neural networks. *Chemometrics and Intelligent Laboratory Systems, 39*(1), 43–62. https://doi.org/10.1016/S0169-7439(97)00061-0.

Tang, F., Barbacioru, C., Wang, Y., Nordman, E., Lee, C., Xu, N., ... Surani, M. A. (2009). mRNA-Seq whole-transcriptome analysis of a single cell. *Nature Methods, 6*(5), 377–382. https://doi.org/10.1038/nmeth.1315.

Tang, Z., Li, Z., Hou, T., Zhang, T., Yang, B., Su, J., & Song, Q. (2023). SiGra: Single-cell spatial elucidation through an image-augmented graph transformer. *Nature Communications, 14*(1), https://doi.org/10.1038/s41467-023-41437-w.

Thorpe, M. (2019). Introduction to optimal transport. *Lecture Notes, 3.*

Thrane, K., Eriksson, H., Maaskola, J., Hansson, J., & Lundeberg, J. (2018). Spatially resolved transcriptomics enables dissection of genetic heterogeneity in stage III cutaneous malignant melanoma. *Cancer Research, 78*(20), 5970–5979. https://doi.org/10.1158/0008-5472.CAN-18-0747.

Tian, L., Chen, F., & Macosko, E. Z. (2023). The expanding vistas of spatial transcriptomics. *Nature Biotechnology, 41*(6), 773–782. https://doi.org/10.1038/s41587-022-01448-2.

Tsien, R. Y. (1998). The green fluorescent protein. *Annual Review of Biochemistry, 67*(1), 509–544. https://doi.org/10.1146/annurev.biochem.67.1.509.

Vaswani, A., Shazeer, N., Parmar, N., Uszkoreit, J., Jones, L., Aidan, ... Polosukhin, I. (2017). Attention is all you need. *arXiv pre-print server* doi:Nonearxiv:1706.03762v7.

Vickovic, S., Eraslan, G., Salmén, F., Klughammer, J., Stenbeck, L., Schapiro, D., ... Ståhl, P. L. (2019). High-definition spatial transcriptomics for in situ tissue profiling. *Nature Methods, 16*(10), 987–990. https://doi.org/10.1038/s41592-019-0548-y.

Wan, X., Xiao, J., Tam, S. S. T., Cai, M., Sugimura, R., Wang, Y., ... Yang, C. (2023). Integrating spatial and single-cell transcriptomics data using deep generative models with SpatialScope. *Nature Communications, 14*(1), https://doi.org/10.1038/s41467-023-43629-w.

Welch, J. D., Kozareva, V., Ferreira, A., Vanderburg, C., Martin, C., & Macosko, E. Z. (2019). Single-cell multi-omic integration compares and contrasts features of brain cell identity. *Cell, 177*(7), 1873–1887.e1817. https://doi.org/10.1016/j.cell.2019.05.006.

Xia, C., Fan, J., Emanuel, G., Hao, J., & Zhuang, X. (2019). Spatial transcriptome profiling by MERFISH reveals subcellular RNA compartmentalization and cell cycle-dependent gene expression. *Proceedings of the National Academy of Sciences USA, 116*(39), 19490–19499. https://doi.org/10.1073/pnas.1912459116.

Xia, C.-R., Cao, Z.-J., Tu, X.-M., & Gao, G. (2023). Spatial-linked alignment tool (SLAT) for aligning heterogenous slices. *Nature Communications, 14*(1), https://doi.org/10.1038/s41467-023-43105-5.

Xu, H., Wang, S., Fang, M., Luo, S., Chen, C., Wan, S., ... Qu, K. (2023). SPACEL: Deep learning-based characterization of spatial transcriptome architectures. *Nature Communications, 14*(1), https://doi.org/10.1038/s41467-023-43220-3.

Yang, J., Qu, Y., Wei, N., Hu, C., Wu, H.-J., & Zheng, X. (2023). *Spider: A flexible and unified framework for simulating spatial transcriptomics data.* Cold Spring Harbor Laboratory. Retrieved from. https://doi.org/10.1101/2023.05.21.541605.

Yu, Q., Jiang, M., & Wu, L. (2022). Spatial transcriptomics technology in cancer research. *Frontiers in Oncology, 12*, 1019111. https://doi.org/10.3389/fonc.2022.1019111.

Zaret, K. S., & Carroll, J. S. (2011). Pioneer transcription factors: Establishing competence for gene expression. *Genes & Development, 25*(21), 2227–2241. https://doi.org/10.1101/gad.176826.111.

Zeira, R., Land, M., Strzalkowski, A., & Raphael, B. J. (2022). Alignment and integration of spatial transcriptomics data. *Nature Methods, 19*(5), 567–575. https://doi.org/10.1038/s41592-022-01459-6.

Zeng, Y., Wei, Z., Yu, W., Yin, R., Yuan, Y., Li, B., ... Yang, Y. (2022). Spatial transcriptomics prediction from histology jointly through transformer and graph neural networks. *Briefings in Bioinformatics, 23*(5), https://doi.org/10.1093/bib/bbac297.

Zhang, D., Schroeder, A., Yan, H., Yang, H., Hu, J., Lee, M. Y. Y., ... Li, M. (2024). Inferring super-resolution tissue architecture by integrating spatial transcriptomics with histology. *Nature Biotechnology.* https://doi.org/10.1038/s41587-023-02019-9.

Zhang, Q., Jiang, S., Schroeder, A., Hu, J., Li, K., Zhang, B., ... Li, M. (2023). Leveraging spatial transcriptomics data to recover cell locations in single-cell RNA-seq with CeLEry. *Nat Commun, 14*(1), 4050. https://doi.org/10.1038/s41467-023-39895-3.

Zhang, X., Wang, X., Shivashankar, G. V., & Uhler, C. (2022). Graph-based autoencoder integrates spatial transcriptomics with chromatin images and identifies joint biomarkers for Alzheimer's disease. *Nature Communications, 13*(1), https://doi.org/10.1038/s41467-022-35233-1.

Zhao, C., Xu, Z., Wang, X., Chen, K., Huang, H., & Chen, W. (2022). *Transformer enables reference free and unsupervised analysis of spatial transcriptomics.* Cold Spring Harbor Laboratory. Retrieved from. https://doi.org/10.1101/10.1101/2022.08.11.503261.

Zhao, E., Stone, M. R., Ren, X., Guenthoer, J., Smythe, K. S., Pulliam, T., ... Gottardo, R. (2021). Spatial transcriptomics at subspot resolution with BayesSpace. *Nature Biotechnology, 1*–10. https://doi.org/10.1038/s41587-021-00935-2.

Zhou, X., Dong, K., & Zhang, S. (2023). Integrating spatial transcriptomics data across different conditions, technologies and developmental stages. *Nature Computational Science, 3*(10), 894–906. https://doi.org/10.1038/s43588-023-00528-w.

Zhu, J., Shang, L., & Zhou, X. (2023). SRTsim: Spatial pattern preserving simulations for spatially resolved transcriptomics. *Genome Biology, 24*(1), https://doi.org/10.1186/s13059-023-02879-z.

Zhuang, X. (2021). Spatially resolved single-cell genomics and transcriptomics by imaging. *Nature Methods, 18*(1), 18–22 doi:10.1038/s41592-020-01037-8.

CHAPTER THREE

Current computational methods for spatial transcriptomics in cancer biology

Jaewoo Mo[a,1], Junseong Bae[b,c,1], Jahanzeb Saqib[a], Dohyun Hwang[d], Yunjung Jin[a], Beomsu Park[a], Jeongbin Park[b,d,e,*], and Junil Kim[a,*]

[a]School of Systems Biomedical Science, Soongsil University, Dongjak-Gu, Seoul, Republic of Korea
[b]Interdisciplinary Program of Genomic Data Science, Pusan National University, Yangsan, Republic of Korea
[c]Graduate School of Medical AI, Pusan National University, Yangsan, Republic of Korea
[d]Department of Information Convergence Engineering, Pusan National University, Yangsan, Republic of Korea
[e]School of Biomedical Convergence Engineering, Pusan National University, Yangsan, Republic of Korea
*Corresponding authors. e-mail address: jeongbin.park@pusan.ac.kr; junilkim@ssu.ac.kr

Contents

1. Introduction	72
2. Overview of spatial transcriptomics techniques	73
2.1 Imaging-based methods	75
2.2 NGS-based methods	77
2.3 Commercial methods	79
2.4 Cell segmentation	82
3. Application of spatial transcriptomics for cancer study	84
3.1 Deconvolution of NGS-based spatial transcriptomics data	84
3.2 Tissue domain analysis	93
3.3 Niche-gene analysis	96
4. Discussion	97
Acknowledgments	99
References	99

Abstract

Cells in multicellular organisms constitute a self-organizing society by interacting with their neighbors. Cancer originates from malfunction of cellular behavior in the context of such a self-organizing system. The identities or characteristics of individual tumor cells can be represented by the hallmark of gene expression or transcriptome, which can be addressed using single-cell dissociation followed by RNA sequencing. However, the dissociation process of single cells results in losing the cellular address in tissue or neighbor information of each tumor cell, which is critical to understanding the malfunctioning cellular behavior in the microenvironment. Spatial transcriptomics

[1] These authors contributed equally as co-first authors.

Advances in Cancer Research, Volume 163
ISSN 0065-230X, https://doi.org/10.1016/bs.acr.2024.06.006
Copyright © 2024 Elsevier Inc. All rights are reserved, including those for text and data mining, AI training, and similar technologies.

technology enables measuring the transcriptome which is tagged by the address within a tissue. However, to understand cellular behavior in a self-organizing society, we need to apply mathematical or statistical methods. Here, we provide a review on current computational methods for spatial transcriptomics in cancer biology.

1. Introduction

Cancer, a complex and often fatal disease, has traditionally been studied through the lens of genetics and molecular biology (Bertram, 2000; Fares, Fares, Khachfe, Salhab, & Fares, 2020). Cancer involves a heterogeneous cellular milieu known as the tumor microenvironment (TME), which encompasses surrounding blood vessels, immune cells, fibroblasts, signaling molecules, and the extracellular matrix (Giraldo et al., 2019). Notably, within the TME, the interactions between tumor and non-tumor cells or spatial configuration of various cell types are critical factors directly influencing tumor growth and invasion into new tissues (de Visser & Joyce, 2023), and its resistance and recurrence (Barker, Paget, Khan, & Harrington, 2015).

The advent of single-cell omics technologies enables cataloging each group of cells in the TMEs across various cancer types (Barkley et al., 2022; Cheng et al., 2021; Hong et al., 2021; Kinker et al., 2020; Luo et al., 2022; Ma et al., 2023; Nofech-Mozes, Soave, Awadalla, & Abelson, 2023; Qian et al., 2020; Qin et al., 2023; Tang et al., 2023; Zheng et al., 2021). However, while these technologies offer high-quality molecular characterization at the individual cell level, they do not allow for the analysis of interactions between neighboring cells as they are completely dissociated during the process (Lafzi, Moutinho, Picelli, & Heyn, 2018; Ziegenhain et al., 2017). Recent advancements in spatial transcriptomics techniques have provided a direct way to study the complexities in TMEs by capturing the spatial abundance of gene expression in the spatial context (Moses & Pachter, 2022).

Among them, techniques capable of profiling gene expression at single-cell resolution are especially valuable for examining cell-cell interactions within the TME. For instance, any fluorescence-based or next-generation sequencing (NGS)-based method with an array size smaller than a single cell (several micrometers) permits the profiling of gene expressions at this high resolution. However, even with single-cell resolution in gene expression profiling, inaccurate cell segmentation can yield flawed results. Watershed segmentation, a traditional algorithm for cell segmentation utilized in conjunction with hematoxylin and eosin (H&E), 4',6-diamidino-2-phenylindole (DAPI), or Poly-A staining images, often suffers from over- or under-segmentation, leading to an excessive number of insignificant small clusters or merged cells, respectively. In response, novel cell segmentation algorithms leveraging machine learning and artificial intelligence have been developed to address these issues.

The growth and progression of cancer occur with the transformation of surrounding components. Configuration surrounding malignant cells called the TME, consists of immune cells, non-immune stromal cells, proteins, blood vessels, intercellular substances, and metabolites (Jin & Jin, 2020). Tumor cells can shape the specific microenvironment around them through the secretion of cytokines, chemokines, and other substances. This leads to the change of neighboring cells, which can play a crucial role in tumor survival and advancement. TME ultimately promotes tumor proliferation and metastasis (Hinshaw & Shevde, 2019).

Many cancer studies using bulk RNA-seq data or scRNA-seq data have provided a great deal of understanding of TME but are insufficient to reveal structural aspects within the TME. Analysis using spatial transcriptome data interacts with traditional analysis methods to provide a spatial understanding of the different types of immune and non-immune stromal cells in the TME (Wang et al., 2023). It can also identify rare or unique cell populations within tumor tissue, including malignant cells (Berglund et al., 2018).

This chapter aims to discuss cutting-edge spatial transcriptomics techniques, computational methods including precise cell segmentation, deconvolution of cell types, tissue domain analysis, and niche analysis and their applications in cancer biology, which facilitate the study of cell-cell interactions and spatial compositions in the TMEs.

2. Overview of spatial transcriptomics techniques

As of today, numerous spatial transcriptomics methods have been developed (Giolai et al., 2019; Vandereyken, Sifrim, Thienpont, & Voet, 2023; Williams, Lee, Asatsuma, Vento-Tormo, & Haque, 2022). However, particularly in the context of cancer research, knowing the names and understanding the underlying principles of all these methods could demand an excessive amount of effort and time. In practice, it is quite common to opt for one of the commercially available methods to profile spatial gene expression. Instead of introducing every spatial transcriptomics method available to date, we focus on several selected techniques here. These have been chosen for their utility in understanding the principles behind the commercial methods commonly used in the field, and it discusses how these techniques have evolved into their current commercial forms.

The currently available spatial transcriptomics techniques can be divided into two main categories: imaging-based methods and NGS-based methods (Fig. 1). On the one hand, imaging-based methods utilize fluorescence microscopy images to localize mRNAs for hundreds or thousands of genes at single-molecule resolution. The methods are further divided into multiplexed fluorescence in situ hybridization (FISH) (Chen, Boettiger, Moffitt, Wang, & Zhuang, 2015; Codeluppi et al., 2018; Eng et al., 2019; Lubeck & Cai, 2012), and in situ sequencing (ISS) (Gyllborg et al., 2020; Ke et al., 2013; Lee et al., 2014; Mitra, Shendure, Olejnik, Edyta Krzymanska, & Church, 2003). On the other hand, NGS-based methods (Chen et al., 2022;

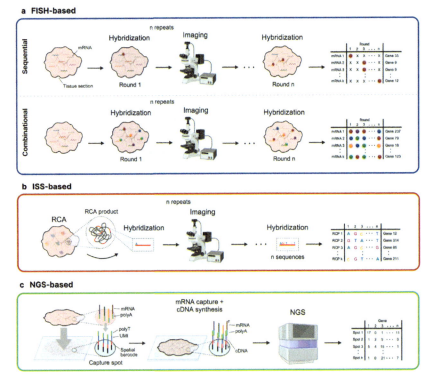

Fig. 1 Diagram for experimental flow of current spatial transcriptomics technologies. Spatial transcriptomics based on (A) fluorescence in situ hybridization (FISH), (B) in situ sequencing (ISS), and (C) next generation sequencing (NGS).

Cho et al., 2021; Rodriques et al., 2019; Russell et al., 2024; Ståhl et al., 2016) employ arrays of spatially barcoded oligonucleotides capable of capturing mRNAs. These mRNAs are sequenced using NGS technologies, demultiplexed based on the spatial barcodes, and quantified at each location. Although they offer comprehensive coverage of the transcriptome, these methods typically provide lower resolution and sensitivity compared to the fluorescence-based methods. Both imaging-based and NGS-based approaches have facilitated the mapping of gene expression within the TME (Walsh & Quail, 2023). In recent years, the development of commercial technologies based on both imaging (RNASCOPE, MERSCOPE, Xenium) and NGS-based methods (Stereo-Seq, Visium) has further increased the accessibility of these spatial transcriptomics techniques.

2.1 Imaging-based methods

Imaging-based methods refer to those based on fluorescence images of tissue at single-molecule resolution, enabling the in situ quantification of each mRNA molecule (Fig. 1A). Initially, this capability was enabled by single-molecule fluorescence in situ hybridization (smFISH), which became a widely used tool for detecting several genes of interest for cell-type identification in situ (Femino, Fay, Fogarty, & Singer, 1998). Later, several approaches were introduced to overcome the limitation of the number of detectable genes by smFISH, referred to as FISH-based methods (Chen et al., 2015; Codeluppi et al., 2018; Eng et al., 2019; Lubeck & Cai, 2012). Additionally, the more recently introduced in situ sequencing (Ke et al., 2013) enabled the sequencing of target molecules in situ, facilitating the identification of cell types in a spatial context. ISS has also given rise to several new techniques based on it, referred to as ISS-based methods (Fig. 1B) (Gyllborg et al., 2020; Ke et al., 2013; Lee et al., 2014; Mitra et al., 2003). In this section, we will review both FISH-based and ISS-based methods.

2.1.1 FISH-based methods

Recent improvements in imaging techniques have made it possible to detect more genes than was possible with conventional smFISH (Raj, van den Bogaard, Rifkin, van Oudenaarden, & Tyagi, 2008). There are two main types of these advanced techniques: (1) sequential FISH methods, which mainly automate the smFISH imaging process to reduce manual work, and (2) combinatorial FISH methods, which use non-specific probes in each round of imaging, so that the unique combination of probes used across all rounds at the same location identifies a single gene. By automating the imaging process and enabling the detection of genes with unique probe combinations, these advanced techniques open an innovative step forward in spatial transcriptomics.

One landmark sequential FISH method is osmFISH (Codeluppi et al., 2018). The name of the method came from "ouroboros" smFISH, which means the cyclic imaging procedure of smFISH based on a special apparatus designed for automation of each imaging round of smFISH so that the whole imaging procedure is cyclic. This enabled imaging of tens of genes of interest with less human effort, which enabled the mapping of cell types in a spatial context. However, it is fundamentally repeated smFISH experiments, implying that the time required for imaging increases proportionally with the number of genes, thereby constraining the number of detectable genes per experiment. This is a limitation of osmFISH on the delineation of sub-cell types, requiring further consideration of genes beyond the primary marker genes.

To surmount the constraint regarding the number of genes, SeqFISH (Eng et al., 2019; Lubeck & Cai, 2012) and MERFISH (Chen et al., 2015) introduced a novel technique called combinatorial barcoding. The principle of this technique is labeling a single mRNA with multiple different fluorescence probes so that the same mRNA is imaged in multiple imaging rounds, and then interpreting the fluorescence signal as binary bits (1 is imaged, 0 is not). After all the imaging rounds, the binary barcode of each mRNA molecule is used to identify the gene. As the gene identity is a combination of binary bits, the number of detectable genes is much more than the number of imaging rounds. Based on this, both methods enabled the detection of hundreds of genes, even up to 10,000 genes as proof of concept by combining a super-resolution microscopy.

2.1.2 ISS-based method

Unlike smFISH, ISS is a relatively new method that amplifies and reads transcripts in situ. ISS employs rolling circle amplification (RCA) to amplify the reverse transcribed complementary DNA (cDNA) from mRNAs, which results in amplicons generated at each location of mRNAs. Then, reading either (1) the short sequence of cDNA, or (2) the barcode sequences of each amplicon, to determine the gene identity of each mRNA. Typically, sequencing-by-ligation (SBL) chemistry, rounds of hybridization of the fluorescence probes followed by the identification of 1–2 bases by imaging, is used to read the sequence. More specifically, the chemistry is further categorized into sequencing by oligonucleotide ligation and detection (SOLiD), in situ sequencing with two-base encoding for error correction (SEDAL), or combinatorial Probe-Anchor Ligation (cPAL). Then the gene identity is revealed by using the sequence read.

Fluorescence in situ sequencing (FISSEQ) was first introduced in 2003 (Mitra et al., 2003), and later improved in 2014 (Lee et al., 2014) for profiling gene expression in situ. FISSEQ is an untargeted method, which relies on cross-linking of the cDNA amplicons after RCA and sequencing them in the cells via SOLiD chemistry. The resulting amplicon is sized at ~600 nm, which has enough resolution to capture the subcellular structure of cells.

In situ sequencing (ISS) developed by Ke et al. (2013) is another approach based on padlock probing (Nilsson et al., 1994). ISS takes a targeted approach to avoid high crowding of amplicons after RCA. Two different targeted methods were introduced, namely "gap-targeted" and "barcode-targeted" sequencing. The gap-targeted sequencing method involves the filling of the gap between the two ends of the probe after the hybridization, which contains the sequence of mRNAs. The barcode-targeted method is based on introducing a unique barcode to identify genes in the probe. After that either the sequence of mRNA (gap-targeted) or the barcode sequences (barcode-targeted) are read by sequence-by-ligation (SBL) chemistry.

Later, an improved version of the ISS method was released by the same group, called HybISS (Gyllborg et al., 2020). HybISS introduces a sequence-by-hybridization (SBH) chemistry, based on a combinatorial barcoding of padlock probes. By introducing SBH chemistry, HybISS achieved a higher signal-to-noise ratio (SNR), detection efficiency, and number of genes to profile.

2.2 NGS-based methods

NGS-based methods utilize the presence of the poly(A) tail contained in mRNA to capture released mRNA from permeabilized tissue in situ, followed by the synthesis of cDNA based on the mRNAs, and subsequently read the sequence through high-throughput NGS (Fig. 1C). This allows for the capture of both the spatial barcode and the sequence of the mRNAs. The arrays used for capturing mRNA comprise oligonucleotides, including poly(T) to capture the poly(A) of mRNAs, and spatial barcodes to identify spatial location. These oligonucleotides are distributed on solid surfaces in various forms: as clusters on glass slides, attached to the surfaces of beads, or contained within microwells. Given that most NGS-based methods are based on poly(T) and can capture the whole transcriptome, they offer a broader scope than imaging-based methods, which have inherent limitations of time and budget.

The term "spatial transcriptomics" is widely used today to describe techniques that are capable of profiling gene expression in a spatial context. However, the term was first used to refer to the method introduced by Ståhl et al. (2016). The "Spatial transcriptomics (ST)", or the Ståhl method introduced a new way to profile gene expression in situ by mounting tissue on a spatially barcoded array and quantifying the captured transcripts via NGS. However, this method has a significant limitation in terms of resolution (with a spot diameter of $100\,\mu m$ and a center-to-center distance of $200\,\mu m$) compared to imaging-based methods.

Slide-seq (Rodriques et al., 2019) introduces DNA-barcoded microparticles called "beads", to achieve a higher spatial resolution to profile gene expression in a spatial context. The procedure involves preparing tissue sections placed on a glass coverslip with barcoded beads, and the bead barcode sequences are determined using SOLiD (sequencing by oligonucleotide ligation and detection) chemistry. The resolution of the method depends on the size of the beads, which are reported as $10\,\mu m$ in diameter in the original publication, which reaches a cellular resolution.

Later, to further increase the capability to capture the transcriptome in the cells, the method was leveraged to tag nuclei, instead of capturing mRNA molecules. The method is called Slide-tags (Russell et al., 2024), which rely on single-cell sequencing assays to conduct multiomics profiling (including snRNA-seq, snATAC-seq, TCR-seq) of nuclei at the single-cell level in a spatial context. Specifically, like Slide-Seq, the method uses oligonucleotides incorporating a photocleavable linker and a spatial barcode attached to the beads, and these beads are spatially indexed through SOLiD sequencing. After mounting tissue on the bead array, the spatial barcodes are photocleaved and associated with nuclei. The tagged nuclei are isolated and captured via a droplet-based microfluidics device to profile the spatial barcodes.

Seq-Scope is a novel technique that uses an Illumina flow cell to pattern an oligonucleotide array on the surface of a flow cell at a submicrometer resolution (Cho et al., 2021). This approach involves two sequencing steps: In the first step, an oligonucleotide library containing the poly(T) sequences and spatial barcodes, named high-definition map coordinate identifier (HDMI) sequences, are amplified on the surface of the flow cell to form spatial clusters with size ~ 0.5–$0.8\,\mu m$ (Bentley et al., 2008). Subsequently, each cluster is sequenced by an Illumina sequencing machine to identify its location on the flow cell and the corresponding HDMI sequence. In the second step, the glass window of the flow cell is removed to reveal the patterned HDMI array, and tissue is mounted on top of it to capture mRNA. The captured mRNA is reverse-transcribed to form the sequencing library containing both the HDMI and cDNA sequences, and then the library is sequenced again to locate the spatial location of mRNA based on the HDMI sequence. Although Seq-Scope is a method that only requires an existing Illumina sequencing machine, disassembling flow cells is not an easy task, prohibiting the widespread adoption of the method. Recently, an open-source experimental and computational resource to easily perform Seq-Scope based on Illumina NovaSeq 6000 S4 flow cell was released, which increased the accessibility of the method [https://www.biorxiv.org/content/10.1101/2023.12.22.572554v1.full].

Stereo-Seq is a method that achieves sub-cellular resolution alongside a field of view (FOV) that can extend to a centimeter or larger (Chen et al., 2022). This technique is based on DNA nanoball (DNB) sequencing technology (Drmanac et al., 2010), which employs a patterned nanoarray for high-efficiency genome sequencing with reduced reagent consumption. Like Seq-Scope, Stereo-Seq involves imaging the patterned nanoarray with DNBs to determine the spatial location of each spot. Unique molecular identifiers (UMIs) and poly(T) sequences are then ligated at each spot to capture mRNAs with poly(A) tails. Following the ligation step, tissue is mounted atop the nanoarray, facilitating the spatial profiling of mRNA abundance. Notably, Stereo-Seq achieves a high spatial resolution on a nanometer scale, characterized by a 220 nm spot size and 500 nm spacing (center-to-center), which permits high-resolution gene expression profiling within a spatial context. As Stereo-Seq utilizes a customized nanoarray, it allows for a flexible FOV as the size of the array is customizable.

2.3 Commercial methods

2.3.1 Imaging-based methods

Among FISH-based methods, current commercialized options include Spatial Genomics' commercialized version of seqFISH known as GenePS, Vizgen's commercialized MERFISH known as MERSCOPE, ACD's RNAScope, and Nanostring's CosMx Spatial Molecular Imager (SMI) (Fig. 2).

RNAScope, launched by ACD in 2011, is a sequential FISH method combining cyclic smFISH and signal amplification to achieve a high detection rate of fluorophores using double Z-shaped probes. The principle of this technique is as follows: when a pair of z probes bind adjacently on

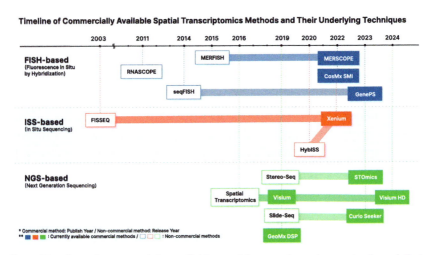

Fig. 2 Timeline of commercially available spatial transcriptomics methods and their underlying techniques.

the target sequence of mRNA, a pre-amplifier hybridizes to the exposed adjacent pre-amplifier binding sites of the Z-shaped probes (each 14-base, together 28-base), followed by the hybridization of multiple amplifiers. These amplifiers have multiple binding sites where many fluorescently labeled probes hybridize, achieving high signal amplification. The method is highly specific, for example, if only one Z-shaped probe hybridizes to the mRNA, the pre-amplifier cannot hybridize to the pre-amplifier binding site of the Z-shaped probe, thus preventing signal amplification.

More recently, in 2022, Nanostring launched the CosMx SMI, another commercially available method based on combinatorial FISH, combined with the capability of spatial protein profiling. This method uses primary probes containing a target binding domain and multiple subdomains of a readout domain, along with fluorescently labeled secondary probes, to perform combinatorial barcoding and signal amplification. The available gene panel scale is at the 1000-plex level. In the case of ISS-based methods, Xenium by 10× Genomics is currently the most used method. Previously there have been two commercial companies, CARTANA and ReadCoor, established based on HybISS and FISSEQ techniques, respectively. Both commercial companies were acquired by 10× Genomics in 2020 and evolved into Xenium, which was launched in late 2022. Xenium uses targeted gene panels for spatial gene expression profiling, with the panel capable of including up to 480 genes as of today.

2.3.2 NGS-based methods

In the case of NGS-based methods, a variety of companies are actively developing and bringing technologies to market. Examples include Visium, VisiumHD, GeoMx Digital Spatial Profiler (DSP), Curio Seeker, and STOmics.

10× Genomics' Visium platform was released in 2019, after the acquisition of Spatial Transcriptomics (ST) in 2018 (Ståhl et al., 2016). Compared to ST, the spatial resolution was improved, having a hexagonal-shaped spot size of 55 μm with a center-to-center distance of 100 μm. Still, the resolution is not high enough to distinguish single cells. As the spot size is bigger than the size of a single cell, gene expressions from multiple cells are mixed. Thus, to understand the proportion of cell types within each spot, various cell-type deconvolution methods have been developed to address this. Such methods are discussed in detail in the "3.1 Deconvolution" section.

Recently, 10× Genomics announced the launch of an improved version of Visium, named VisiumHD. Unlike Visium, VisiumHD contains oligonucleotides capable of capturing mRNAs in 2×2 μm square spots aligned without gaps. Notably, VisiumHD is not based on Poly(A) capture – it is a targeted method with a fixed gene panel, which makes designing a custom gene panel will not be easy.

STOmics is another NGS-based high-resolution spatial transcriptomics platform released in April 2023, developed by BGI Genomics based on Stereo-Seq (Chen et al., 2022). By commercializing Stereo-Seq, a variety of choices of field-of-view (FOV) are now available, ranging from 1 cm × 1 cm to 13 cm × 13 cm.

Curio Seeker is a commercialized product by Curio Bioscience, launched in 2023, bringing Slide-seq technology to the market. The platform is based on Slide-seqV2 (Stickels et al., 2021), which significantly improved upon the original Slide-seq by boosting RNA capture efficiency approximately tenfold through advancements in array indexing and library preparation methods. The product is available in two versions, with capture areas of 3 mm × 3 mm and 10 mm × 10 mm, each containing beads of 10 μm in diameter that can capture 1–2 cells.

GeoMx DSP is a product of Nanostring, launched in 2019. GeoMx DSP uses a microdissection-based approach, which makes it slightly different from the other methods. It employs photocleavable targeted probes attached to transcripts and allows the optical selection of regions of interest (ROI) via UV light illumination. GeoMx DSP is a targeted method, but the available gene panel contains around 20,000 genes (over 18,000 for humans, over 20,000 for mice) covering almost the whole transcriptome.

2.3.3 FFPE sample capability

Many spatial transcriptomics methods are primarily tested with fresh frozen tissues, as these tissues preserve the high-quality molecules present within them. However, particularly in the study of human disease samples, obtaining suitable fresh frozen tissue can be very challenging. In the case of cancer research, many tissues are archived as formalin-fixed, paraffin-embedded (FFPE) tissue blocks in hospitals, biobanks, and biomedical research laboratories. Often, FFPE samples represent the only available tissue type for study, highlighting a critical need for spatial transcriptomics techniques that can effectively work with these preserved samples.

However, in the case of the analysis of FFPE tissues, often there is a considerable amount of degradation of mRNAs within the tissue. This degradation often affects the poly(A) tail at the mRNA's $3'$ end, posing a challenge for the methods that rely on poly(T) sequences to capture mRNAs (Lin et al., 2019). To address this issue, targeted approaches such as Visium FFPE and Visium HD have been introduced which capture mRNAs with an array but do not sequence the mRNAs via NGS. Including these, many commercial spatial transcriptomics methods are capable of profiling spatial gene expression in FFPE tissue, including imaging-based methods (MERSCOPE, RNAScope, CosMx SMI, Xenium), and NGS-based methods (GeoMx DSP, Visium, VisiumHD, STOmics Stereo-seq FFPE).

2.4 Cell segmentation

The cell segmentation process aims to identify and delineate individual cells within spatially resolved datasets obtained from imaging-based spatial transcriptomics methods. The primary goal of cell segmentation is to convert raw spatial data, typically represented as a gene-spot matrix detailing a single molecule of mRNA and its localization within a tissue, into a gene-cell matrix. This transformation assigns detected gene expression signals to specific cells, offering insights into the unique transcriptional profiles characterizing individual cells within the tissue.

ClusterMap is a tool for cell segmentation using imaging-based spatial transcriptomics data (He et al., 2021). ClusterMap leverages density peak clustering (DPC) to pinpoint cluster centers based on RNA spot density and distance within the joint physical and neighborhood gene composition (P-NGC) space. The initial phase involves extracting P-NGC coordinates for mRNA spots, capturing both spatial mRNA spot locations and their neighboring genes. Following this, these coordinates undergo projection into P-NGC spaces, seamlessly integrating them computationally. And the

DPC algorithm is engaged to cluster mRNA within this integrated P-NGC space. Subsequently, each mRNA spot finds assignment to a distinct cluster, thereby signifying one cell. This innovative method shows remarkable precision in cell identification, achieving cellular resolution in a variety of biological samples and excelling in cell segmentation. Even when extending coverage to 3D volumetric data from thick tissue blocks, it shows consistent performance despite irregular boundaries, varying densities, and heterogeneous shapes and sizes.

Baysor is a probabilistic model-based cell segmentation tool, that facilitates segmentation performance with an optimized approach that considers transcriptome composition, cell size, and shape (Petukhov et al., 2022). This model works holistically to detect cell locations and assign molecules to each cell. It generates a cell-by-gene expression matrix from the initial image data, converting the spot-by-gene matrix. Baysor uses a new concept: neighborhood composition vectors (NCVs). The concept of NCVs introduces an approach to analyzing spatial data without the need for traditional segmentation processes. Here, NCVs harness the spatial relationships between molecules by examining the k nearest neighbors for each molecule. These vectors effectively visualize gene distribution in the spatial context, enabling pattern identification through clustering or lower-dimensional embeddings. Another important aspect of Baysor is its general approach to statistical labeling of spatial data in transcriptomics. This method, driven by Markov Random Field (MRF) priors and an expectation-maximization (EM) algorithm, addresses label assignment challenges like cell segmentation and differentiating signal from background molecules. Its versatility in using observable data and prior information from scRNA-seq profiles reflects its adaptability across diverse spatial labeling problems, despite requiring tailored adaptations for each challenge.

To study the role and diversity of tumor-associated macrophages (TAMs) in PDAC, Carroni et al. integrated scRNAseq, Visium, and molecular cartography (single molecule combinatorial FISH) (Caronni et al., 2023). The authors found that IL-1β+ TAMs have a critical role in PDAC based on scRNAseq data of PDAC patients and mouse models. Tumor-intrinsic IL-1β response signature positive (T1RS+) cancer cells were determined through the analysis of single-cell spatial gene expression patterns. These patterns were intricately calculated through segmentation using Baysor on spatial images, providing detailed insights into the interactions between IL-1β+ TAMs and T1RS+ cancer cells. The analysis revealed high co-expression of the IL-1β target gene CXCL1, which is strongly expressed in T1RS+ cancer cells, along with tumor marker (KRT19), macrophage marker (CD6), IL-1β (IL1A, IL1B, THBS1), and PGE2 (PTGS2, PTGER1, which induces the IL-1β+ TAM). Marker expression in IL-1β+ TAMs was found to have a stronger correlation with CXCL1 expression than markers of other TAM subtypes. Additionally, T1RS+ PDAC cells and IL-1β+ TAM cells were significantly and selectively enriched in each other's spatial neighborhoods. These findings indicate that T1RS+ PDAC cells and IL-1β+ TAM cells are closely located to each other, suggesting local interaction between them.

SCS is a method that combines sequencing data and staining data to perform cell segmentation (Chen, Li, & Bar-Joseph, 2023). Until recently, array-based spatial transcriptomics technologies did not have subcellular resolution. However, methods with subcellular resolution, such as StereoSeq and Seq-Scope, have emerged, and the challenge of analyzing data obtained from these methods is the task of assigning multiple spots to a single cell. SCS improves the accuracy of this task by combining sequencing data and staining data produced by these high-resolution array-based methods, enabling more cells to be identified and more accurate cell size estimation. Examples of combining data include Stereo-seq data with nucleus staining images, or Seq-scope data with H&E staining images. It performs cell segmentation in three steps, the first of which is to identify cell nuclei from the staining image using the Watershed algorithm. Second, the transformer model uses an attention mechanism to infer whether each spot is part of the cell or extracellular matrix based on the cell center, using the gene expression data of neighboring spots as input. Finally, for spots that are determined to be part of a cell, the cell boundary is drawn by tracking the gradient flow from spots to the cell center.

3. Application of spatial transcriptomics for cancer study

3.1 Deconvolution of NGS-based spatial transcriptomics data

Among various spatial transcriptomics platforms, 10× Visium has been used in most of the publications of spatial transcriptomics for cancer due to low cost and easy availability (Moses & Pachter, 2022). However, in these publications, interactions between cell types cannot be investigated at a single cell level since the 10× Visium platform cannot provide high-resolution transcriptomes. High-resolution spots in spatial transcriptomics such as Slide-seqV2 (Stickels et al., 2021) also cannot be treated as single cells since cells are irregularly distributed. To analyze these averaged transcriptomes of 2–30 cells, they borrowed deconvolution algorithms providing insights on cellular compositions depending on spatial context. In this section, we summarize the deconvolution algorithms utilized in various cancer studies (Table 1).

3.1.1 Stereoscope

Stereoscope is one of the pioneering deconvolution tools for spatial tran-scriptomics. It assumes that both spatial and single-cell data follow a negative binomial distribution (Andersson et al., 2020). Based on this assumption, stereoscope first infers cell type specific parameters using maximum likelihood estimation (MLE) and then deconvolves each spot of the spatial data using maximum a posteriori (MAP) estimation. Andersson et al. (2020) applied stereoscope to the ST datasets obtained from mouse brain and developmental heart, estimating spatial arrangement of cell type proportions, which are similar to what was previously reported.

In another study from the same research group, Andersson et al. (2021) employed the stereoscope to analyze the spatial distribution and co-loca-lization patterns of cell types using ST/Visium datasets of 36 sections obtained from HER2+ breast cancer and scRNAseq datasets from the same cancer type. Based on the deconvolution results, Andersson et al. first confirmed spatial enrichment or depletion of cell types in each region which was annotated by pathologists. For example, B cells, T cells, and myeloid cells were enriched in immune infiltrated regions and depleted in invasive cancer regions. Next, colocalization analysis between sub-types of myeloid cells and sub-types of T cells were investigated, identifying that IFIT1+ T cells and CXCL10+ macrophages were closely located. The expression of CXCL10 and IFIT1 is linked to type 1 interferon signaling, and a strong concentration of type 1 interferon signaling was detected in this area. Given the anti-tumor immune function of type 1 interferon signaling within the tumor (Zitvogel, Galluzzi, Kepp, Smyth, & Kroemer, 2015) and the presence of cancer treatments that induce the activation of Type 1 interferon (Parker, Rautela, & Hertzog, 2016), spatial colocaliza-tion patterns between IFIT1+ T cells and CXCL10+ macrophage could be associated with patient prognosis.

Wu et al. (2021) also applied stereoscope for studying heterotypic interactions in the TMEs of breast cancer. The deconvolution results on 10× visium data of triple negative breast cancer (TNBC) samples identified

Table 1 Examples of application of spatial transcriptomics for cancer study.

Tool	Data platform	Organ	Cancer type	References
SpatialDe	ST	Prostate	Prostate adenocarcinoma	Wang et al. (2020)
Stereoscope	10× Visium	Breast	Luminal, HER2+, Triple negative breast cancer	Wu et al. (2021)
Stereoscope	ST	Breast	HER2+ breast cancer	Andersson et al. (2021)
BayesSpace	ST, 10× Visium	Skin, Breast, Ovary	Melanoma, HER2+ breast cancer, ovarian adenocarcinoma	Zhao et al. (2021)
NMF	10× Visium	Ovary, Uterus, Breast, Pancreas, Gastrointestine, Liver	Ovarian cancer, Uterine corpus endometrial carcinoma, Breast cancer, Pancreatic ductal adenocarcinoma, Gastrointestinal stromal tumor, Liver hepatocellular carcinoma	Barkley et al. (2022)
Vesalius	10× Visium	Breast	Breast cancer	Martin et al. (2022)
RCTD	10× Visium	Pancreas	Pancreatic ductal adenocarcinoma	Cui Zhou et al. (2022)
SpatialDecon	Digital spatial profiling	Lung	Non-small cell lung cancer	Zhang et al. (2022)
CARD	ST	Pancreas	Pancreatic ductal adenocarcinoma	Chen et al. (2022)

cell2location	10× Visium	Kidney	Renal cell carcinoma	Li et al. (2022)
cell2location	10× Visium	Colon	Colorectal cancer	Ozato et al. (2023)
cell2location	10× Visium	Nasopharynx	Nasopharyngeal carcinoma	Gong et al. (2023)
SpatialDecon	Digital spatial profiling	Esophagus	Esophageal squamous precancerous lesions, Esophageal squamous cell carcinoma	Liu et al. (2023)
RCTD	10× Visium	Brain	Recurrent glioblastoma, brain metastases	Sun et al. (2023)
CellNeighborEx	Slide-seq	Liver	Mouse liver cancer	Kim et al. (2023)
Baysor	10× Visium	Pancreas	Pancreatic ductal adenocarcinoma	Caronni et al. (2023)
RCTD, CARD	10× Visium	Skin	Acral melanoma	Wei et al. (2023)
cell2location	10× Visium	Pancreas	Pancreatic ductal adenocarcinoma	Kim et al. (2024)
BayesSpace	10× Visium	Colon	Colorectal cancer	Valdeolivas et al. (2024)
Niche-DE	10× Visium	Liver	Liver metastases of colorectal cancer	Mason et al. (2024)

the distribution of TMEs. In TNBC, myofibroblast-like CAFs (mCAFs) were enriched in invasive cancer regions, while inflammatory CAFs (iCAFs) were distributed across stroma and tumor-infiltrating lymphocytes (TILs)-aggregate regions, and invasive cancer, which was also discovered in Andersson et al. (2021). Both studies demonstrated that iCAFs colocalized with multiple lymphocyte groups, including CD4+/CD8+ T cells and memory/naïve B cells, whereas mCAFs were spatially positively correlated with CD8+ T cells, suggesting that mCAFs are associated with invasive cancers characterized by an immune inflamed phenotype.

3.1.2 SpatialDecon

SpatialDecon is another deconvolution method using log-normal regression with modeling background expression count (Danaher et al., 2022). In the paper of SpatialDecon, the authors showed that SpatialDecon outperformed other approaches based on least-squares regression. For the comparison, they used expression data of varying proportions of two cell lines. The authors also suggested that SpatialDecon is suitable for deconvolution of cancer and immune cells in tumor tissues since they constructed reference cell type profiles based on various tissue types.

With a non-small cell lung cancer (NSCLC) dataset, SpatialDecon suggested that the cell abundance profiles in the TMEs of primary tumors and metastases were clearly distinguished (Zhang et al., 2022). The TMEs of lung primary tumor regions were more enriched in T cells, B/Plasma cells than those of brain metastatic regions. This study also found that the degree of activation of T and B cells and the production of antibodies by cytokines and chemokines were more reduced in the TMEs of brain metastatic regions, suggesting that the immune response is more reduced in the brain metastatic environment.

Liu et al. (2023) studied the progression mechanisms from esophageal squamous precancerous lesions (ESPL) to esophageal squamous cell carcinoma (ESCC) using spatial transcriptomics data obtained from different status of ESPL and ESCC samples. SpatialDecon revealed the differential immune cell infiltration across four pathological stages; normal epithelia (NE), low-grade intraepithelial neoplasia (LGIN), high-grade intraepithelial neoplasia (HGIN), and ESCC. For example, the results showed an increase in fibroblasts and macrophages during the transition from ESPL to ESCC. Progressing from NE to ESPL, the proportion of M1 macrophages with anti-tumor effects decreased, while the proportion of M2 macrophages with cancer-promoting functions increased. The increment of fibroblasts and macrophages across the ESCC progression can be interpreted that these two cell types are known to be involved in tumor cell proliferation and immune regulation (Kalluri & Zeisberg, 2006; Salmaninejad et al., 2019; Zhang et al., 2019).

3.1.3 Cell2location

Cell2location is a Bayesian model mapping cell type in spatial transcriptomics data by employing a negative binomial regression to estimate reference single cell expression signatures (Kleshchevnikov et al., 2022). It achieved high sensitivity in the estimation of cell abundances by modeling technological sensitivity, gene-specific additive shift, and per-location sensitivity. Owing to the high sensitivity of cell2location, it already has been applied in many cancer researchs (Gong et al., 2023; Kim et al., 2024; Li et al., 2022; Ozato et al., 2023).

In a study of renal cell carcinoma (RCC), analyzing 10× visium data with using cell2location, revealed that malignant cells with epithelial-mesenchymal transition (EMT) characteristics were found in greater numbers at the tumor-normal boundary (the edge of the tumor) compared to the center of the tumor (Li et al., 2022). This observation reflects that tumor cells in EMT are highly invasive and have metastatic status.

In a study of nasopharyngeal carcinoma (NPC) (Gong et al., 2023), Gong et al. found that CD70-CD27 interactions are enhanced between NPC cells and CD4+ regulatory T cells (Tregs) using cell-cell interaction analysis in scRNAseq. To further investigate the exact role of CD70-CD27 signaling, Gong et al. deconvolved 10× visium data using cell2location, identifying co-localization of CD70+ NPC cells and Tregs. This finding indicates that CD70+ NPC cells play an important role in the CD70-CD27 co-stimulatory pathway, which has been shown to facilitate Treg proliferation in other studies (Claus et al., 2012; Yang, Novak, Ziesmer, Witzig, & Ansell, 2007). The authors also confirmed the role of CD70+ NPC cells using knockout experiments.

Ozato et al. (2023) also used cell2location on 10× visium data in a study of colorectal cancer (CRC) using a public scRNAseq data as a reference (Lee et al., 2020). Cell2location enables them to identify differential spatial distributions between CRC tissue regions. CRC cells in the invasion front region showed high co-localization with SPP1+ macrophages, which is associated with M2 phenotype or tumor immunosuppression and promotion of tumor progression (Capote et al., 2016; Li et al., 2021; Wang & Denhardt, 2008). Further analysis of spatial distribution identified that

HLA-G is expressed in the invasive front, which is known to promote M2 macrophage polarization. Further analysis of the spatial distribution confirmed that HLA-G, known to promote M2 macrophage polarization, was expressed at the invasive front (Lin & Yan, 2021; Zhang, He, Wang, & Liao, 2017). The authors validated this finding in spatial transcriptomics using HLA-G knock-out mouse experiments.

In a study on the characterization of tumor epithelial cells and cancer associated fibroblasts in the tumor microenvironment of PDAC, Kim et al. identified the spatial distribution of cancer and fibroblast subtypes in the tissue using cell2location on paired scRNAseq and 10× visium datasets (Kim et al., 2024). Spatial neighborhood enrichment analysis between cell types identified that a specific cancer subtype (VGLL1+) was spatially correlated with the two major cancer subtypes, classical (TRIM54+) and basal-like (KRT6A+), which were not adjacent to each other. This finding of the bridging role of VGLL1+ cancer cells was predicted by their model of pancreatic cancer dynamics based on scRNAseq analysis. Kim et al. also found TME cell types that are spatially proximal to cancer cells and those that are not. Among the fibroblast subtypes, myoblastic cancer associated fibroblast (LRRC15+) exhibited high proximity to cancer cells, while inflammatory cancer associated fibroblast (SFRP1+) exhibited low proximity to cancer spots.

3.1.4 RCTD

RCTD (Robust Cell Type Decomposition) (Cable et al., 2022) is another probabilistic model-based deconvolution method, specifically designed for analyzing Slide-seq data, a high-resolution (10 μM scale) spatial transcriptomics platform. RCTD used MLE to infer cell type compositions in each spot. Like cell2location, RCTD model includes pixel-specific effect, gene-specific platform random effect, and gene-specific random overdispersion. However, RCTD calculates the log likelihood of the model based on the observation of Poisson-lognormal mixture in count data instead of a linear function.

Sun et al. used RCTD to predict the spatial distribution of cell types in 10× Visium data of recurrent glioblastoma (rGBM) and brain metastases (BrM), which is altered by the immune checkpoint blockade (ICB) treatment (Sun et al., 2023). Based on the deconvolution results, Sun et al. investigated the cell types proximal to tumor cells and found that the tumor-adjacent terminally exhausted T cells consistently increased by ICB treatment in both rGBM and BrM samples. Comprehensive investigation of the neighborhood analysis between all sub-cell types, identified that MRC1+ macrophages also known as CD206+ or M2 macrophages were enriched with a blood vasculature signature in BrM samples without treating ICB. The MRC1+ macrophages are known to be involved in promoting angiogenesis and maintaining the blood-brain barrier in other studies (Harney et al., 2015; He et al., 2012, 2016). However, in BrM treated with ICB, there was a significant reduction in the co-localization of MRC1+ macrophages and vascular concentrated spots. Conversely, in rGBM, the neighborhood fractions of vascular cells and MRC1+ macrophages were increased by ICB treatment. These findings suggest that the spatial distribution of immune subtypes varies based on the application of ICB treatment and whether the tumor is metastatic or primary.

Zhou et al. studied the TMEs in untreated and chemo resistant PDAC using multi-omics atlas including scRNAseq, bulk proteogenomics, spatial transcriptomics with 10× Visium platform, and cellular imaging (Cui Zhou et al., 2022). Dissection of the TMEs using RCTD on 10× Visium data observed spatially and transcriptionally distinct tumor subpopulations even within the same patient sample. Mapping of tumor subtypes distinguished in scRNA-seq data to spatial transcriptome slides identified a subgroup that clustered separately from other tumor subpopulations and exhibited distinct morphology. Gene set enrichment analysis showed that this sub-population upregulated genes associated with the fucosylation, hydroxylation and HIF pathways and the basal-like tumor subtype (BTNL8, AGR3 and LYZ). This spatially and transcriptionally distinct population had different H&E morphology and was likely more aggressive, indicating the heterogeneity of PDAC tumor cells.

3.1.5 CARD

The deconvolution algorithms including stereoscope, spatialDecon, RCTD and cell2location deconvolved spatial transcriptomics data by assuming that the expression profile of each spot can be treated as independent bulk RNAseq data. To incorporate spatial localization information, Ma et al. developed NMF (non-negative matrix factorization)-based algorithm, CARD (conditional autoregressive-based deconvolution), which accommodates the spatial correlation structure (Ma & Zhou, 2022). This strategy enhances the accuracy and robustness of deconvolution, even when using a scRNA-seq reference that is not perfectly matched. Ma et al. also showed that CARD allows to impute cell type compositions by modeling spatial correlation. Ma et al. applied CARD to a ST dataset from a pancreatic ductal adenocarcinoma (PDAC) sample (Moncada et al., 2020), displaying clear regional separations, which were not able to be captured in the other tools.

Using CARD, Chen et al. analyzed the same ST dataset of PDAC to identify metastatic gene signatures. CARD enables the spatial location of a metastatic tumor cell type exhibiting (Chen et al., 2022). The metastatic tumor cell types were found to preferentially distribute in the ductal epithelium and cancerous regions of the pancreas and differentially expressed genes such as TFF1, TFF2, and HMGA1, which are associated with PDAC cell migration (Arumugam et al., 2011; Guppy et al., 2012; Kajioka et al., 2021). This is consistent with previous reports of inflammation promoting PDAC metastasis (Khalafalla & Khan, 2017; Padoan, Plebani, & Basso, 2019; Steele et al., 2013).

To investigate the environmental system of primary tumor tissue according to the presence or absence of lymph node (LN) metastasis in acral melanoma (AM), Wei et al. (2023) used both CARD and RCTD on 10× Visium data of LN+ AM and LN- AM samples. The deconvolution analysis identified heterogeneity of different CNV levels and differences in immune regions depending on the presence or absence of secondary tumor. The tumor regions with high CNV levels were highly heterogeneous among patients. In contrast, the non-tumor regions mostly had subtypes shared among patients. These results reflect the high heterogeneity among AM patients in tumor regions. In addition, immune cells were predominantly clustered in non-tumor regions compared to tumor regions. Notably, fewer immune cells were penetrated in the local cellular environment of LN+ AM than LN- AM. These observations suggest that immune activity was downregulated in AM patients with LN metastases compared to patients without LN metastases.

3.1.6 Other methods

Barkley et al. investigated the composition and interactions between tumor cell states and TMEs across a pan-cancer scale (Barkley et al., 2022). Barkley et al. deconvolved 10× Visium data of 6 cancer types with pancancer scRNAseq datasets across 15 cancer types using non-negative linear least squares (NNLS) regression. The 6 cancer types of 10× Visium data include ovarian cancer (OVCA), breast cancer (BRCA), uterine corpus endometrial cancer (UCEC), pancreatic ductal adenocarcinoma (PDAC), gastrointestinal stromal cancer (GIST) and liver hepatocellular carcinoma cancer (LIHC). The deconvolution enables delineated 10× Visium spots into 'Malignant', 'Normal' and 'Both' spots, providing also differential characteristics in these three types of spots. 'Both' spots containing combinations of malignant and normal cells exhibited a lower percentage of endothelial cells compared to 'Normal' spots (with no malignant cells) and displayed higher levels of neutrophils. They also found higher M1/M2 ratios in 'Both' spots compared to 'Normal' spots in four cancer types including three gynecological cancer types (OVCA, BRCA, UCEC) and PDAC. This difference was not found in the other two cancer types (GIST and LIHC). This indicates a robust activity of anti-cancer macrophages in close vicinity to tumor cells. To further utilize these cell type compositions, Barkley et al. calculated the correlations between cell type compositions and cancer cell characteristics within 'Malignant' spots. This correlation analysis exhibited a positive correlation between the proximity of macrophages and T cells with interferon response signatures. And EMT gene signatures were positively correlated with fibroblasts and endothelial cells and negative correlated with other malignant cells, aligning with the observation that fibroblasts and endothelial cells are enriched at the border region between tumors and surrounding normal tissue and interact with cancer-associated fibroblasts (Puram et al., 2017). Collectively, these findings suggest that the cancer cell state is influenced by stromal and immune cells.

3.2 Tissue domain analysis

Both deconvolution and cell segmentation methods usually operate independently of spatial coordinates, focusing on each single cell. Typically, the spatial information used in the downstream analysis following these processes. In contrast to these two strategies, tissue domain analysis focuses on identifying spatial features such as anatomical structures within a tissue. Its objective is to uncover spatial features, employing algorithms that group transcriptomically similar spots or cells impartially to unveil spatial gene expression patterns. Examples of these tissue domain analysis methods include SpatialDE (Svensson, Teichmann, & Stegle, 2018), SpaGCN (Hu et al., 2021), SPARK (Sun, Zhu, & Zhou, 2020), BayesSpace (Zhao et al., 2021), Vesalius (Martin, Kim, Lövkvist, Hong, & Won, 2022), and (Table 1).

3.2.1 SpatialDE

SpatialDE is a tool designed to identify genes demonstrating spatial patterns of expression variation across tissues or spatially variable (SV) genes (Svensson et al., 2018). This method utilizes Gaussian process regression to dissect expression variability into spatial and nonspatial components for each gene. This decomposition involves employing two random effect terms: a spatial variance term that characterizes gene expression covariance based on pairwise distances between samples, and a noise term representing non-spatial variability. This process implements automatic expression histology (AEH), a spatial gene-clustering approach facilitating the creation of expression-based tissue histology. Based on the expression patterns of the marker gene set, AEH can be performed to identify tissue domains. The observed gene expression information is used to assign genes to individual patterns, and eventually histological pattern expression information is obtained.

To investigate the spatial expression heterogeneity within prostate cancer tissues, Berglund et al. analyzed spatial transcriptomics data from a prostate cancer tissue (Berglund et al., 2018). Using factor analysis, the authors identified differential gene expression patterns in normal glands, stromal, PIN, cancerous or inflamed regions. Wang et al. revisited this dataset using SpatialDE to investigate metabolic gene expression patterns and malignant cell-specific metabolic vulnerabilities (Wang, Ma, & Ruzzo, 2020). The metabolic pathway enrichment analysis of these SV genes displayed considerable heterogeneity in domains such as oxidative phosphorylation, arachidonic (i.e., eicosanoid) and fatty acid metabolism, and reactive oxygen species. These findings suggest that tumor cells in different areas of the same primary tumor biopsy of the prostate may have developed varied survival strategies.

3.2.2 BayesSpace

BayesSpace (Zhao et al., 2021) is a computational method designed to enhance the resolution of spatial transcriptomic data and facilitate the clustering analysis of such data. BayesSpace algorithm addresses the limitations of spatial transcriptomics, such as the relatively low resolution of the technology, which may not fully capture the cellular complexity of tissues. Traditional methods may not use spatial information efficiently, leading to a loss of potential insights. BayesSpace utilizes information from spatial neighborhoods to improve the resolution of data and to perform clustering analysis, thereby revealing tissue structure and transcriptional heterogeneity that might be undetectable at the original resolution or through histological analysis alone.

The workflow of BayesSpace begins with the preprocessing of spatial transcriptomic (ST) or Visium data by log transformation and normalization using library size, then PCA is performed on the top 2000 HVGs. The algorithm employs a fully Bayesian model with a Markov random field. This spatial statistics method is adapted from techniques used in image analysis and microarray data. It uses a t-distributed error model that is robust against outliers in clusters, which may be driven by technical artifacts. The algorithm utilizes Markov chain Monte Carlo (MCMC) for parameter estimation and performs spatial clustering to infer regions within the tissue that have similar expression profiles. This step models a low-dimensional representation of the gene expression matrix and uses a spatial prior to encourage neighboring spots to belong to the same cluster.

BayesSpace enhances the resolution of the clustering map by segmenting each spot into sub-spots. This is achieved by leveraging spatial information using the Potts model spatial prior. Specifically, each ST spot is divided into nine sub-spots, and each Visium spot into six sub-spots. BayesSpace is adaptable to other platforms where spots are arranged on a lattice, and it may be extended to jointly cluster spots from multiple samples with appropriate data normalizations.

In a study of the spatial characteristics and heterogeneity of the consensus molecular subtypes (CMS) classification of CRC, BayesSpace analysis, which uses the neighborhood structure of $10\times$ Visium data to increase gene expression resolution, identified biological diversity within specific CRC subgroups (Valdeolivas et al., 2024). CMS is a criterion proposed by the CRC subtyping consortium that categorizes CRC into four groups (Thanki et al., 2017). Using deconvolution of spatial transcriptome data (cell2location) and pathologist annotations, we spatially determined CMS and identified dominant CMS groups for each patient sample. Among them, we identified regions of differential expression of genes, biological processes, and pathway activities when subclustering tumor-annotated spots in patient samples representing the dominant CMS2 group. The WNT and VEGF pathways associated with CMS2 showed a consistent distribution of activity across tumor subclusters compared to the activity of the EGFR and MAPK pathways. Subcluster1 showed higher activity of the TGF-β pathway compared to the other clusters, indicating that this region has a higher cancer proliferation and metastatic ability compared to other regions, revealing heterogeneity among intratumor regions.

3.2.3 Vesalius

Vesalius is developed for characterizing tissue architecture from spatial transcriptomics data by embedding the transcriptome into RGB color code

(Martin et al., 2022). Using image analysis based on the embedded color dimension, vesalius enables clustering of tissue territories followed by downstream analysis including differential expression analysis based on different territories, territory borders, and different morphologies. Using vesalius, Martin et al. recovers tissue territories in the 10× Visium dataset of a breast cancer sample.

3.3 Niche-gene analysis

Tissue domain analysis provides delineation of global patterns of spatial gene expression. However, cells' expression profiles include expression variations altered by neighboring information. Here, we introduced recent approaches of identifying such expression variations from spatial transcriptomics data (Table 1).

3.3.1 NicheSVM

NicheSVM (Kim et al., 2023), was originally designed to analyze the physically interacting cell sequencing (PIC-seq) data of mouse embryonic development. NicheSVM borrowed the power of support vector machines (SVMs) for the classification of spots. It first trains multi-class SVMs with artificial combinations of multiple cells' expression. With the trained SVMs, Visium spots can be classified into the combinations of cell types. Finally, genes expressed in specific combinations of malignant cells or TMEs are identified by comparing the spatial data with the artificial spots.

Saqib et al. (2023) applied NicheSVM to integrate scRNA-seq data with spatial transcriptomics data obtained from a pan-cancer study (Barkley et al., 2022). The NicheSVM pipeline was utilized to discern niche-specific genes within five different cancer types (BRCA, GIST, LIHC, OVCA, and UCEC). Through the application of the NicheSVM algorithm, it was able to distinguish niche-specific genes that were separate from the cell-type-specific markers. These niche-specific genes exhibited unique expression profiles that were not shared with the traditional markers, indicating their distinct roles within the TME, which was related to cytokine-mediated signaling and antigen processing, which are critical processes in the immune response and cancer progression. Furthermore, niche-specific genes are more spatially correlated with each other than with cell-type markers. This higher spatial cross-correlation indicates that these genes have similar expression patterns and are more localized, suggesting a coordinated role in cell-cell communication within the TMEs.

3.3.2 CellNeighborEx

Kim et al. (2023) developed CellNeighborEX for identifying gene expression alterations induced by cell–cell interactions from spatial transcriptomics data. This method can be applied to diverse data types, including image-based NGS-based spatial transcriptomics data, without restricting its analysis to ligand–receptor co-expression. This method categorizes cells (or spots) into two groups—homotypic and heterotypic—using Delaunay triangulation for image-based data and RCTD deconvolution for Slide-seq data. Subsequently, it compares the transcriptomes of these groups using appropriate statistical tests, providing neighbor-specific genes. To find the origin of the neighbor-specific genes, CellNeighborEx employs a linear regression model that leverages spot deconvolution results and investigates scRNA-seq data to accurately attribute gene expression changes to specific cell types. CellNeighborEx was applied to a Slide-seq data obtained from a mouse liver cancer sample, providing 42 up-regulated genes in 10 heterotypic pairs and 3 down-regulated genes in 2 heterotypic pairs including F13a1 which is up-regulated in monocytes when they contact tumor cells.

3.3.3 Niche-DE and Niche-LR

Mason et al. (2024) developed niche-DE offering niche-differential expression analysis, suitable for both image-based and NGS-based data, with a notable advantage in low-resolution settings like 10× Visium. It employs a negative binomial regression model to analyze gene expression counts in spots, based on deconvolution results. The model incorporates separate terms to account for the inherent cell-type composition of spots and the dynamic gene expression changes triggered by neighborhood effects. With the same framework, niche-LR was developed for analyzing ligand receptor signaling mechanisms. Mason et al. applied niche-DE to five 10× Visium data obtained from liver metastasis samples of colorectal cancer. Niche-DE identified GAL3 which are predicted to be expressed in tumor cells when they are enriched with fibroblasts-niche. This finding of niche-induced expression was validated using a high-resolution spatial transcriptomics platform CODEX.

4. Discussion

In this chapter, we provided an up-to-date overview on (1) cutting edge technologies of spatially resolved transcriptomics, (2) computational algorithms for analyzing spatial transcriptomics, and (3) their applications in cancer studies. The spatial transcriptomics can be broadly classified into image-based and NGS-based methods, each with its own strengths and weaknesses. Image-based methods can detect a single molecule, which is considered the highest resolution for the purpose of spatial transcriptomics, but they require inclusion in a predetermined set of nucleotide sequences. By expanding the number of detectable genes, most of the critical genes in cancer biology can be incorporated into a cancer panel, offering a plausible solution within limited budgets. However, imaging-based methods with a panel of hundreds or thousands of genes are not suitable for novel discoveries of interactions between cancer cells and TME. In contrast, NGS-based methods read sequences at the single nucleotide level, enabling novel discovery including the detection of spatially resolved variants detection (Chen et al., 2023) and the involvement of microbiome in cancer development (Galeano Niño et al., 2022). However, current NGS-based spatial transcriptomics methods are limited by low resolution or detect only a limited portion of nucleotide molecules due to loss in various experimental processes such as extraction of mRNA molecules, synthesis of cDNA, or barcoding gaps.

Various computational methods have been developed to compensate for such technological limitations of those two categories of spatial transcriptomics. The limitation of gene panels in imaging-based methods can be addressed by prioritization of gene panel selection (Yafi et al., 2024) or prediction of unmeasured gene expression from the measured gene expression. The undetected gene expression can be inferred by modeling the relationship between genes in scRNAseq data. Several methods have been proposed to address this problem using ensemble learning (ENGEP) (Yang & Zhang, 2023) or deep generative models (SpatialScope) (Wan et al., 2023).

Despite the low resolution, 10× Visium is still broadly used in many studies (Moses & Pachter, 2022). To overcome the low-resolution problem, researchers can deconvolute each spot or generate high resolution spatial transcriptomics data based on scRNAseq. As we introduced in Section 3.1, cell2location is the most widely used deconvolution method in cancer biology (Kleshchevnikov et al., 2022). For the generation of high-resolution data, several methods have been suggested such as CytoSPACE (Vahid et al., 2023) and DeepSpaCE (Monjo, Koido, Nagasawa, Suzuki, & Kamatani, 2022). CytoSPACE achieved superior performance by assignment of single cell expression on the spot by solving a convex optimization problem.

Most of these computational methods to fill the technological gaps were developed within two years from early 2023 to this year (2024). This suggests that the field of spatial transcriptomics still has room for innovative approaches to excavate novel findings from this rapidly accumulating spatially resolved transcriptomics from cancer samples.

Acknowledgments

This work was supported by the National Research Foundation of Korea (NRF) grant funded by the Korea government (MSIT) to J.K. (2021M3H9A2096988) and J.P. (No. RS-2023-00278658), and Pusan National University Research Grant, 2022. The diagrams in Fig. 1 were created using BioRender.

References

Andersson, A., Bergenstråhle, J., Asp, M., Bergenstråhle, L., Jurek, A., Fernández Navarro, J., & Lundeberg, J. (2020). Single-cell and spatial transcriptomics enables probabilistic inference of cell type topography. *Communications Biology, 3*(1), 565. https://doi.org/10.1038/s42003-020-01247-y.

Andersson, A., Larsson, L., Stenbeck, L., Salmén, F., Ehinger, A., Wu, S. Z., ... Lundeberg, J. (2021). Spatial deconvolution of HER2-positive breast cancer delineates tumor-associated cell type interactions. *Nature Communications, 12*(1), 6012. https://doi.org/10.1038/s41467-021-26271-2.

Arumugam, T., Brandt, W., Ramachandran, V., Moore, T. T., Wang, H., May, F. E., ... Logsdon, C. D. (2011). Trefoil factor 1 stimulates both pancreatic cancer and stellate cells and increases metastasis. *Pancreas, 40*(6), 815–822. https://doi.org/10.1097/MPA.0b013e31821f6927.

Barker, H. E., Paget, J. T., Khan, A. A., & Harrington, K. J. (2015). The tumour microenvironment after radiotherapy: Mechanisms of resistance and recurrence. *Nature Reviews. Cancer, 15*(7), 409–425. https://doi.org/10.1038/nrc3958.

Barkley, D., Moncada, R., Pour, M., Liberman, D. A., Dryg, I., Werba, G., ... Yanai, I. (2022). Cancer cell states recur across tumor types and form specific interactions with the tumor microenvironment. *Nature Genetics, 54*(8), 1192–1201. https://doi.org/10.1038/s41588-022-01141-9.

Bentley, D. R., Balasubramanian, S., Swerdlow, H. P., Smith, G. P., Milton, J., Brown, C. G., ... Smith, A. J. (2008). Accurate whole human genome sequencing using reversible terminator chemistry. *Nature, 456*(7218), 53–59. https://doi.org/10.1038/nature07517.

Berglund, E., Maaskola, J., Schultz, N., Friedrich, S., Marklund, M., Bergenstråhle, J., ... Lundeberg, J. (2018). Spatial maps of prostate cancer transcriptomes reveal an unexplored landscape of heterogeneity. *Nature Communications, 9*(1), 2419. https://doi.org/10.1038/s41467-018-04724-5.

Bertram, J. S. (2000). The molecular biology of cancer. *Molecular Aspects of Medicine, 21*(6), 167–223. https://doi.org/10.1016/s0098-2997(00)00007-8.

Cable, D. M., Murray, E., Zou, L. S., Goeva, A., Macosko, E. Z., Chen, F., & Irizarry, R. A. (2022). Robust decomposition of cell type mixtures in spatial transcriptomics. *Nature Biotechnology, 40*(4), 517–526. https://doi.org/10.1038/s41587-021-00830-w.

Capote, J., Kramerova, I., Martinez, L., Vetrone, S., Barton, E. R., Sweeney, H. L., ... Spencer, M. J. (2016). Osteopontin ablation ameliorates muscular dystrophy by shifting macrophages to a pro-regenerative phenotype. *The Journal of Cell Biology, 213*(2), 275–288. https://doi.org/10.1083/jcb.201510086.

Caronni, N., La Terza, F., Vittoria, F. M., Barbiera, G., Mezzanzanica, L., Cuzzola, V., ... Ostuni, R. (2023). IL-1β(+) macrophages fuel pathogenic inflammation in pancreatic cancer. *Nature, 623*(7986), 415–422. https://doi.org/10.1038/s41586-023-06685-2.

Chen, A., Liao, S., Cheng, M., Ma, K., Wu, L., Lai, Y., ... Wang, J. (2022). Spatiotemporal transcriptomic atlas of mouse organogenesis using DNA nanoball-patterned arrays. *Cell, 185*(10), 1777–1792.e1721. https://doi.org/10.1016/j.cell.2022.04.003.

Chen, H., Li, D., & Bar-Joseph, Z. (2023). SCS: Cell segmentation for high-resolution spatial transcriptomics. *Nature Methods, 20*(8), 1237–1243. https://doi.org/10.1038/s41592-023-01939-3.

Chen, K. H., Boettiger, A. N., Moffitt, J. R., Wang, S., & Zhuang, X. (2015). RNA imaging. Spatially resolved, highly multiplexed RNA profiling in single cells. *Science (New York, N. Y.), 348*(6233), aaa6090. https://doi.org/10.1126/science.aaa6090.

Chen, L., Chang, D., Tandukar, B., Deivendran, D., Pozniak, J., Cruz-Pacheco, N., ... Shain, A. H. (2023). STmut: A framework for visualizing somatic alterations in spatial transcriptomics data of cancer. *Genome Biology, 24*(1), 273. https://doi.org/10.1186/s13059-023-03121-6.

Chen, S., Zhou, S., Huang, Y. E., Yuan, M., Lei, W., Chen, J., ... Jiang, W. (2022). Estimating metastatic risk of pancreatic ductal adenocarcinoma at single-cell resolution. *International Journal of Molecular Sciences, 23*(23), https://doi.org/10.3390/ijms232315020.

Cheng, S., Li, Z., Gao, R., Xing, B., Gao, Y., Yang, Y., ... Zhang, Z. (2021). A pan-cancer single-cell transcriptional atlas of tumor infiltrating myeloid cells. *Cell, 184*(3), 792–809.e723. https://doi.org/10.1016/j.cell.2021.01.010.

Cho, C. S., Xi, J., Si, Y., Park, S. R., Hsu, J. E., Kim, M., ... Lee, J. H. (2021). Microscopic examination of spatial transcriptome using Seq-Scope. *Cell, 184*(13), 3559–3572.e3522. https://doi.org/10.1016/j.cell.2021.05.010.

Claus, C., Riether, C., Schürch, C., Matter, M. S., Hilmenyuk, T., & Ochsenbein, A. F. (2012). CD27 signaling increases the frequency of regulatory T cells and promotes tumor growth. *Cancer Research, 72*(14), 3664–3676. https://doi.org/10.1158/0008-5472.Can-11-2791.

Codeluppi, S., Borm, L. E., Zeisel, A., La Manno, G., van Lunteren, J. A., Svensson, C. I., & Linnarsson, S. (2018). Spatial organization of the somatosensory cortex revealed by osmFISH. *Nature Methods, 15*(11), 932–935. https://doi.org/10.1038/s41592-018-0175-z.

Cui Zhou, D., Jayasinghe, R. G., Chen, S., Herndon, J. M., Iglesia, M. D., Navale, P., ... Ding, L. (2022). Spatially restricted drivers and transitional cell populations cooperate with the microenvironment in untreated and chemo-resistant pancreatic cancer. *Nature Genetics, 54*(9), 1390–1405. https://doi.org/10.1038/s41588-022-01157-1.

Danaher, P., Kim, Y., Nelson, B., Griswold, M., Yang, Z., Piazza, E., & Beechem, J. M. (2022). Advances in mixed cell deconvolution enable quantification of cell types in spatial transcriptomic data. *Nature Communications, 13*(1), 385. https://doi.org/10.1038/s41467-022-28020-5.

De Visser, K. E., & Joyce, J. A. (2023). The evolving tumor microenvironment: From cancer initiation to metastatic outgrowth. *Cancer Cell, 41*(3), 374–403. https://doi.org/10.1016/j.ccell.2023.02.016.

Drmanac, R., Sparks, A. B., Callow, M. J., Halpern, A. L., Burns, N. L., Kermani, B. G., ... Reid, C. A. (2010). Human genome sequencing using unchained base reads on self-assembling DNA nanoarrays. *Science (New York, N. Y.), 327*(5961), 78–81. https://doi.org/10.1126/science.1181498.

Eng, C. L., Lawson, M., Zhu, Q., Dries, R., Koulena, N., Takei, Y., ... Cai, L. (2019). Transcriptome-scale super-resolved imaging in tissues by RNA seqFISH. *Nature, 568*(7751), 235–239. https://doi.org/10.1038/s41586-019-1049-y.

Fares, J., Fares, M. Y., Khachfe, H. H., Salhab, H. A., & Fares, Y. (2020). Molecular principles of metastasis: A hallmark of cancer revisited. *Signal Transduction and Targeted Therapy, 5*(1), 28. https://doi.org/10.1038/s41392-020-0134-x.

Femino, A. M., Fay, F. S., Fogarty, K., & Singer, R. H. (1998). Visualization of single RNA transcripts in situ. *Science (New York, N. Y.), 280*(5363), 585–590. https://doi.org/10.1126/science.280.5363.585.

Galeano Niño, J. L., Wu, H., LaCourse, K. D., Kempchinsky, A. G., Baryiames, A., Barber, B., ... Bullman, S. (2022). Effect of the intratumoral microbiota on spatial and cellular heterogeneity in cancer. *Nature, 611*(7937), 810–817. https://doi.org/10.1038/s41586-022-05435-0.

Giolai, M., Verweij, W., Lister, A., Heavens, D., Macaulay, I., & Clark, M. D. (2019). Spatially resolved transcriptomics reveals plant host responses to pathogens. *Plant Methods, 15*, 114. https://doi.org/10.1186/s13007-019-0498-5.

Giraldo, N. A., Sanchez-Salas, R., Peske, J. D., Vano, Y., Becht, E., Petitprez, F., ... Sautès-Fridman, C. (2019). The clinical role of the TME in solid cancer. *British Journal of Cancer, 120*(1), 45–53. https://doi.org/10.1038/s41416-018-0327-z.

Gong, L., Luo, J., Zhang, Y., Yang, Y., Li, S., Fang, X., ... Guan, X. Y. (2023). Nasopharyngeal carcinoma cells promote regulatory T cell development and suppressive activity via CD70-CD27 interaction. *Nature Communications, 14*(1), 1912. https://doi.org/10.1038/s41467-023-37614-6.

Guppy, N. J., El-Bahrawy, M. E., Kocher, H. M., Fritsch, K., Qureshi, Y. A., Poulsom, R., ... Alison, M. R. (2012). Trefoil factor family peptides in normal and diseased human pancreas. *Pancreas, 41*(6), 888–896. https://doi.org/10.1097/MPA.0b013e31823c9ec5.

Gyllborg, D., Langseth, C. M., Qian, X., Choi, E., Salas, S. M., Hilscher, M. M., ... Nilsson, M. (2020). Hybridization-based in situ sequencing (HybISS) for spatially resolved transcriptomics in human and mouse brain tissue. *Nucleic Acids Research, 48*(19), e112. https://doi.org/10.1093/nar/gkaa792.

Harney, A. S., Arwert, E. N., Entenberg, D., Wang, Y., Guo, P., Qian, B. Z., ... Condeelis, J. S. (2015). Real-time imaging reveals local, transient vascular permeability, and tumor cell intravasation stimulated by TIE2hi macrophage-derived VEGFA. *Cancer Discovery, 5*(9), 932–943. https://doi.org/10.1158/2159-8290.Cd-15-0012.

He, H., Mack, J. J., Güç, E., Warren, C. M., Squadrito, M. L., Kilarski, W. W., ... Iruela-Arispe, M. L. (2016). Perivascular macrophages limit permeability. *Arteriosclerosis, Thrombosis, and Vascular Biology, 36*(11), 2203–2212. https://doi.org/10.1161/atvbaha.116.307592.

He, H., Xu, J., Warren, C. M., Duan, D., Li, X., Wu, L., & Iruela-Arispe, M. L. (2012). Endothelial cells provide an instructive niche for the differentiation and functional polarization of M2-like macrophages. *Blood, 120*(15), 3152–3162. https://doi.org/10.1182/blood-2012-04-422758.

He, Y., Tang, X., Huang, J., Ren, J., Zhou, H., Chen, K., ... Wang, X. (2021). ClusterMap for multi-scale clustering analysis of spatial gene expression. *Nature Communications, 12*(1), 5909. https://doi.org/10.1038/s41467-021-26044-x.

Hinshaw, D. C., & Shevde, L. A. (2019). The tumor microenvironment innately modulates cancer progression. *Cancer Research, 79*(18), 4557–4566. https://doi.org/10.1158/0008-5472.Can-18-3962.

Hong, F., Meng, Q., Zhang, W., Zheng, R., Li, X., Cheng, T., ... Gao, X. (2021). Single-cell analysis of the pan-cancer immune microenvironment and scTIME portal. *Cancer Immunology Research, 9*(8), 939–951. https://doi.org/10.1158/2326-6066.Cir-20-1026.

Hu, J., Li, X., Coleman, K., Schroeder, A., Ma, N., Irwin, D. J., ... Li, M. (2021). SpaGCN: Integrating gene expression, spatial location and histology to identify spatial domains and spatially variable genes by graph convolutional network. *Nature Methods, 18*(11), 1342–1351. https://doi.org/10.1038/s41592-021-01255-8.

Jin, M. Z., & Jin, W. L. (2020). The updated landscape of tumor microenvironment and drug repurposing. *Signal Transduction and Targeted Therapy, 5*(1), 166. https://doi.org/10.1038/s41392-020-00280-x.

Kajioka, H., Kagawa, S., Ito, A., Yoshimoto, M., Sakamoto, S., Kikuchi, S., ... Fujiwara, T. (2021). Targeting neutrophil extracellular traps with thrombomodulin prevents pancreatic cancer metastasis. *Cancer Letters, 497*, 1–13. https://doi.org/10.1016/j.canlet.2020.10.015.

Kalluri, R., & Zeisberg, M. (2006). Fibroblasts in cancer. *Nature Reviews. Cancer, 6*(5), 392–401. https://doi.org/10.1038/nrc1877.

Ke, R., Mignardi, M., Pacureanu, A., Svedlund, J., Botling, J., Wählby, C., & Nilsson, M. (2013). In situ sequencing for RNA analysis in preserved tissue and cells. *Nature Methods, 10*(9), 857–860. https://doi.org/10.1038/nmeth.2563.

Khalafalla, F. G., & Khan, M. W. (2017). Inflammation and epithelial-mesenchymal transition in pancreatic ductal adenocarcinoma: Fighting against multiple opponents. 1179064417709287 *Cancer Growth Metastasis, 10*. https://doi.org/10.1177/1179064417709287.

Kim, H., Kumar, A., Lovkvist, C., Palma, A. M., Martin, P., Kim, J., ... Won, K. J. (2023). CellNeighborEX: Deciphering neighbor-dependent gene expression from spatial transcriptomics data. *Molecular Systems Biology, 19*(11), e11670. https://doi.org/10.15252/msb.202311670.

Kim, J., Rothova, M. M., Madan, E., Rhee, S., Weng, G., Palma, A. M., ... Won, K. J. (2023). Neighbor-specific gene expression revealed from physically interacting cells during mouse embryonic development. *Proceedings of the National Academy of Sciences of the United States of America, 120*(2), e2205371120. https://doi.org/10.1073/pnas.2205371120.

Kim, S., Leem, G., Choi, J., Koh, Y., Lee, S., Nam, S. H., ... Park, J. E. (2024). Integrative analysis of spatial and single-cell transcriptome data from human pancreatic cancer reveals an intermediate cancer cell population associated with poor prognosis. *Genome Medicine, 16*(1), 20. https://doi.org/10.1186/s13073-024-01287-7.

Kinker, G. S., Greenwald, A. C., Tal, R., Orlova, Z., Cuoco, M. S., McFarland, J. M., ... Tirosh, I. (2020). Pan-cancer single-cell RNA-seq identifies recurring programs of cellular heterogeneity. *Nature Genetics, 52*(11), 1208–1218. https://doi.org/10.1038/s41588-020-00726-6.

Kleshchevnikov, V., Shmatko, A., Dann, E., Aivazidis, A., King, H. W., Li, T., ... Bayraktar, O. A. (2022). Cell2location maps fine-grained cell types in spatial transcriptomics. *Nature Biotechnology, 40*(5), 661–671. https://doi.org/10.1038/s41587-021-01139-4.

Lafzi, A., Moutinho, C., Picelli, S., & Heyn, H. (2018). Tutorial: Guidelines for the experimental design of single-cell RNA sequencing studies. *Nature Protocols, 13*(12), 2742–2757. https://doi.org/10.1038/s41596-018-0073-y.

Lee, H. O., Hong, Y., Etlioglu, H. E., Cho, Y. B., Pomella, V., Van den Bosch, B., ... Park, W. Y. (2020). Lineage-dependent gene expression programs influence the immune landscape of colorectal cancer. *Nature Genetics, 52*(6), 594–603. https://doi.org/10.1038/s41588-020-0636-z.

Lee, J. H., Daugharthy, E. R., Scheiman, J., Kalhor, R., Yang, J. L., Ferrante, T. C., ... Church, G. M. (2014). Highly multiplexed subcellular RNA sequencing in situ. *Science (New York, N. Y.), 343*(6177), 1360–1363. https://doi.org/10.1126/science.1250212.

Li, R., Ferdinand, J. R., Loudon, K. W., Bowyer, G. S., Laidlaw, S., Muyas, F., ... Mitchell, T. J. (2022). Mapping single-cell transcriptomes in the intra-tumoral and associated territories of kidney cancer. *Cancer Cell, 40*(12), 1583–1599.e1510. https://doi.org/10.1016/j.ccell.2022.11.001.

Li, Y., Liu, H., Zhao, Y., Yue, D., Chen, C., Li, C., ... Wang, C. (2021). Tumor-associated macrophages (TAMs)-derived osteopontin (OPN) upregulates PD-L1 expression and predicts poor prognosis in non-small cell lung cancer (NSCLC). *Thoracic Cancer, 12*(20), 2698–2709. https://doi.org/10.1111/1759-7714.14108.

Lin, A., & Yan, W. H. (2021). HLA-G/ILTs targeted solid cancer immunotherapy: Opportunities and challenges. *Frontiers in Immunology, 12*, 698677. https://doi.org/10.3389/fimmu.2021.698677.

Lin, X., Qiu, L., Song, X., Hou, J., Chen, W., & Zhao, J. (2019). A comparative analysis of RNA sequencing methods with ribosome RNA depletion for degraded and low-input total RNA from formalin-fixed and paraffin-embedded samples. *BMC Genomics, 20*(1), 831. https://doi.org/10.1186/s12864-019-6166-3.

Liu, X., Zhao, S., Wang, K., Zhou, L., Jiang, M., Gao, Y., ... Dong, Z. (2023). Spatial transcriptomics analysis of esophageal squamous precancerous lesions and their progression to esophageal cancer. *Nature Communications, 14*(1), 4779. https://doi.org/10.1038/s41467-023-40343-5.

Lubeck, E., & Cai, L. (2012). Single-cell systems biology by super-resolution imaging and combinatorial labeling. *Nature Methods, 9*(7), 743–748. https://doi.org/10.1038/nmeth.2069.

Luo, H., Xia, X., Huang, L. B., An, H., Cao, M., Kim, G. D., ... Xu, H. (2022). Pan-cancer single-cell analysis reveals the heterogeneity and plasticity of cancer-associated fibroblasts in the tumor microenvironment. *Nature Communications, 13*(1), 6619. https://doi.org/10.1038/s41467-022-34395-2.

Ma, C., Yang, C., Peng, A., Sun, T., Ji, X., Mi, J., ... Feng, Q. (2023). Pan-cancer spatially resolved single-cell analysis reveals the crosstalk between cancer-associated fibroblasts and tumor microenvironment. *Molecular Cancer, 22*(1), 170. https://doi.org/10.1186/s12943-023-01876-x.

Ma, Y., & Zhou, X. (2022). Spatially informed cell-type deconvolution for spatial transcriptomics. *Nature Biotechnology, 40*(9), 1349–1359. https://doi.org/10.1038/s41587-022-01273-7.

Martin, P. C. N., Kim, H., Lövkvist, C., Hong, B. W., & Won, K. J. (2022). Vesalius: High-resolution in silico anatomization of spatial transcriptomic data using image analysis. *Molecular Systems Biology, 18*(9), e11080. https://doi.org/10.15252/msb.202211080.

Mason, K., Sathe, A., Hess, P. R., Rong, J., Wu, C. Y., Furth, E., ... Zhang, N. (2024). Niche-DE: Niche-differential gene expression analysis in spatial transcriptomics data identifies context-dependent cell-cell interactions. *Genome Biology, 25*(1), 14. https://doi.org/10.1186/s13059-023-03159-6.

Mitra, R. D., Shendure, J., Olejnik, J., Edyta Krzymanska, O., & Church, G. M. (2003). Fluorescent in situ sequencing on polymerase colonies. *Analytical Biochemistry, 320*(1), 55–65. https://doi.org/10.1016/s0003-2697(03)00291-4.

Moncada, R., Barkley, D., Wagner, F., Chiodin, M., Devlin, J. C., Baron, M., ... Yanai, I. (2020). Integrating microarray-based spatial transcriptomics and single-cell RNA-seq reveals tissue architecture in pancreatic ductal adenocarcinomas. *Nature Biotechnology, 38*(3), 333–342. https://doi.org/10.1038/s41587-019-0392-8.

Monjo, T., Koido, M., Nagasawa, S., Suzuki, Y., & Kamatani, Y. (2022). Efficient prediction of a spatial transcriptomics profile better characterizes breast cancer tissue sections without costly experimentation. *Scientific Reports, 12*(1), 4133. https://doi.org/10.1038/s41598-022-07685-4.

Moses, L., & Pachter, L. (2022). Museum of spatial transcriptomics. *Nature Methods, 19*(5), 534–546. https://doi.org/10.1038/s41592-022-01409-2.

Nilsson, M., Malmgren, H., Samiotaki, M., Kwiatkowski, M., Chowdhary, B. P., & Landegren, U. (1994). Padlock probes: Circularizing oligonucleotides for localized DNA detection. *Science (New York, N. Y.), 265*(5181), 2085–2088. https://doi.org/10.1126/science.7522346.

Nofech-Mozes, I., Soave, D., Awadalla, P., & Abelson, S. (2023). Pan-cancer classification of single cells in the tumour microenvironment. *Nature Communications, 14*(1), 1615. https://doi.org/10.1038/s41467-023-37353-8.

Ozato, Y., Kojima, Y., Kobayashi, Y., Hisamatsu, Y., Toshima, T., Yonemura, Y., ... Mimori, K. (2023). Spatial and single-cell transcriptomics decipher the cellular environment containing HLA-G+ cancer cells and SPP1+ macrophages in colorectal cancer. *Cell Reports, 42*(1), 111929. https://doi.org/10.1016/j.celrep.2022.111929.

Padoan, A., Plebani, M., & Basso, D. (2019). Inflammation and pancreatic cancer: Focus on metabolism, cytokines, and immunity. *International Journal of Molecular Sciences, 20*(3), https://doi.org/10.3390/ijms20030676.

Parker, B. S., Rautela, J., & Hertzog, P. J. (2016). Antitumour actions of interferons: Implications for cancer therapy. *Nature Reviews. Cancer, 16*(3), 131–144. https://doi.org/10.1038/nrc.2016.14.

Petukhov, V., Xu, R. J., Soldatov, R. A., Cadinu, P., Khodosevich, K., Moffitt, J. R., & Kharchenko, P. V. (2022). Cell segmentation in imaging-based spatial transcriptomics. *Nature Biotechnology, 40*(3), 345–354. https://doi.org/10.1038/s41587-021-01044-w.

Puram, S. V., Tirosh, I., Parikh, A. S., Patel, A. P., Yizhak, K., Gillespie, S., ... Bernstein, B. E. (2017). Single-cell transcriptomic analysis of primary and metastatic tumor ecosystems in head and neck cancer. e1624 *Cell, 171*(7), 1611–1624. https://doi.org/10.1016/j.cell.2017.10.044.

Qian, J., Olbrecht, S., Boeckx, B., Vos, H., Laoui, D., Etlioglu, E., ... Lambrechts, D. (2020). A pan-cancer blueprint of the heterogeneous tumor microenvironment revealed by single-cell profiling. *Cell Research, 30*(9), 745–762. https://doi.org/10.1038/s41422-020-0355-0.

Qin, Y., Liu, Y., Xiang, X., Long, X., Chen, Z., Huang, X., ... Li, W. (2023). Cuproptosis correlates with immunosuppressive tumor microenvironment based on pan-cancer multiomics and single-cell sequencing analysis. *Molecular Cancer, 22*(1), 59. https://doi.org/10.1186/s12943-023-01752-8.

Raj, A., van den Bogaard, P., Rifkin, S. A., van Oudenaarden, A., & Tyagi, S. (2008). Imaging individual mRNA molecules using multiple singly labeled probes. *Nature Methods, 5*(10), 877–879. https://doi.org/10.1038/nmeth.1253.

Rodriques, S. G., Stickels, R. R., Goeva, A., Martin, C. A., Murray, E., Vanderburg, C. R., ... Macosko, E. Z. (2019). Slide-seq: A scalable technology for measuring genome-wide expression at high spatial resolution. *Science (New York, N. Y.), 363*(6434), 1463–1467. https://doi.org/10.1126/science.aaw1219.

Russell, A. J. C., Weir, J. A., Nadaf, N. M., Shabet, M., Kumar, V., Kambhampati, S., ... Chen, F. (2024). Slide-tags enables single-nucleus barcoding for multimodal spatial genomics. *Nature, 625*(7993), 101–109. https://doi.org/10.1038/s41586-023-06837-4.

Salmaninejad, A., Valilou, S. F., Soltani, A., Ahmadi, S., Abarghan, Y. J., Rosengren, R. J., & Sahebkar, A. (2019). Tumor-associated macrophages: Role in cancer development and therapeutic implications. *Cellular Oncology (Dordr), 42*(5), 591–608. https://doi.org/10.1007/s13402-019-00453-z.

Saqib, J., Park, B., Jin, Y., Seo, J., Mo, J., & Kim, J. (2023). Identification of niche-specific gene signatures between malignant tumor microenvironments by integrating single cell and spatial transcriptomics data. *Genes (Basel), 14*(11), https://doi.org/10.3390/genes14112033.

Ståhl, P. L., Salmén, F., Vickovic, S., Lundmark, A., Navarro, J. F., Magnusson, J., ... Frisén, J. (2016). Visualization and analysis of gene expression in tissue sections by spatial transcriptomics. *Science (New York, N. Y.), 353*(6294), 78–82. https://doi.org/10.1126/science.aaf2403.

Steele, C. W., Jamieson, N. B., Evans, T. R., McKay, C. J., Sansom, O. J., Morton, J. P., & Carter, C. R. (2013). Exploiting inflammation for therapeutic gain in pancreatic cancer. *British Journal of Cancer, 108*(5), 997–1003. https://doi.org/10.1038/bjc.2013.24.

Stickels, R. R., Murray, E., Kumar, P., Li, J., Marshall, J. L., Di Bella, D. J., ... Chen, F. (2021). Highly sensitive spatial transcriptomics at near-cellular resolution with Slide-seqV2. *Nature Biotechnology, 39*(3), 313–319. https://doi.org/10.1038/s41587-020-0739-1.

Sun, L., Kienzler, J. C., Reynoso, J. G., Lee, A., Shiuan, E., Li, S., ... Prins, R. M. (2023). Immune checkpoint blockade induces distinct alterations in the microenvironments of primary and metastatic brain tumors. *The Journal of Clinical Investigation, 133*(17), https://doi.org/10.1172/jci169314.

Sun, S., Zhu, J., & Zhou, X. (2020). Statistical analysis of spatial expression patterns for spatially resolved transcriptomic studies. *Nature Methods, 17*(2), 193–200. https://doi.org/10.1038/s41592-019-0701-7.

Svensson, V., Teichmann, S. A., & Stegle, O. (2018). SpatialDE: Identification of spatially variable genes. *Nature Methods, 15*(5), 343–346. https://doi.org/10.1038/nmeth.4636.

Tang, F., Li, J., Qi, L., Liu, D., Bo, Y., Qin, S., ... Zhang, Z. (2023). A pan-cancer single-cell panorama of human natural killer cells. *Cell, 186*(19), 4235–4251.e4220. https://doi.org/10.1016/j.cell.2023.07.034.

Thanki, K., Nicholls, M. E., Gajjar, A., Senagore, A. J., Qiu, S., Szabo, C., ... Chao, C. (2017). Consensus molecular subtypes of colorectal cancer and their clinical implications. *International Biological and Biomedical Journal, 3*(3), 105–111.

Vahid, M. R., Brown, E. L., Steen, C. B., Zhang, W., Jeon, H. S., Kang, M., ... Newman, A. M. (2023). High-resolution alignment of single-cell and spatial transcriptomes with CytoSPACE. *Nature Biotechnology, 41*(11), 1543–1548. https://doi.org/10.1038/s41587-023-01697-9.

Valdeolivas, A., Amberg, B., Giroud, N., Richardson, M., Gálvez, E. J. C., Badillo, S., ... Hahn, K. (2024). Profiling the heterogeneity of colorectal cancer consensus molecular subtypes using spatial transcriptomics. *npj Precision Oncology, 8*(1), 10. https://doi.org/10.1038/s41698-023-00488-4.

Vandereyken, K., Sifrim, A., Thienpont, B., & Voet, T. (2023). Methods and applications for single-cell and spatial multi-omics. *Nature Reviews. Genetics, 24*(8), 494–515. https://doi.org/10.1038/s41576-023-00580-2.

Walsh, L. A., & Quail, D. F. (2023). Decoding the tumor microenvironment with spatial technologies. *Nature Immunology, 24*(12), 1982–1993. https://doi.org/10.1038/s41590-023-01678-9.

Wan, X., Xiao, J., Tam, S. S. T., Cai, M., Sugimura, R., Wang, Y., ... Yang, C. (2023). Integrating spatial and single-cell transcriptomics data using deep generative models with SpatialScope. *Nature Communications, 14*(1), 7848. https://doi.org/10.1038/s41467-023-43629-w.

Wang, K. X., & Denhardt, D. T. (2008). Osteopontin: Role in immune regulation and stress responses. *Cytokine & Growth Factor Reviews, 19*(5-6), 333–345. https://doi.org/10.1016/j.cytogfr.2008.08.001.

Wang, Q., Zhi, Y., Zi, M., Mo, Y., Wang, Y., Liao, Q., ... Xiong, W. (2023). Spatially resolved transcriptomics technology facilitates cancer research. *Advanced Science (Weinh), 10*(30), e2302558. https://doi.org/10.1002/advs.202302558.

Wang, Y., Ma, S., & Ruzzo, W. L. (2020). Spatial modeling of prostate cancer metabolic gene expression reveals extensive heterogeneity and selective vulnerabilities. *Scientific Reports, 10*(1), 3490. https://doi.org/10.1038/s41598-020-60384-w.

Wei, C., Sun, W., Shen, K., Zhong, J., Liu, W., Gao, Z., ... Gu, J. (2023). Delineating the early dissemination mechanisms of acral melanoma by integrating single-cell and spatial transcriptomic analyses. *Nature Communications, 14*(1), 8119. https://doi.org/10.1038/s41467-023-43980-y.

Williams, C. G., Lee, H. J., Asatsuma, T., Vento-Tormo, R., & Haque, A. (2022). An introduction to spatial transcriptomics for biomedical research. *Genome Medicine, 14*(1), 68. https://doi.org/10.1186/s13073-022-01075-1.

Wu, S. Z., Al-Eryani, G., Roden, D. L., Junankar, S., Harvey, K., Andersson, A., ... Swarbrick, A. (2021). A single-cell and spatially resolved atlas of human breast cancers. *Nature Genetics, 53*(9), 1334–1347. https://doi.org/10.1038/s41588-021-00911-1.

Yafi, M. A., Hisham, M. H. H., Grisanti, F., Martin, J. F., Rahman, A., & Samee, M. A. H. (2024). scGIST: Gene panel design for spatial transcriptomics with prioritized gene sets. *Genome Biology, 25*(1), 57. https://doi.org/10.1186/s13059-024-03185-y.

Yang, S. T., & Zhang, X. F. (2023). ENGEP: Advancing spatial transcriptomics with accurate unmeasured gene expression prediction. *Genome Biology, 24*(1), 293. https://doi.org/10.1186/s13059-023-03139-w.

Yang, Z. Z., Novak, A. J., Ziesmer, S. C., Witzig, T. E., & Ansell, S. M. (2007). CD70+ non-Hodgkin lymphoma B cells induce Foxp3 expression and regulatory function in intratumoral CD4+CD25 T cells. *Blood, 110*(7), 2537–2544. https://doi.org/10.1182/blood-2007-03-082578.

Zhang, Q., Abdo, R., Iosef, C., Kaneko, T., Cecchini, M., Han, V. K., & Li, S. S. (2022). The spatial transcriptomic landscape of non-small cell lung cancer brain metastasis. *Nature Communications, 13*(1), 5983. https://doi.org/10.1038/s41467-022-33365-y.

Zhang, R., Qi, F., Zhao, F., Li, G., Shao, S., Zhang, X., ... Feng, Y. (2019). Cancer-associated fibroblasts enhance tumor-associated macrophages enrichment and suppress NK cells function in colorectal cancer. *Cell Death Dis, 10*(4), 273. https://doi.org/10.1038/s41419-019-1435-2.

Zhang, Y. H., He, M., Wang, Y., & Liao, A. H. (2017). Modulators of the balance between M1 and M2 macrophages during pregnancy. *Frontiers in Immunology, 8*, 120. https://doi.org/10.3389/fimmu.2017.00120.

Zhao, E., Stone, M. R., Ren, X., Guenthoer, J., Smythe, K. S., Pulliam, T., ... Gottardo, R. (2021). Spatial transcriptomics at subspot resolution with BayesSpace. *Nature Biotechnology, 39*(11), 1375–1384. https://doi.org/10.1038/s41587-021-00935-2.

Zheng, L., Qin, S., Si, W., Wang, A., Xing, B., Gao, R., ... Zhang, Z. (2021). Pan-cancer single-cell landscape of tumor-infiltrating T cells. abe6474 *Science (New York, N. Y.), 374*(6574), https://doi.org/10.1126/science.abe6474.

Ziegenhain, C., Vieth, B., Parekh, S., Reinius, B., Guillaumet-Adkins, A., Smets, M., ... Enard, W. (2017). Comparative analysis of single-cell RNA sequencing methods. e634 *Molecular Cell, 65*(4), 631–643. https://doi.org/10.1016/j.molcel.2017.01.023.

Zitvogel, L., Galluzzi, L., Kepp, O., Smyth, M. J., & Kroemer, G. (2015). Type I interferons in anticancer immunity. *Nature Reviews. Immunology, 15*(7), 405–414. https://doi.org/10.1038/nri3845.

CHAPTER FOUR

Applications of spatial transcriptomics and artificial intelligence to develop integrated management of pancreatic cancer

Rishabh Maurya[a,b,1], Isha Chug[b,1], Vignesh Vudatha[a,b], and António M. Palma[b,c,d,*]

[a]Department of Surgery, Virginia Commonwealth University, School of Medicine, Richmond, VA, United States
[b]Massey Comprehensive Cancer Center, Virginia Commonwealth University, Richmond, VA, United States
[c]VCU Institute of Molecular Medicine, Department of Human and Molecular Genetics, Virginia Commonwealth University, School of Medicine, Richmond, VA, United States
[d]Department of Human and Molecular Genetics, Virginia Commonwealth University, School of Medicine, Richmond, VA, United States
*Corresponding author. e-mail address: antonio.palma@vcuhealth.org

Contents

1. Introduction	108
2. Overview of cancer	109
3. Spatial transcriptomics	111
4. Artificial intelligence	118
5. Spatial transcriptomics contribution to understanding the heterogeneity of pancreatic cancer tumors	121
6. The use of Artificial Intelligence to diagnose and define pancreatic cancer therapy	126
7. Conclusions	130
Acknowledgments	131
References	131

Abstract

Cancer is a complex disease intrinsically associated with cellular processes and gene expression. With the development of techniques such as single-cell sequencing and sequential fluorescence in situ hybridization (seqFISH), it was possible to map the location of cells based on their gene expression with more precision. Moreover, in recent years, many tools have been developed to analyze these extensive datasets by integrating machine learning and artificial intelligence in a comprehensive manner.

[1] These authors contributed equally to the first author position.

Advances in Cancer Research, Volume 163
ISSN 0065-230X, https://doi.org/10.1016/bs.acr.2024.06.007
Copyright © 2024 Elsevier Inc. All rights are reserved, including those for text and data mining, AI training, and similar technologies.

107

Since these tools analyze sequencing data, they offer the chance to analyze any tissue regardless of its origin. By applying this to cancer settings, spatial transcriptomic analysis based on artificial intelligence may help us understand cell-cell communications within the tumor microenvironment. Another advantage of this analysis is the identification of new biomarkers and therapeutic targets.

The integration of such analysis with other omics data and with routine exams such as magnetic resonance imaging can help physicians with the earlier diagnosis of tumors as well as establish a more personalized treatment for pancreatic cancer patients. In this review, we give an overview description of pancreatic cancer, describe how spatial transcriptomics and artificial intelligence have been used to study pancreatic cancer and provide examples of how integrating these tools may help physicians manage pancreatic cancer in a more personalized approach.

1. Introduction

Cancer is defined as a disease in which cells divide uncontrollably to invade surrounding tissues, causing failure of the affected organs and ultimately leading to death. The definition of cancer is very simple, but the biology behind it is immensely complex. Cancer arises from genetic changes within a cell that allow it to pass by quality controls unchecked, thus leading to increased cell proliferation, resistance to apoptosis, and other processes described as the hallmarks of cancer (Hanahan & Weinberg, 2011).

Our body is composed of trillions of cells, which are grouped in tissues and, at a larger level, in organs. Within a tissue, different cell types with specific functions establish a communicational microenvironment to maintain organ homeostasis. Thus, it is logical that a cancer developing in a particular organ will have a different biology depending on the type of cell it arises from. Adding to this complexity, each cell of our body has around 20,000 genes, which form a network of different signaling pathways to keep cellular functions. There are genes whose expression can confer specificity to a cell, whereas others are expressed regardless of the cell type. Among those are the gatekeeping genes, which play an essential role in checking cellular processes such as proliferation, apoptosis, cell cycle, and DNA repair. Alterations in the expression of these gatekeeping genes may constitute an initial step toward the formation of a tumor, characterized by the accumulation of genetic alterations such as mutations, deletions, and amplifications. Within this line of thought, it is logical to assume that a tumor will behave differently depending on the genes affected within the tumor cell. Though tumor origin results from a single clone, its uncontrolled growth leads to the accumulation of more genetic alterations in a random fashion, and by the time tumors are detectable, they already show clonal heterogeneity.

All this complexity helps us to understand why patients respond differently to the same treatment and supports the idea that cancer patients need a more personalized therapeutic strategy that fits better within the biological characteristics of their cancer. In this context, the development and analysis of -omics data such as genomic, transcriptomic, proteomic, and metabolomic, among others, pose a significant advance to understanding the biology of cancer but also to discover new biomarkers and therapeutic targets that will help in the earlier detection and the development of more effective treatments for cancer. For example, using spatial transcriptomics (ST) allows us to analyze, identify, and locate cells within a tissue based on their gene expression profiles, thus helping us understand the cellular communications established within the tissue microenvironment. The analysis of such extensive datasets requires the development of tools capable of processing and accurately identifying the outcomes in a comprehensive manner. In this context, many tools have been developed over the last few years based on machine learning (ML) and artificial intelligence (AI) that aim to help understand the biology behind cancer and assist physicians in decision-making regarding the diagnosis and treatment of cancer.

This review will focus on how integrating ST analysis with AI may help manage pancreatic cancer.

2. Overview of cancer

Cancer is one of the leading causes of death globally, with millions of new cases diagnosed annually. In 2020, there were 19 million new cancer cases and nearly 10 million deaths worldwide (Sung et al., 2021). Pancreatic ductal adenocarcinoma (PDAC) is one of the most lethal cancer types, with a 5-year survival rate of 11%. By 2022, there were over 60,000 new PDAC cases and nearly 50,000 deaths in the United States alone (Siegel, Miller, Fuchs, & Jemal, 2022). Characterized by uncontrolled and abnormal cell growth, cancerous cells invade the surrounding tissues and spread to distant organs, leading to severe morbidity and mortality. The irregular activity of cancerous cells stems from genetic mutations that accumulate over time, which disrupt the balance between cell proliferation and programmed cell death (Hanahan & Weinberg, 2011).

A prominent aspect of cancer is its pronounced heterogeneity, occurring between tumors of different origins and within tumors arising from the same tissue. Heterogeneity refers to the diverse cellular identities, states, and behaviors within and across tumors stemming from genetic, epigenetic, and microenvironmental differences that drive distinct functional phenotypes. This heterogeneity contributes to cancer complexity, evasion, and mortality (McGranahan & Swanton, 2017). Tumor heterogeneity is one of the major obstacles in curing cancer, as it enables therapeutic resistance and disease recurrence (Yu, Jiang, & Wu, 2022). Pancreatic tumors exhibit pronounced heterogeneity, with multiple direct clones driving their progression and metastasis (Maddipati & Stanger, 2015). Unraveling and targeting heterogeneity is critical for developing more effective personalized cancer therapies.

In addition to cancer cells, the tumor microenvironment contains diverse populations of stromal, immune, and other host cells that modulate cancer behaviors (Hanahan & Weinberg, 2011). The tumor microenvironment typically comprises diverse cell types, including cancer cells, fibroblasts, various immune cell populations, endothelial cells, and numerous accessory cells (Quail & Joyce, 2013). These cells interact through direct contact, paracrine signaling, and matrix remodeling to modulate tumor progression. The specific identities, quantities, activation states, and spatial distributions of these cells dictate processes such as angiogenesis, invasion, metastasis, and immunotherapy response (Tan & Naylor, 2022). Additionally, different cells and regions of a heterogeneous tumor may exhibit distinct ecological niches, such as modification of the environment, regulation of immune-system checkpoints, absorption of nutrients, and creation of hypoxic or arid zones with unique microenvironments (Yuan, 2016). A prime example of specialized tumor environments shaping cancer progression is seen in PDAC, which exhibits characteristic desmoplastic stroma surrounding cancer cells. Pancreatic stellate cells within this fibrotic tissue promote invasive properties and stem cell phenotypes in cancer cells through signaling pathways. They also facilitate immune evasion by recruiting suppressor cells. Together, these stromal interactions induce therapeutic resistance and support recurrence and metastasis (Hamada, Masamune, & Shimosegawa, 2013). To unravel cancer biology mechanisms and improve therapeutic strategies, the composition and spatial structure of the tumor microenvironment must be delineated. In general, capturing multicellular complexity and spatial context is essential to identifying critical cell interactions, signaling networks, and microenvironment dynamics driving cancer progression.

However, traditional bulk tumor profiling overlooks spatial heterogeneity and dimensions by homogenizing the tissue (Zhao & Rosen, 2022). Single-cell RNA sequencing enables fine-grained cellular decomposition but loses spatial arrangements through dissociation (Satija, Farrell, Gennert, Schier, & Regev, 2015). Revolutionary spatial profiling technologies, such as multiplexed ion beam imaging (Angelo et al., 2014), imaging mass cytometry (Giesen et al., 2014), and ST (Stahl et al., 2016), allow quantitative mapping of numerous targets while retaining spatial context. Moreover, emerging ST technologies will enable the quantification of gene expression while preserving spatial coordinates within intact tissues (Rodriques et al., 2019). These approaches leverage sequencing or imaging to capture gene expression data that are mapped to spatial spots; in particular, sequencing techniques provide information regarding the whole transcriptome, while imaging approaches lack full coverage (Wan et al., 2023). By enabling high-resolution mapping of diverse cell populations in situ, these emerging spatial profiling technologies have the potential to reveal critical aspects of the microenvironment structure and dynamics that drive cancer progression and therapeutic response.

In parallel, advances in AI and ML have provided critical tools for deriving insights from complex, multidimensional spatial omics datasets. AI methods enable the automated segmentation of diverse cell populations in situ based on spatial expression patterns (Goltsev et al., 2018). ML techniques can also integrate transcripts with histology to classify cell types and reconstruct tissue architectures from spatial molecular data (Le et al., 2022). Deep-learning approaches show promise for inferring cell–cell interactions and signaling networks from spatial proximity and expression mapping (Wang et al., 2023). The application of AI techniques provides an exciting new paradigm for mining the immense complexity of spatial omics data to uncover new knowledge on cancer biology. Although challenges remain, integrated AI analysis could elucidate crucial tumor microenvironment dynamics, interactions, and therapeutic targets mapped using emerging spatial profiling technologies.

3. Spatial transcriptomics

ST provides a transformative approach compared to traditional bulk tissue profiling techniques, such as microarray or RNA sequencing, which destroy spatial context through tissue dissociation (Satija et al., 2015). ST

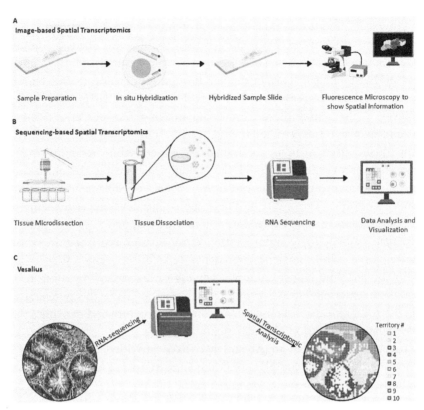

Fig. 1 Methods for spatial transcriptomics. (A) Fluorescence in-situ hybridization is the basis of many image-based ST techniques. A paraffin-embedded section of tissue is permeabilized and labeled with probes. The in-situ hybridization occurs, and the probes can be visualized through a fluorescence microscope. (B) Laser-captured microdissection is a common sequencing-based ST method. A laser beam is used to dissect a selected area of a tissue sample, which is then dissociated and lysed. After, the sample is sequenced to give insight into spatial localization. (C) Vesalius is a ST tool that produces high-resolution ST data from tissue samples. Vesalius is able to identity tissue territories that form the various structures, and layers observed in tissue samples stained with hematoxylin and eosin.

preserves native spatial gene expression patterns within intact tissue sections (Zhang et al., 2022), enabling direct visualization and mapping of distinct cell populations while retaining information about localization, tissue structure, and cellular interactions critical to understanding cancer biology (Stahl et al., 2016). There are two main approaches to ST technologies: imaging-based and sequencing-based (Fig. 1). Imaging-based ST relies on directly visualizing RNA transcripts in tissue context through methods like in situ hybridization and in situ sequencing, providing subcellular resolution but limited throughput (Fig. 1A). Sequencing-based ST involves initially tagging or isolating RNA transcripts from spatial regions prior to sequencing through techniques such as laser capture microdissection to isolate defined tissue regions and in situ barcoding to tag RNAs with spatial barcodes (Fig. 1B). These methods enable higher throughput transcriptome-wide analysis but sacrifice imaging resolution (Yu et al., 2022). The utility of ST in revealing multi-scale biology based on spatial context extends across diverse research areas, including cancer biology. For example, spatial techniques in reproductive biology and neuroscience uncover developmental gradients and neighborhood-specific pathology. Likewise, in cancer, spatial approaches reveal tumor-wide expression patterns, reflecting disease morphology and therapeutic response (Williams, Lee, Asatsuma, Vento-Tormo, & Haque, 2022). These applications are particularly relevant in highly heterogeneous cancers, including PDAC. PDAC exhibits pronounced inter- and intra-tumor heterogeneity, containing multiple distinct clones that drive progression. Within the realm of PDAC, ST approaches have immense potential for unraveling this complexity by mapping heterogeneous clones, microenvironment interactions, and spatially delineated biology (Maddipati & Stanger, 2015). Just as spatial techniques elucidate developmental and pathological patterns in other fields, they can provide critical insights into the drivers of heterogeneity and progression in cancers, such as PDAC.

Various technologies have been developed for targeted or transcriptome-wide spatial profiling at different resolutions. Imaging-based ST provides single-cell resolution, enabling precise mapping of individual RNA transcripts within cells. Utilizing direct fluorescent labeling of probes hybridized to target mRNAs, these methods allow for highly efficient and specific detection. The sequential barcoding strategies also make these approaches robust and reproducible across experiments. Additionally, imaging techniques offer subcellular spatial resolution, as mRNAs can be localized to distinct compartments within a cell. The images generated provide a permanent record of the spatial transcriptome, allowing re-examination and spatial pattern analysis (Wang et al., 2018).

Surfacing in the late 1990s, single-molecule fluorescence in situ hybridization (smFISH) is a quantitative method for visualizing individual RNA molecules within fixed cells. It enabled the transition from qualitative to quantitative imaging of gene expression by fluorescence microscopy. This technique uses fluorescently labeled DNA probes designed to hybridize to target RNA sequences, allowing direct visualization of single

RNA transcripts as distinct fluorescent spots. smFISH provides spatial gene expression information at single molecule sensitivity without amplification steps that could potentially introduce bias. Rather, it produces a native snapshot of RNA distributions within individual cells (Williams et al., 2022). As smFISH was initially limited to only imaging a select number of genes at a time, sequential barcoding approaches were developed to dramatically increase multiplexing capabilities (Chen, Boettiger, Moffitt, Wang, & Zhuang, 2015).

Sequential fluorescence in situ hybridization (seqFISH) uses iterative imaging and labeling to map specific RNA transcripts with high subcellular precision and efficiency. Through utilizing a limited set of fluorophores, a unique color barcode is generated on each mRNA molecule during sequential rounds of hybridization. After the sequential barcoding process, the cells are imaged. The color barcode on each mRNA transcript is decoded computationally to identify which RNA it represents, allowing for highly multiplexed profiling of many RNA species in single cells while preserving spatial context (Stahl et al., 2016). However, the optical density of mRNA transcripts within cells causes overcrowding, preventing the seqFISH technique, in particular, from profiling the entire genome. A combination of the seqFISH technique with super-resolution microscopy, seqFISH+, was developed to assist in overcoming issues regarding overcrowding. While current super-resolution microscopies are restricted in detection, seqFISH+ enables highly multiplexed imaging of gene expression, enabling up to 10,000 genes to be multiplexed (Eng et al., 2019). ST array-based approaches, techniques that utilize reverse transcription, integrate sequencing with tissue positioning for mapping major cell types and zones within a heterogeneous tumor. Tested in breast cancer samples, ST has been applied to reveal intra-tumor heterogeneity and somatic mutation expression patterns while generating high-quality sequencing data (Stahl et al., 2016).

Emerging multicolor imaging methods, such as multiplexed error-robust FISH (MERFISH), permit the parallel quantification and visualization of numerous distinct RNA molecules within individual cells by using successive rounds of imaging with fluorescent probes that bind to specific RNA sequences (Chen et al., 2015). Chemical cleavage, rather than photobleaching, removes the fluorescence between imaging cycles (Moffitt et al., 2016). Additionally, the imaging capacity of MERFISH enabled the characterization of the spatial distribution of hundreds to thousands of RNA transcripts inside the cell (Chen, et al., 2015). Within intact tumor samples embedded in paraffin and fixed with formalin,

MERFISH categorizes diverse cancer cell subpopulations and immune infiltrates based on their gene expression profiles, pinpointing their locations within the tumor architecture. Segmentation and transcriptional profiling of each cell facilitate intricate characterization of varied gene activity patterns in the assorted cell types. Spatial gene expression landscapes provide insights into tumor heterogeneity and the interplay between malignant and immune compartments. MERFISH enabled the construction of cell atlases with transcriptomic and protein signatures within these preserved clinical cancer samples, empowering detailed decoding of cancer and immune cell interrelationships within the tumor microenvironment pertinent to cancer biology (Moffitt et al., 2016).

Unlike targeted imaging methods, sequencing-based ST takes an unbiased approach by profiling all polyadenylated RNAs in a manner similar to single-cell RNA sequencing. This allows the discovery of genome-wide expression patterns and differential gene activity across spatial regions, enabling the identification of novel biological mechanisms (Chen, You, Hardillo, & Chien, 2023). There are two categories of sequencing-based ST technologies: microdissection-based and array-based (Williams et al., 2022).

Digital Spatial Profiling (DSP) is an innovative microdissection-based mechanism that is used for highly multiplexed profiling of proteins and RNAs in a spatial context within formalin-fixed, paraffin-embedded tissue samples. Its key innovation is using photocleavable oligos on affinity reagents, allowing specific release of molecular tags for quantification when light is projected onto user-defined regions of interest. This provides single-cell resolution protein profiling and sensitive RNA detection (Merritt et al., 2020). Another microdissection-based technology is laser capture microscopy (LCM), which involves extracting specific cells or regions from tissue sections using laser beams. In fact, LCM is the most widely adopted microdissection method across diverse biological disciplines, such as oncology, neuroscience, immunology, and botany (Moses & Pachter, 2022). LCM enables precise excision of specific cell populations or regions from heterogeneous tumor tissue. By allowing downstream profiling to be linked to these spatially defined samples, LCM empowers the discovery of localized biomarkers and molecular heterogeneity. For instance, LCM coupled with mass spectrometry reveals proteomic signatures associated with invasion and stromal interactions. Additionally, LCM enables epigenetic analysis to characterize DNA methylation patterns tied to different states of circulating tumor cells; integration with other epigenetic assays can provide insights into heterogeneous regulation mechanisms within tumors (Lee, Kim, Lee, Lee, & Kwon, 2024). Overall, LCM facilitates multifaceted characterization of the cancer proteome and epigenome with spatial context intact.

Continued improvements in spot density, tissue processing, and sequencing have enabled expanding throughput and resolution (Rodriques et al., 2019). Array-based ST techniques involve arrayed capture probes on a slide surface to fix mRNA in tissue sections for subsequent sequencing and mapping of transcripts back to spatial coordinates (Williams et al., 2022). An example of an array-based technology is Slide-seq, which performs ST on glass slides using barcoded arrays, enabling precise mapping of the tumor microenvironment. Slide-seq is able to analyze spatial gene expression in frozen tissue with high resolution across large volumes, integrating readily with single-cell RNA sequencing datasets. Through this integration, Slide-seq allows the identification of localized gene expression patterns in normal and diseased tissues (Rodriques et al., 2019). The original Slide-seq protocol had limitations in transcript detection sensitivity, constraining potential applications. To address this, substantial improvements were made to the barcoded bead synthesis, array sequencing protocol, and cDNA processing, leading to a new high-sensitivity approach: Slide-seqV2. This approach uses the development of a new mono-base encoding scheme for spatial barcode indexing, avoiding reliance on inefficient barcode format conversion. The advancements made to develop Slide-seqV2 provide a 10-fold increase in sensitivity along with more efficient gene expression recovery per sequencing read. This expanded analytical capability allowed for the identification of spatially localized gene expression and in situ characterization of developmental trajectories, increasing its utility for probing transcriptional heterogeneity (Stickels et al., 2021). These technologies empower spatially resolved study of cancer heterogeneity and microenvironment cell interactions with customizable resolution.

While ST captures gene expression data and provides insight into tissue organization and microenvironmental influences, high-dimensional datasets often present computational challenges for analysis and interpretation (Asp, Bergenstrahle, & Lundeberg, 2020). Computational tools are critical to fully realize the potential of ST. By developing specialized algorithms, effective computational analysis can help extract key biological insights from spatial information.

CellNeighborEx is a computational tool that enables the analysis of ST data generated by either sequencing or imaging methods to identify genes differentially expressed due to direct cell–cell interactions. For image-based data, such as seqFISH, CellNeighborEx categorizes cells by their immediate neighbors using algorithms and then compares heterotypic neighbors to homotypic neighbors. For sequencing-based data, such as Slide-seq, this tool compares the expression profiles of heterotypic spots containing transcripts from cell types to homotypic spots of single cell types. In both cases, CellNeighborEx reveals neighbor-dependent gene regulation specifically due to cell contact, while also determining the source cell type of differential expression. This provides an unbiased approach to studying how cell-cell interactions within intact tissues influence gene expression, capturing effects distinct from ligand-receptor analysis. The niche-specific gene expression changes revealed by CellNeighborEx give key insights into cellular communication during development and disease from both imaging- and sequencing-based ST data (Kim et al., 2023).

Vesalius is another innovative computational tool that leverages image analysis techniques to isolate tissue territories effectively from ST data. It represents transcriptomes as colors to build images, overcoming challenges in identifying heterogeneous spatial domains (Fig. 1C). Through territory isolation, Vesalius enhances the analysis of high-resolution image-based or sequencing-based ST data, uncovering finer tissue organization and spatial gene expression patterns. Vesalius outperformed existing computational tools, such as BayesSpace, STAGATE, SpaGCN, SEDR, Giotto, and Seurat, in recovering distinct territories across cell heterogeneity and resolutions. It reveals territory-specific expression profiles, reflecting the importance of spatial context. Vesalius complements current approaches to provide biological insights, annotate cell types by location, and investigate borders between neighboring tissue regions. By bridging ST and computational innovation, Vesalius supports the analysis of datasets to highlight tissue organization and the microenvironmental influences on gene regulation (Martin, Kim, Lovkvist, Hong, & Won, 2022).

While ST shows potential for highlighting tumor heterogeneity, this technique faces hurdles such as high costs and technical limitations in detection sensitivity (Asp, et al., 2020), transcript coverage (Moses & Pachter, 2022), and formalin-fixed paraffin-embedded sample compatibility (Merritt et al., 2020). Additionally, bioinformatic tools tailored for analyzing ST data are lacking, as most RNA-seq analysis software does not account for spatial relationships (Yu et al., 2022). Developing algorithms for spatial spot deconvolution, cell identification from subcellular data, and multi-platform integration would expand analytical capabilities. However, the larger constraint of ST is that it provides only a single-omics view of the tumor microenvironment (Yu et al., 2022). Emerging spatial multi-omics techniques aim to overcome this by integrating proteomic, genomic, and transcriptomic profiles while retaining spatial context (Stickels et al., 2021). Early applications in cancer research demonstrate the value of this approach for obtaining a multidimensional understanding of intratumor heterogeneity at the protein, RNA, and DNA levels. By mapping diverse biomolecules to spatial coordinates, these technologies promise to provide unparalleled characterization of the tumor ecosystem. Advances in spatial multi-omics are poised to yield profound insights into the interconnected regulatory networks driving cancer progression and therapeutic response (Yu et al., 2022).

4. Artificial intelligence

AI and ML are transforming the analysis of multidimensional spatial omics datasets to gain new insights into cancer biology. Whereas traditional computational approaches rely heavily on statistical testing, AI provides more versatile and scalable tools to uncover complex patterns, spatial modeling, and simulation using multidimensional data (Zhang, Shi, & Wang, 2023).

Deep-learning techniques, especially convolutional neural networks (CNNs), have shown great promise for extracting features and patterns from spatially structured biological data, including histology slides, multiplexed tissue images, and medical scans (Zidane et al., 2023). CNNs are well-suited for processing grid-structured data, a characteristic of images. Through stacked convolutional layers, CNNs automatically learn hierarchical feature presentations from pixel values. This allows CNNs to effectively recognize patterns and objects in images (Fig. 2). By training on labeled data, CNNs develop specialized feature detectors tailored to problems, such as medical image diagnosis (Yamashita, Nishio, Do, & Togashi, 2018). A major application of CNNs has been in cancer imaging, where they can identify malignant lesions and tumors with high accuracy. For example, CNNs have been applied to mammogram images for breast cancer screening (Kooi et al., 2017) and CT scans for PDAC (Chen et al., 2023), outperforming human radiologists (McKinney et al., 2020) and technologies such as Computer-Aided Detection (CAD) systems that were widely used within hospital settings (Kooi et al., 2017). Beyond cancer detection, CNNs have been applied to segment and classify cells morphologically from histology images, enabling subcellular compartment analysis as well as the categorization of

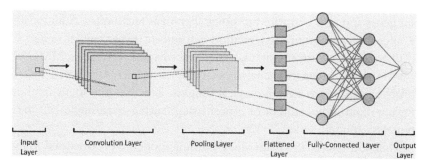

Fig. 2 Convolutional Neural Networks. CNNs are a deep-learning technique that utilizes stacked specialized convolution and pooling layers to learn spatial visual features at increasing levels of abstraction, enabling the recognition and classification of complex patterns. The pooling layer is reshaped into a flattened layer, preparing the data to enter the fully-connected layer. The fully-connected layer is a dense layer that integrates the convolutional features across the spatial dimensions to allow for classification. As the complexity of the input increases, the CNN may contain additional layers to ensure proper detection of detailed patterns.

cells into clinically relevant subtypes (Huang et al., 2022). Some deep-learning models also estimate tumor purity, immune infiltration, and other microenvironment features directly from digitized histopathology slides, showing the potential for assisting in diagnoses and enabling optimal personalized immunotherapies (Saltz et al., 2018).

Other AI techniques include generative adversarial networks (GANs), which can synthesize realistic spatial distributions of cells consistent with input spatial omics data (Lan et al., 2020). A novel deep-learning technique was developed, called single-cell multi-modal generative adversarial network (scMMGAN), to integrate heterogeneous single-cell data, such as ST, single-cell RNA sequencing, and proteomics. scMMGAN is based on GANs and uses specialized losses, including a new correspondence loss based on diffusion geometry (Amodio et al., 2022). Conventional GANs are comprised of generator and discriminator networks trained in an adversarial manner. They are used for digital image processing, medical image processing, medical informatics, and applications in -omic data (Lan et al., 2020). scMMGAN enables more meaningful integration and mapping between different modalities while preserving intrinsic data structure. Through experiments on multi-modal breast cancer data, scMMGAN demonstrates the ability to combine the strengths of each modality, infer spatial information, and discover novel correlations missed by individual datasets. Quantitative evaluations also show that scMMGAN better preserves signals from genes and cells compared to other alignment techniques. Additionally, the uncertainty quantification method can inform experimental design by identifying genes with unique vs. shared information. Overall, scMMGAN allows in-depth integrative analysis of complex cancer data across diverse modalities, leading to new biological insights. This has important implications for multi-omics studies of cancer using cutting-edge single-cell measurements (Amodio et al., 2022).

While just two of the many deep-learning models were discussed in this review, other AI techniques that provide insight into spatial data include physics-based neural networks, topology-aware neural networks, and graph neural networks (Zidane et al., 2023). Though numerous studies demonstrate the potential of deep-learning and neural networks in explaining cancer biology and advancing care, there are important challenges and ethical concerns that need to be addressed. There is a need for extensive real-world testing and validation before establishing an AI system in clinical settings (Topol, 2019). The IBM Watson for Oncology algorithm was developed to recommend treatments and drugs for cancer patients. However, it was found to have significant flaws, as it would suggest treatments detrimental to the patient's health. Rushing to deploy unproven AI systems into healthcare facilities compromises patient safety on a wide scale. Additionally, the "black box" nature of deep-learning algorithms makes it difficult to understand the outputs; physicians may not be able to justify the recommended treatments. Installing black-box AI systems removes the understanding of why certain therapies are recommended at a physician level and a patient level, leading to ethical problems, especially in the field of oncology, where shared decision-making is essential (Topol, 2019). Additional critical concerns surrounding the use of AI and ML in healthcare revolve around bias and fairness. Flawed data and design choices can propagate biases leading to inequitable model performance across patient subgroups. There are also many ethical issues around informed consent, privacy, accountability, and appropriate use of sensitive patient data (McCradden, Joshi, Mazwi, & Anderson, 2020). Hence, establishing governance frameworks and best practices for access, transparency, and maintaining responsibility are crucial. Moving forward, priorities for leveraging AI to improve cancer therapy include building predictive models to match patients with optimal treatments based on molecular profiling, designing intelligent clinical decision support systems to recommend evidence-based therapeutic strategies, and using deep learning on large biological datasets to uncover new therapeutic targets and drug combinations (Topol, 2019).

5. Spatial transcriptomics contribution to understanding the heterogeneity of pancreatic cancer tumors

Traditional diagnostic approaches, such as imaging and histological samples, are limited in detecting early-stage PDAC and micrometastases or peritoneal lesions (Wang, Li, Wang, & Ding, 2021). However, ST can potentially enhance the diagnosis of PDAC by offering significant insights into the spatial arrangement of cells and their gene expression patterns within the tumor microenvironment, thus improving the accuracy of diagnosis. For instance, endoscopic ultrasound-guided fine needle aspiration (EUS-FNA) is a standard procedure for histological diagnosis in PDAC, aiding in identifying specific genetic alterations like KRAS and molecular markers such as CEA and CA19-9 (Kitano, Minaga, Hatamaru, & Ashida, 2022; Wang et al., 2023). ST can complement this approach by providing spatially resolved gene expression data. This data can contribute to understanding the various molecular pancreatic cancer types and their tumor microenvironment while developing novel therapeutic strategies based on precision medicine approaches (Bazzichetto et al., 2020; George et al., 2024).

A recent paper by George et al. highlights the importance of integrating a precision medicine approach based on PDAC TME subtypes into treatment decisions (George et al., 2024). The study shows an association between PDAC TME subtypes and treatment responses to neoadjuvant chemotherapy regimens, finding that specific subtypes of the PDAC TME respond better to FOLFIRINOX and Gemcitabine with Nab-paclitaxel (George et al., 2024). To classify the PDAC TME, they performed a transcriptomic analysis and identified four distinct subtypes: immune enriched (IE), immune enriched-fibrotic (IE/F), fibrotic (F), and immune-depleted (D) (George et al., 2024). Then, they emphasized the distribution of KRAS-mutated PDAC among various TME subtypes and the presence of subclonal mutations in KRAS. Interestingly, KRAS mutations were evenly distributed across these subtypes; however, 64% of patients with KRAS WT tumors had an IE/F TME subtype (George et al., 2024). Furthermore, PDAC patients with lung metastases harbored a distinct IE TME and tended to have a favorable prognosis compared to the liver (George et al., 2024). Patients with IE and IE/F subtypes also showed better responses to immunotherapy and a more favorable prognosis than those with F and D TMEs subtypes. These findings suggest that FOLFIRINOX chemotherapy may be effective for patients with IE or IE/F

TMEs, while Gemcitabine and Nab-paclitaxel may benefit patients with F or D TMEs (George et al., 2024). The study results show significant implications for utilizing immunotherapeutic strategies based on the TME subtype approach that will enhance the advancement of precision-based medicine and the potential for biomarker-driven patient selection in personalized immunotherapeutic approaches (Table 1).

Promising results were observed using multi-omics approaches to identify biomarkers for predicting how esophageal cancer will respond to neoadjuvant therapy. For example, a study published by Yang et al. identified several biomarkers using ST, including genetic variations such as single-nucleotide polymorphisms (SNPs) in the genes ERCC1 and XRCC1, radiomic feature from CT imaging, miRNA ratios, and spatial metabolite data data (Yang, Guan, Bronk, & Zhao, 2024). Reduced expression of miR-350-5p in esophageal adenocarcinoma can be a potential biomarker predictor of non-responders to neoadjuvant chemotherapy (Bibby, Reynolds, & Maher, 2015). Additionally, chemoradiation resistance in esophageal cancer is associated with SNPs in genes such as ERCC1 and XRCC1. In pancreatic cancer, miR-350-5p acts as a tumor suppressor by targeting MUC1 (Trehoux et al., 2015), and the expression of ERCC1 impacts the response to FOLFIRINOX treatment in metastatic pancreatic cancer (Strippoli et al., 2016). Thus, integrating multi-omics data can help clinicians personalize treatment plans considering each patient's molecular complexity and treatment outcomes in pancreatic cancer.

Multi-omics analysis can also detect RNA modifications such as N6-methyl adenosine (m6A), which was identified as a potential biomarker to diagnose pancreatic cancer. The m6A RNA modification is crucial in regulating gene expression by affecting RNA splicing, degradation, translocation efficiency, stability, decay, and RNA-protein interactions. Consequently, this modification impacts various biological processes such as cancer cell proliferation, differentiation, apoptosis, invasion, and metastasis (Li, Wang, Liu, Wang, & Ni, 2021; Xu, 2020; Zhou et al., 2020).

Applying clustering and signal enrichment analysis to transcriptomic data can contribute to predicting drug responses. For example, single-sample Gene Set Enrichment Analysis (ssGSEA) and K-means clustering identified gene signatures and classification scores essential to distinguish small cell lung cancer (SCLC) tumors based on their molecular properties (Nemes et al., 2024). By performing a preprocessing step, it was possible to cluster the gene datasets to predict drug responses, generating a model capable of describing how the different SCLC subtypes identified by the

Table 1 The transcriptomic profiling of pancreatic ductal adenocarcinoma (PDAC) has enabled the comprehensive categorization of tumor microenvironment (TME) subtypes.

TME subtype	Description	Prognosis and response to therapy	Metastatic pattern	Clinical implications
Immune Enriched (IE)	High immune cell infiltration, low fibrosis	Favorable prognosis; Potential response to immunotherapy	Lung Metastases	Potential for immunotherapy; Consideration for neoadjuvant therapy
Immune Enriched–Fibrotic (IE/F)	High immune cell infiltration with fibrosis	Favorable prognosis; Potential response to immunotherapy	–	Consideration for neoadjuvant therapy
Fibrotic (F)	High fibrosis, low immune cell infiltration	Less favorable prognosis; Potential response to specific chemotherapy regimens	–	Considered for tailored chemotherapy
Immune Depleted (D)	Low immune cell infiltration, low fibrosis	Less favorable prognosis; Potential for resistance to immunotherapy	Liver Metastases	Consideration for alternative therapeutic strategies

This categorization outlines the prognostic implication, response to therapy, metastatic patterns, and clinical implications. Personalized therapeutic treatment based on the PDAC TME subtype can improve the outcomes and personalized therapeutic strategies in precision medicine.

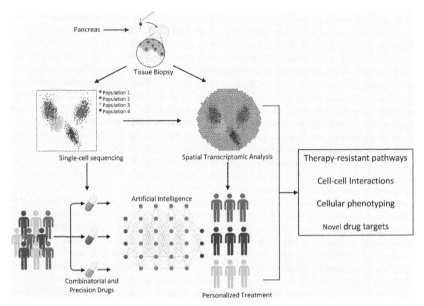

Fig. 3 Spatial transcriptomic analysis can contribute to cancer diagnosis and treatment. The integrative approach for precision medicine in pancreatic tissue biopsy analysis utilizing single cell-sequencing and ST analysis, combined with AI, can help identify therapy-resistant pathways, cell-cell interaction, cellular phenotyping, and novel drug targets. By utilizing this comprehensive strategy, it is possible to develop combinatorial and precision drugs that are tailored to individual patients' treatment approaches to improve patient outcomes.

ssGSEA analysis will respond to therapy. Thus, applying this type of analysis to SCLC patient datasets can help tailor the treatment strategy that best fits the characteristics of the tumor, which may lead to better healthcare outcomes by giving the appropriate treatment and avoiding unnecessary toxicity. Furthermore, it is possible to use the same analysis principles as used by ssGESA in datasets for types of cancers, such as PDAC, to compare genomics traits, gene expression, and molecular subtyping to predict the drug response in pancreatic cancer patients (Fig. 3).

Integrating ST with single-cell RNA sequencing (scRNA-seq) can offer a promising approach for diagnosing, analyzing heterogeneity, and identifying potential therapeutic targets for PDAC. Moncada et al. developed a multimodal interaction analysis (MIA) by combining scRNA-seq and a microarray-based ST method (Visium) to analyze the heterogeneity of the TME of PDAC (Moncada et al., 2020). MIA helped determine the potential functional roles of different cell states and subpopulations by examining their spatial distribution in a particular tissue

(Moncada et al., 2020). This methodology can be especially valuable in tumor analysis, where the composition of different tumor subclasses can vary among individuals. By identifying the subpopulation composition and spatial localization of these subclasses in a given patient, we may be able to provide valuable prognostic information for personalized cancer treatment.

Similarly, Hwang et al. used single-nucleus RNA-seq and whole-transcriptome to construct a high-resolution molecular landscape of PDAC to investigate the cellular dynamics within PDAC patients undergoing neoadjuvant treatment (Hwang et al., 2022). This study analyzed 43 primary tumor samples that either received neoadjuvant therapy or were untreated. The analysis revealed significant modification in the cellular population and their spatial arrangement in TME after neoadjuvant treatment. These changes included alternation in immune cell infiltration, remodeling of the stromal compartment, and modification in tumor cell plasticity. The study revealed recurrent expression patterns in malignant cells and fibroblasts, which shed light on the intra- and intertumoral heterogeneity in PDAC (Hwang et al., 2022).

Furthermore, using single-cell and ST, the study identified a neural-like malignant cell program enriched after chemotherapy associated with poor prognosis; this association was validated by analyzing external datasets, strengthening the reliability of linking this program to aggressive disease. By reproducing findings across independent patient cohorts, the researchers avoided false positive results and demonstrated the neural-like program's connection to shorter survival times in pancreatic cancer (Hwang et al., 2022). The study also revealed spatially defined receptor-ligand interactions between cells that were increased in residual PDAC after neoadjuvant therapy. Mapping these spatial intercellular networks could help identify new therapeutic targets to improve treatment response (Hwang et al., 2022). The findings of this study provide insights into the complex molecular mechanisms that determine the response of pancreatic cancer to treatment, suggesting encouraging opportunities for developing novel therapeutic approaches to effectively target specific subpopulations of cells or spatially defined regions within the tumor.

Despite the immense potential of ST in diagnosing pancreatic cancer, it faces challenges that need to be addressed. However, researchers are actively working towards overcoming these limitations by leveraging advanced technologies, adopting immunotherapeutic approaches, and emphasizing the significance of timely interventions and precision medicine strategies. By doing so, they aim to enhance the accuracy and efficacy of ST-based diagnostic methods for pancreatic cancer.

6. The use of Artificial Intelligence to diagnose and define pancreatic cancer therapy

ST and AI tools can be applied to effectively diagnose cancer by analyzing complex data and identifying patterns that may not be easily recognizable to the human eye. Previous studies have demonstrated a correlation between gene expression patterns and histological image features, suggesting the possibility of predicting gene expression based on histological samples (Badea & Stanescu, 2020). However, current techniques only utilize a fraction of the extensive cellular information in high-resolution histology images. Zhang et al. recently developed the iStart (Inferring Super-Resolution Tissues Architecture) tool that utilizes AI to combine high-resolution histology images and ST data. This tool can predict spatial gene expression and quickly examine gene activity in histology images (Zhang et al., 2024). iStart can automatically detect multicellular structures correlated with better immunotherapy responses, such as tertiary lymphoid – clusters of anti-tumor immune cells found in non-lymphoid tissue, including solid tumors (Zhang et al., 2024). This feature is especially significant in pancreatic cancer, where the immune cell infiltration and tumor microenvironment play a crucial role in cancer progression and treatment response. iStart can help determine which patients diagnosed with pancreatic cancer would benefit from immunotherapy. The paper also highlights iStart's efficacy in various cancer types, including prostate cancer, kidney cancer, breast cancer, and colorectal cancer. The study demonstrates iStart's superior efficiency by comparing it to X-fuse, a deep generative model to estimate super-resolution gene expression at the pixel level using hematoxylin and eosin-stained (H&E) histology images (Hu et al., 2023; Zhang et al., 2024). The results show that iStart is more accurate and less time-consuming than XFuse. Further research using iStart to study pancreatic cancer can accurately describe the spatial gene expression patterns and tissue structure. It can also help identify vital multicellular structures within the tumor microenvironment.

Additionally, a recent study by Wang et al. showed the importance of understanding colorectal cancer (CRC) and the potential application of ML in cancer research (Wang et al., 2024). This study focuses on the development of a prognostic index for predicting outcomes and identifying immunotherapeutic responses in patients with CRC by using ML techniques based on the cytotoxic T lymphocyte evasion genes (GERGs), where it aims to uncover molecular signatures linked to the immune invasion and chemotherapy response through the use of ML algorithms to find potential personalized therapeutic targets for CRC (Wang et al., 2024). Moreover, FOLFOX and FOLFIRI combine chemotherapies commonly used as first-line treatment for late-stage CRC (Giacchetti et al., 2000), where FOLFOX and FOLFIRI have an efficacy of 59% and 39%, respectively. However, while these therapies can potentially prolong overall survival, unfortunately, many patients do not experience long-term benefits (Neugut et al., 2019). A study by Amniouel et al. identified specific markers that can accurately predict the effectiveness of 5-FU-based chemotherapy (FOLFOX and FOLFIRI) on patients with CRC by analyzing the datasets retrieved from Gene Expression Omnibus (GEO) using ML models. The author determined that about 28.6% of patients who did not respond well to their initial therapy could benefit from an alternative treatment approach (Amniouel & Jafri, 2023). Similarly, both studies can use ML methods combined with the analysis of gene expression data, which can potentially identify molecular signatures that can predict the efficacy of specific chemotherapy treatments in patients with pancreatic cancer. Ultimately, this approach will lead to personalized treatment plans that consider the molecular characteristics of the tumor and the patient's genetic background, aiding in selecting the most effective chemotherapy regimen.

Another study by Mao et al. investigated on the molecular mechanisms that cause breast capsular contracture, a common complication developed after breast capsular prothesis contracture (Chicco, Ahmadi, & Cheng, 2021; Mao, Hou, Fu, & Luan, 2024). In this study, the researcher used RNA sequencing technology and ML algorithms to identify the hub genes that correlate with the progression of breast capsular contracture to predict potential mechanisms that can lead to more accurate diagnosis and treatment (Mao et al., 2024). This study found that the PBKAR2B gene could potentially serve as a new diagnostic biomarker to detect breast capsular contracture, a complication of breast augmentation surgery. The researchers suggest that PBKAR2B may be used in a novel grading system to identify the progression of this condition, which appears to be linked to the cCMP-PKG, PI3k-Akt, and thyroid signaling pathways (Mao et al., 2024). Similar concepts can be applied to pancreatic cancer to identify the hub genes and signaling pathways involved in the molecular mechanisms of pancreatic cancer progression.

Previous studies have shown that cell-cell communication networks (CCNs) have a significant impact on the development of the tumor immune microenvironment (TIME) and can affect the response to immune checkpoint inhibitors (ICIs) in cancer patients (Wei, Duffy, & Allison, 2018). Lee et al. demonstrated that ML based on CCNs extracted from bulk tumor transcriptome data could be used to predict the efficiency of ICN treatment for various types of cancer. The model developed achieved a median AUROC (area under the receiver operating characteristic curve) score of 0.79 (Lee et al., 2024). This study analyzed samples of over 700 ICI patients who received ICIs using the ML algorithm. The results showed that communication strength between various types of cancer within the TIME plays a significant role in determining the response to ICIs (Lee et al., 2024). For example, programmed cell death ligands (PD-L1) are produced by antigen-presenting cells (APCs) to control the activation of T cells, while tumors are expressed by the same gene used for immune invasion. This indicates that the cooperation between T cells and APCs is crucial for patients responding to ICI therapy (Escors et al., 2018; Lee et al., 2024; Suresh & O'Donnell, 2021). Their findings highlight that targeting specific cell-cell communication pathways can play a role in the recruitment and activation within TIME and could be a viable approach to improving the effectiveness of ICIs. This similar approach using the transcriptome and ML can be adapted to examine CNNs in pancreatic cancer patients, which can help predict the efficacy of ICI treatment.

Moreover, AI-powered approaches have revolutionized the early detection and intervention capabilities in PDAC. A recent article by the *Mayo Clinic* highlighted how Dr. Goenka's team utilized a vast imaging dataset comprising over 3000 patient CT scans to train an automated AI model. This dataset is one of the largest and most diverse used for such research, with 64% of CT scans sourced from external institutions, highlighting the diversity of the training dataset. The model demonstrated an average accuracy of 92%, accurately classifying 88% of CT scans with cancer and 94% of control CT scans. Despite being trained on CTs with larger tumors, the model could detect pancreatic cancer in pre-diagnostic CTs taken between 3 and 36 months (median 475 days) before the cancer's clinical diagnosis. In these scenarios, the model reported an accuracy of 84%, an AUROC curve of 0.91, a sensitivity of 75%, and a specificity of 90% (Korfiatis et al., 2023). These findings underscore the model's potential in early pancreatic cancer detection and intervention. In addition, AI-assisted endoscopic ultrasound has shown promising preliminary results, indicating its potential as a tool for the computer-aided diagnosis of pancreatic cancer (Dumitrescu et al., 2022). Furthermore, Using various imaging techniques, along with advancements in AI-driven diagnosis based

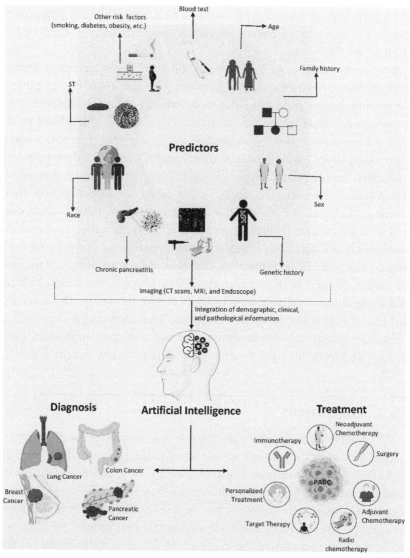

Fig. 4 Integration of patient data by Artificial Intelligence may guide in determining cancer diagnosis and therapy. The comprehensive integration of demographic, clinical, and pathological data to predict pancreatic cancer may use AI algorithms to help clinicians in cancer diagnosis and treatment strategies. The process involves collecting diverse patient data, including demographics, clinical history, and pathological characteristics, which are then analyzed using advanced AI techniques such as ML and deep learning to identify predictive factors and patterns indicative of pancreatic cancer. The AI model accurately diagnoses the presence of cancer and provides insights into optimal treatment approaches tailored to individual patients. This integrative approach enhances the accuracy and efficiency of pancreatic cancer diagnosis and enables personalized treatment strategies, ultimately improving patient outcomes and quality of care.

on cytopathology and serological markers, AI models have the potential to enhance the accuracy of pancreatic cancer diagnosis (Fig. 4) (Hameed & Krishnan, 2022).

Apart from imaging, AI has also been applied to genetic characterization and prediction of carrier status of germline pathogenic variants in cancer-predisposing genes, indicating the potential for AI to contribute to a comprehensive approach to pancreatic cancer diagnosis (Mizukami et al., 2020). AI has shown promise in grading pancreatic cancer in pathological images using deep learning convolutional neural networks, demonstrating its potential in enhancing diagnostic accuracy (Sehmi, Fauzi, Ahmad, & Chan, 2022). However, the development of AI for detecting and characterizing PDAC is challenging due to the scarcity of PDAC data compared to other cancers (Hellström et al., 2023). Additionally, the use of AI to analyze clinical data from electronic health records has not yet been fully explored, particularly in improving the early detection of pancreatic and other types of cancers (Kenner et al., 2021). Overall, challenges exist in establishing a more reliable classification model for accurately detecting pancreatic cancer using designed algorithms. Thus, applying AI approaches to diagnose pancreatic cancer shows promise and could transform early detection and individualized treatment, leading to better patient results.

7. Conclusions

The development of tools capable of integrating and analyzing multiple omics data is improving our knowledge of cancer biology. ML and AI-based analysis of such complex datasets show promising results by identifying new biomarkers and therapeutic targets and helping us better understand cellular interactions within the tumor microenvironment (Yang et al., 2024). The application of such tools is still developing, and there is a long road ahead to improve the accuracy of the predictions based on this type of analysis. Every year, multiple tools are developed and presented to address the complexity of tumor biology, making it hard for the end users to understand which tools to use. That is one reason why healthcare professionals are not using AI as a standard practice nowadays. Thus, the use of AI tools requires standardization.

Another reason is that AI requires powerful processing computers and currently applies complex algorithms that could be more user-friendly. Additionally, these tools still require further testing and must be simplified so that healthcare professionals can comprehensively obtain the results. Moreover, adopting AI-based tools will require healthcare professionals to pass through a training process to become familiar with such tools so that AI can help manage cancer diagnoses and treatments.

It is undeniable that AI will be able to help diagnose and treat cancer patients by contributing to the establishment of strategies focused on the biological characteristics of each patient's tumors. For example, AI can predict cases in which a cancer may be resistant to chemotherapy (Strippoli et al., 2016) or treatment with ICIs (Jiang et al., 2018; Lee et al., 2024; Raskov, Orhan, Christensen, & Gogenur, 2021), thus preventing patients from the toxicity of a treatment from which they will not benefit. Integrating ST analysis with routine exams such as MRIs (Virostko, 2020) and CT scans (Korfiatis et al., 2023; Mukherjee et al., 2022) can help diagnose pancreatic cancer.

An advantage of ST analysis by AI tools is their versatility since the analysis of transcriptomic profiles can be obtained regardless of the type of cancer. Still, it will always reflect the characteristic features of the tumor analyzed, thus contributing to a patient-centered approach to diagnosis and therapy.

Ultimately, the application of AI in the management of pancreatic cancer needs to overcome the challenges described here and must be used as a tool to help healthcare professionals make better decisions and not as a replacement for such professionals.

All authors declare that they have no conflict of interests.

Acknowledgments

This study was supported by Fundação para a Ciência e a Tecnologia Grant 2020.05319. BD to A.M.P. All figures are original and were prepared using BioRender.

References

Amniouel, S., & Jafri, M. S. (2023). High-accuracy prediction of colorectal cancer chemotherapy efficacy using machine learning applied to gene expression data. *Frontiers in Physiology, 14*, 1272206. https://doi.org/10.3389/fphys.2023.1272206.

Amodio, M., Youlten, S. E., Venkat, A., San Juan, B. P., Chaffer, C. L., & Krishnaswamy, S. (2022). Single-cell multi-modal GAN reveals spatial patterns in single-cell data from triple-negative breast cancer. *Patterns (N Y), 3*(9), 100577. https://doi.org/10.1016/j.patter.2022.100577.

Angelo, M., Bendall, S. C., Finck, R., Hale, M. B., Hitzman, C., Borowsky, A. D., et al. (2014). Multiplexed ion beam imaging of human breast tumors. *Nature Medicine, 20*(4), 436–442. https://doi.org/10.1038/nm.3488.

Asp, M., Bergenstrahle, J., & Lundeberg, J. (2020). Spatially resolved transcriptomes-next generation tools for tissue exploration. *Bioessays: News and Reviews in Molecular, Cellular and Developmental Biology, 42*(10), e1900221. https://doi.org/10.1002/bies.201900221.

Badea, L., & Stanescu, E. (2020). Identifying transcriptomic correlates of histology using deep learning. *PLoS One, 15*(11), e0242858. https://doi.org/10.1371/journal.pone.0242858.

Bazzichetto, C., Luchini, C., Conciatori, F., Vaccaro, V., Di Cello, I., Mattiolo, P., et al. (2020). Morphologic and molecular landscape of pancreatic cancer variants as the basis of new therapeutic strategies for precision oncology. *International Journal of Molecular Sciences, 21*(22), https://doi.org/10.3390/ijms21228841.

Bibby, B. A., Reynolds, J. V., & Maher, S. G. (2015). MicroRNA-330-5p as a putative modulator of neoadjuvant chemoradiotherapy sensitivity in oesophageal adenocarcinoma. *PLoS One, 10*(7), e0134180. https://doi.org/10.1371/journal.pone.0134180.

Chen, K. H., Boettiger, A. N., Moffitt, J. R., Wang, S., & Zhuang, X. (2015). RNA imaging. Spatially resolved, highly multiplexed RNA profiling in single cells. *Science (New York, N. Y.), 348*(6233), aaa6090. https://doi.org/10.1126/science.aaa6090.

Chen, P. T., Wu, T., Wang, P., Chang, D., Liu, K. L., Wu, M. S., et al. (2023). Pancreatic cancer detection on CT scans with deep learning: A nationwide population-based study. *Radiology, 306*(1), 172–182. https://doi.org/10.1148/radiol.220152.

Chen, T. Y., You, L., Hardillo, J. A. U., & Chien, M. P. (2023). Spatial transcriptomic technologies. *Cells, 12*(16), https://doi.org/10.3390/cells12162042.

Chicco, M., Ahmadi, A. R., & Cheng, H. T. (2021). Systematic review and meta-analysis of complications following mastectomy and prosthetic reconstruction in patients with and without prior breast augmentation. *Aesthetic Surgery Journal, 41*(7), NP763–NP770. https://doi.org/10.1093/asj/sjab028.

Dumitrescu, E. A., Ungureanu, B. S., Cazacu, I. M., Florescu, L. M., Streba, L., Croitoru, V. M., et al. (2022). Diagnostic value of artificial intelligence-assisted endoscopic ultrasound for pancreatic cancer: A systematic review and meta-analysis. *Diagnostics.* https://doi.org/10.3390/diagnostics12020309.

Eng, C. L., Lawson, M., Zhu, Q., Dries, R., Koulena, N., Takei, Y., et al. (2019). Transcriptome-scale super-resolved imaging in tissues by RNA seqFISH. *Nature, 568*(7751), 235–239. https://doi.org/10.1038/s41586-019-1049-y.

Escors, D., Gato-Canas, M., Zuazo, M., Arasanz, H., Garcia-Granda, M. J., Vera, R., et al. (2018). The intracellular signalosome of PD-L1 in cancer cells. *Signal Transduction and Targeted Therapy, 3*, 26. https://doi.org/10.1038/s41392-018-0022-9.

George, B., Kudryashova, O., Kravets, A., Thalji, S., Malarkannan, S., Kurzrock, R., et al. (2024). Transcriptomic-based microenvironment classification reveals precision medicine strategies for PDAC. *Gastroenterology.* https://doi.org/10.1053/j.gastro.2024.01.028.

Giacchetti, S., Perpoint, B., Zidani, R., Le Bail, N., Faggiuolo, R., Focan, C., et al. (2000). Phase III multicenter randomized trial of oxaliplatin added to chronomodulated fluorouracil-leucovorin as first-line treatment of metastatic colorectal cancer. *Journal of Clinical Oncology: Official Journal of the American Society of Clinical Oncology, 18*(1), 136–147. https://doi.org/10.1200/JCO.2000.18.1.136.

Giesen, C., Wang, H. A., Schapiro, D., Zivanovic, N., Jacobs, A., Hattendorf, B., et al. (2014). Highly multiplexed imaging of tumor tissues with subcellular resolution by mass cytometry. *Nature Methods, 11*(4), 417–422. https://doi.org/10.1038/nmeth.2869.

Goltsev, Y., Samusik, N., Kennedy-Darling, J., Bhate, S., Hale, M., Vazquez, G., et al. (2018). Deep profiling of mouse splenic architecture with CODEX multiplexed imaging. e915 *Cell, 174*(4), 968–981. https://doi.org/10.1016/j.cell.2018.07.010.

Hamada, S., Masamune, A., & Shimosegawa, T. (2013). Alteration of pancreatic cancer cell functions by tumor-stromal cell interaction. *Frontiers in Physiology, 4*, 318. https://doi.org/10.3389/fphys.2013.00318.

Hameed, B. S., & Krishnan, U. M. (2022). Artificial intelligence-driven diagnosis of pancreatic cancer. *Cancers.* https://doi.org/10.3390/cancers14215382.

Hanahan, D., & Weinberg, R. A. (2011). Hallmarks of cancer: The next generation. *Cell, 144*(5), 646–674. https://doi.org/10.1016/j.cell.2011.02.013.

Hellström, T.A.E., Viviers, C.G.A., Ramaekers, M., Tasios, N., Nederend, J., & Luyer, M. D.P., et al. (2023). Clinical segmentation for improved pancreatic ductal adenocarcinoma detection and segmentation.

Hu, J., Coleman, K., Zhang, D., Lee, E. B., Kadara, H., Wang, L., et al. (2023). Deciphering tumor ecosystems at super resolution from spatial transcriptomics with TESLA. *Cell Systems, 14*(5), 404–417.e404. https://doi.org/10.1016/j.cels.2023.03.008.

Huang, B., Huang, H., Zhang, S., Zhang, D., Shi, Q., Liu, J., et al. (2022). Artificial intelligence in pancreatic cancer. *Theranostics, 12*(16), 6931–6954. https://doi.org/10.7150/thno.77949.

Hwang, W. L., Jagadeesh, K. A., Guo, J. A., Hoffman, H. I., Yadollahpour, P., Reeves, J. W., et al. (2022). Single-nucleus and spatial transcriptome profiling of pancreatic cancer identifies multicellular dynamics associated with neoadjuvant treatment. *Nature Genetics, 54*(8), 1178–1191. https://doi.org/10.1038/s41588-022-01134-8.

Jiang, P., Gu, S., Pan, D., Fu, J., Sahu, A., Hu, X., et al. (2018). Signatures of T cell dysfunction and exclusion predict cancer immunotherapy response. *Nature Medicine, 24*(10), 1550–1558. https://doi.org/10.1038/s41591-018-0136-1.

Kenner, B., Abrams, N., Chari, S. T., Field, B. F., Goldberg, A., Hoos, W., et al. (2021). Early detection of pancreatic cancer. *Pancreas.* https://doi.org/10.1097/mpa.0000000000001882.

Kim, H., Kumar, A., Lovkvist, C., Palma, A. M., Martin, P., Kim, J., et al. (2023). CellNeighborEX: Deciphering neighbor-dependent gene expression from spatial transcriptomics data. *Molecular Systems Biology, 19*(11), e11670. https://doi.org/10.15252/msb.202311670.

Kitano, M., Minaga, K., Hatamaru, K., & Ashida, R. (2022). Clinical dilemma of endoscopic ultrasound-guided fine needle aspiration for resectable pancreatic body and tail cancer. *Digestive Endoscopy, 34*(2), 307–316. https://doi.org/10.1111/den.14120.

Kooi, T., Litjens, G., van Ginneken, B., Gubern-Merida, A., Sanchez, C. I., Mann, R., et al. (2017). Large scale deep learning for computer aided detection of mammographic lesions. *Medical Image Analysis, 35*, 303–312. https://doi.org/10.1016/j.media.2016.07.007.

Korfiatis, P., Suman, G., Patnam, N. G., Trivedi, K. H., Karbhari, A., Mukherjee, S., et al. (2023). Automated artificial intelligence model trained on a large data set can detect pancreas cancer on diagnostic computed tomography scans as well as visually occult preinvasive cancer on prediagnostic computed tomography scans. *Gastroenterology, 165*(6), 1533–1546.e1534. https://doi.org/10.1053/j.gastro.2023.08.034.

Lan, L., You, L., Zhang, Z., Fan, Z., Zhao, W., Zeng, N., et al. (2020). Generative adversarial networks and its applications in biomedical informatics. *Frontiers in Public Health, 8*, 164. https://doi.org/10.3389/fpubh.2020.00164.

Le, H., Peng, B., Uy, J., Carrillo, D., Zhang, Y., Aevermann, B. D., et al. (2022). Machine learning for cell type classification from single nucleus RNA sequencing data. *PLoS One, 17*(9), e0275070. https://doi.org/10.1371/journal.pone.0275070.

Lee, J., Kim, D., Kong, J., Ha, D., Kim, I., Park, M., et al. (2024). Cell-cell communication network-based interpretable machine learning predicts cancer patient response to immune checkpoint inhibitors. eadj0785 *Science Advances, 10*(5), https://doi.org/10.1126/sciadv.adj0785.

Lee, S., Kim, G., Lee, J., Lee, A. C., & Kwon, S. (2024). Mapping cancer biology in space: Applications and perspectives on spatial omics for oncology. *Molecular Cancer, 23*(1), 26. https://doi.org/10.1186/s12943-024-01941-z.

Li, J., Wang, F., Liu, Y., Wang, H., & Ni, B. (2021). N(6)-methyladenosine (m(6)A) in pancreatic cancer: Regulatory mechanisms and future direction. *International Journal of Biological Sciences, 17*(9), 2323–2335. https://doi.org/10.7150/ijbs.60115.

Maddipati, R., & Stanger, B. Z. (2015). Pancreatic cancer metastases harbor evidence of polyclonality. *Cancer Discovery, 5*(10), 1086–1097. https://doi.org/10.1158/2159-8290.CD-15-0120.

Mao, Y., Hou, X., Fu, S., & Luan, J. (2024). Transcriptomic and machine learning analyses identify hub genes of metabolism and host immune response that are associated with the progression of breast capsular contracture. *Genes & Diseases, 11*(3), 101087. https://doi.org/10.1016/j.gendis.2023.101087.

Martin, P. C. N., Kim, H., Lovkvist, C., Hong, B. W., & Won, K. J. (2022). Vesalius: High-resolution in silico anatomization of spatial transcriptomic data using image analysis. e11080 *Molecular Systems Biology, 18*(9), https://doi.org/10.15252/msb.202211080.

McCradden, M. D., Joshi, S., Mazwi, M., & Anderson, J. A. (2020). Ethical limitations of algorithmic fairness solutions in health care machine learning. *Lancet Digital Health, 2*(5), e221–e223. https://doi.org/10.1016/S2589-7500(20)30065-0.

McGranahan, N., & Swanton, C. (2017). Clonal heterogeneity and tumor evolution: Past, present, and the future. *Cell, 168*(4), 613–628. https://doi.org/10.1016/j.cell.2017.01.018.

McKinney, S. M., Sieniek, M., Godbole, V., Godwin, J., Antropova, N., Ashrafian, H., et al. (2020). International evaluation of an AI system for breast cancer screening. *Nature, 577*(7788), 89–94. https://doi.org/10.1038/s41586-019-1799-6.

Merritt, C. R., Ong, G. T., Church, S. E., Barker, K., Danaher, P., Geiss, G., et al. (2020). Multiplex digital spatial profiling of proteins and RNA in fixed tissue. *Nature Biotechnology, 38*(5), 586–599. https://doi.org/10.1038/s41587-020-0472-9.

Mizukami, K., Iwasaki, Y., Kawakami, E., Hirata, M., Kamatani, Y., Matsuda, K., et al. (2020). Genetic characterization of pancreatic cancer patients and prediction of carrier status of germline pathogenic variants in cancer-predisposing genes. *Ebiomedicine.* https://doi.org/10.1016/j.ebiom.2020.103033.

Moffitt, J. R., Hao, J., Wang, G., Chen, K. H., Babcock, H. P., & Zhuang, X. (2016). High-throughput single-cell gene-expression profiling with multiplexed error-robust fluorescence in situ hybridization. *Proceedings of the National Academy of Sciences of the United States of America, 113*(39), 11046–11051. https://doi.org/10.1073/pnas.1612826113.

Moncada, R., Barkley, D., Wagner, F., Chiodin, M., Devlin, J. C., Baron, M., et al. (2020). Integrating microarray-based spatial transcriptomics and single-cell RNA-seq reveals tissue architecture in pancreatic ductal adenocarcinomas. *Nature Biotechnology, 38*(3), 333–342. https://doi.org/10.1038/s41587-019-0392-8.

Moses, L., & Pachter, L. (2022). Museum of spatial transcriptomics. *Nature Methods, 19*(5), 534–546. https://doi.org/10.1038/s41592-022-01409-2.

Mukherjee, S., Patra, A., Khasawneh, H., Korfiatis, P., Rajamohan, N., Suman, G., et al. (2022). Radiomics-based machine-learning models can detect pancreatic cancer on pre-diagnostic computed tomography scans at a substantial lead time before clinical diagnosis. *Gastroenterology, 163*(5), 1435–1446.e1433. https://doi.org/10.1053/j.gastro.2022.06.066.

Nemes, K., Beno, A., Topolcsanyi, P., Mago, E., Fur, G. M., & Pongor, L. S. (2024). Predicting drug response of small cell lung cancer cell lines based on enrichment analysis of complex gene signatures. *Journal of Biotechnology.* https://doi.org/10.1016/j.jbiotec.2024.01.010.

Neugut, A. I., Lin, A., Raab, G. T., Hillyer, G. C., Keller, D., O'Neil, D. S., et al. (2019). FOLFOX and FOLFIRI use in stage IV colon cancer: Analysis of SEER-medicare data. *Clinical Colorectal Cancer, 18*(2), 133–140. https://doi.org/10.1016/j.clcc.2019.01.005.

Quail, D. F., & Joyce, J. A. (2013). Microenvironmental regulation of tumor progression and metastasis. *Nature Medicine, 19*(11), 1423–1437. https://doi.org/10.1038/nm.3394.

Raskov, H., Orhan, A., Christensen, J. P., & Gogenur, I. (2021). Cytotoxic CD8(+) T cells in cancer and cancer immunotherapy. *British Journal of Cancer, 124*(2), 359–367. https://doi.org/10.1038/s41416-020-01048-4.

Rodriques, S. G., Stickels, R. R., Goeva, A., Martin, C. A., Murray, E., Vanderburg, C. R., et al. (2019). Slide-seq: A scalable technology for measuring genome-wide expression at high spatial resolution. *Science (New York, N. Y.), 363*(6434), 1463–1467. https://doi.org/10.1126/science.aaw1219.

Saltz, J., Gupta, R., Hou, L., Kurc, T., Singh, P., Nguyen, V., et al. (2018). Spatial organization and molecular correlation of tumor-infiltrating lymphocytes using deep learning on pathology images. *Cell Reports, 23*(1), 181–193.e187. https://doi.org/10.1016/j.celrep.2018.03.086.

Satija, R., Farrell, J. A., Gennert, D., Schier, A. F., & Regev, A. (2015). Spatial reconstruction of single-cell gene expression data. *Nature Biotechnology, 33*(5), 495–502. https://doi.org/10.1038/nbt.3192.

Sehmi, M. N. M., Fauzi, M. F. A., Ahmad, W. J. W., & Chan, E. W. L. (2022). Pancreatic cancer grading in pathological images using deep learning convolutional neural networks. *F1000 research*. https://doi.org/10.12688/f1000research.73161.2.

Siegel, R. L., Miller, K. D., Fuchs, H. E., & Jemal, A. (2022). Cancer statistics, 2022. *CA: A Cancer Journal for Clinicians, 72*(1), 7–33. https://doi.org/10.3322/caac.21708.

Stahl, P. L., Salmen, F., Vickovic, S., Lundmark, A., Navarro, J. F., Magnusson, J., et al. (2016). Visualization and analysis of gene expression in tissue sections by spatial transcriptomics. *Science (New York, N. Y.), 353*(6294), 78–82. https://doi.org/10.1126/science.aaf2403.

Stickels, R. R., Murray, E., Kumar, P., Li, J., Marshall, J. L., Di Bella, D. J., et al. (2021). Highly sensitive spatial transcriptomics at near-cellular resolution with Slide-seqV2. *Nature Biotechnology, 39*(3), 313–319. https://doi.org/10.1038/s41587-020-0739-1.

Strippoli, A., Rossi, S., Martini, M., Basso, M., D'Argento, E., Schinzari, G., et al. (2016). ERCC1 expression affects outcome in metastatic pancreatic carcinoma treated with FOLFIRINOX: A single institution analysis. *Oncotarget, 7*(23), 35159–35168. https://doi.org/10.18632/oncotarget.9063.

Sung, H., Ferlay, J., Siegel, R. L., Laversanne, M., Soerjomataram, I., Jemal, A., et al. (2021). Global cancer statistics 2020: GLOBOCAN estimates of incidence and mortality worldwide for 36 cancers in 185 countries. *CA: A Cancer Journal for Clinicians, 71*(3), 209–249. https://doi.org/10.3322/caac.21660.

Suresh, S., & O'Donnell, K. A. (2021). Translational control of immune evasion in cancer. *Trends Cancer, 7*(7), 580–582. https://doi.org/10.1016/j.trecan.2021.04.002.

Tan, K., & Naylor, M. J. (2022). Tumour microenvironment-immune cell interactions influencing breast cancer heterogeneity and disease progression. *Frontiers in Oncology, 12*, 876451. https://doi.org/10.3389/fonc.2022.876451.

Topol, E. J. (2019). High-performance medicine: The convergence of human and artificial intelligence. *Nature Medicine, 25*(1), 44–56. https://doi.org/10.1038/s41591-018-0300-7.

Trehoux, S., Lahdaoui, F., Delpu, Y., Renaud, F., Leteurtre, E., Torrisani, J., et al. (2015). Micro-RNAs miR-29a and miR-330-5p function as tumor suppressors by targeting the MUC1 mucin in pancreatic cancer cells. *Biochimica et Biophysica Acta, 1853*(10 Pt A), 2392–2403. https://doi.org/10.1016/j.bbamcr.2015.05.033.

Virostko, J. (2020). Quantitative magnetic resonance imaging of the pancreas of individuals with diabetes. *Front Endocrinol (Lausanne), 11*, 592349. https://doi.org/10.3389/fendo.2020.592349.

Wan, X., Xiao, J., Tam, S. S. T., Cai, M., Sugimura, R., Wang, Y., et al. (2023). Integrating spatial and single-cell transcriptomics data using deep generative models with SpatialScope. *Nature Communications, 14*(1), 7848. https://doi.org/10.1038/s41467-023-43629-w.

Wang, N., Li, X., Wang, R., & Ding, Z. (2021). Spatial transcriptomics and proteomics technologies for deconvoluting the tumor microenvironment. *Biotechnology Journal, 16*(9), e2100041. https://doi.org/10.1002/biot.202100041.

Wang, Q., Sabanovic, B., Awada, A., Reina, C., Aicher, A., Tang, J., et al. (2023). Single-cell omics: A new perspective for early detection of pancreatic cancer? *European Journal of Cancer, 190*, 112940. https://doi.org/10.1016/j.ejca.2023.112940.

Wang, S., Rong, R., Zhou, Q., Yang, D. M., Zhang, X., Zhan, X., et al. (2023). Deep learning of cell spatial organizations identifies clinically relevant insights in tissue images. *Nat Commun, 14*(1), 7872. https://doi.org/10.1038/s41467-023-43172-8.

Wang, X., Allen, W. E., Wright, M. A., Sylwestrak, E. L., Samusik, N., Vesuna, S., et al. (2018). Three-dimensional intact-tissue sequencing of single-cell transcriptional states. *Science (New York, N. Y.), 361*(6400), https://doi.org/10.1126/science.aat5691.

Wang, X., Chan, S., Chen, J., Xu, Y., Dai, L., Han, Q., et al. (2024). Robust machine-learning based prognostic index using cytotoxic T lymphocyte evasion genes highlights potential therapeutic targets in colorectal cancer. *Cancer Cell International, 24*(1), 52. https://doi.org/10.1186/s12935-024-03239-y.

Wei, S. C., Duffy, C. R., & Allison, J. P. (2018). Fundamental mechanisms of immune checkpoint blockade therapy. *Cancer Discovery, 8*(9), 1069–1086. https://doi.org/10.1158/2159-8290.CD-18-0367.

Williams, C. G., Lee, H. J., Asatsuma, T., Vento-Tormo, R., & Haque, A. (2022). An introduction to spatial transcriptomics for biomedical research. *Genome Medicine, 14*(1), 68. https://doi.org/10.1186/s13073-022-01075-1.

Xu, R. (2020). Association of N6-methyladenosine with viruses and virally induced diseases. *Frontiers in Bioscience-Scholar.* https://doi.org/10.2741/4852.

Yamashita, R., Nishio, M., Do, R. K. G., & Togashi, K. (2018). Convolutional neural networks: An overview and application in radiology. *Insights Imaging, 9*(4), 611–629. https://doi.org/10.1007/s13244-018-0639-9.

Yang, Z., Guan, F., Bronk, L., & Zhao, L. (2024). Multi-omics approaches for biomarker discovery in predicting the response of esophageal cancer to neoadjuvant therapy: A multidimensional perspective. *Pharmacology & Therapeutics, 254*, 108591. https://doi.org/10.1016/j.pharmthera.2024.108591.

Yu, Q., Jiang, M., & Wu, L. (2022). Spatial transcriptomics technology in cancer research. *Frontiers in Oncology, 12*, 1019111. https://doi.org/10.3389/fonc.2022.1019111.

Yuan, Y. (2016). Spatial heterogeneity in the tumor microenvironment. *Cold Spring Harbor Perspectives in Medicine, 6*(8), https://doi.org/10.1101/cshperspect.a026583.

Zhang, B., Shi, H., & Wang, H. (2023). Machine learning and AI in cancer prognosis, prediction, and treatment selection: A critical approach. *Journal of Multidisciplinary Healthcare, 16*, 1779–1791. https://doi.org/10.2147/JMDH.S410301.

Zhang, D., Schroeder, A., Yan, H., Yang, H., Hu, J., Lee, M. Y. Y., et al. (2024). Inferring super-resolution tissue architecture by integrating spatial transcriptomics with histology. *Nature Biotechnology.* https://doi.org/10.1038/s41587-023-02019-9.

Zhang, L., Chen, D., Song, D., Liu, X., Zhang, Y., Xu, X., et al. (2022). Clinical and translational values of spatial transcriptomics. *Signal Transduction and Targeted Therapy, 7*(1), 111. https://doi.org/10.1038/s41392-022-00960-w.

Zhao, N., & Rosen, J. M. (2022). Breast cancer heterogeneity through the lens of single-cell analysis and spatial pathologies. *Seminars in Cancer Biology, 82*, 3–10. https://doi.org/10.1016/j.semcancer.2021.07.010.

Zhou, Z., Lv, J., Yu, H., Han, J., Yang, X., Feng, D., et al. (2020). Mechanism of RNA modification N6-methyladenosine in human cancer. *Molecular Cancer.* https://doi.org/10.1186/s12943-020-01216-3.

Zidane, M., Makky, A., Bruhns, M., Rochwarger, A., Babaei, S., Claassen, M., et al. (2023). A review on deep learning applications in highly multiplexed tissue imaging data analysis. *Frontiers in Bioinformatics, 3*, 1159381. https://doi.org/10.3389/fbinf.2023.1159381.

CHAPTER FIVE

Advancing cancer therapeutics: Integrating scalable 3D cancer models, extracellular vesicles, and omics for enhanced therapy efficacy

Pedro P. Gonçalves[a,b], Cláudia L. da Silva[a,b], and Nuno Bernardes[a,b,*]

[a]Department of Bioengineering and iBB - Institute for Bioengineering and Biosciences at Instituto Superior Técnico, Universidade de Lisboa, Lisboa, Portugal
[b]Associate Laboratory i4HB - Institute for Health and Bioeconomy at Instituto Superior Técnico, Universidade de Lisboa, Lisboa, Portugal
*Corresponding author. e-mail address: nuno.bernardes@tecnico.ulisboa.pt

Contents

1. Introduction	138
2. Beyond 2D: exploring the potential of 3D models in cancer research	140
2.1 Leveraging spheroids as reliable preclinical cancer models	141
2.2 Scalable strategies for tumor spheroid production: bioreactor systems for robust preclinical modeling	143
2.3 Advances and challenges in cancer organoid versus spheroid culture	153
3. Leveraging omics technologies for robust validation and application of 3D cancer models	159
4. Beyond cells: extracellular vesicles as drug delivery systems for targeted cancer therapy	162
4.1 From production to therapeutic application: omics in EV research	166
5. Conclusions and future perspectives	169
Acknowledgments	169
References	169
Further reading	185

Abstract

Cancer remains as one of the highest challenges to human health. However, anticancer drugs exhibit one of the highest attrition rates compared to other therapeutic interventions. In part, this can be attributed to a prevalent use of in vitro models with limited recapitulative potential of the in vivo settings. Three dimensional (3D) models, such as tumor spheroids and organoids, offer many research opportunities to address the urgent need in developing models capable to more accurately mimic cancer biology and drug resistance profiles. However, their wide adoption in high-

Advances in Cancer Research, Volume 163
ISSN 0065-230X, https://doi.org/10.1016/bs.acr.2024.07.001
Copyright © 2024 Elsevier Inc. All rights are reserved, including those for text and data mining, AI training, and similar technologies.

throughput pre-clinical studies is dependent on scalable manufacturing to support large-scale therapeutic drug screenings and multi-omic approaches for their comprehensive cellular and molecular characterization.

Extracellular vesicles (EVs), which have been emerging as promising drug delivery systems (DDS), stand to significantly benefit from such screenings conducted in realistic cancer models. Furthermore, the integration of these nanomedicines with 3D cancer models and omics profiling holds the potential to deepen our understanding of EV-mediated anticancer effects.

In this chapter, we provide an overview of the existing 3D models used in cancer research, namely spheroids and organoids, the innovations in their scalable production and discuss how omics can facilitate the implementation of these models at different stages of drug testing. We also explore how EVs can advance drug delivery in cancer therapies and how the synergy between 3D cancer models and omics approaches can benefit in this process.

1. Introduction

The increasing number of cancer diagnoses worldwide are alarming. In 2022, 20 M new cancer cases and almost 10 M deaths caused by cancer were reported (World Health Organization, 2024), making this disease a leading cause of death worldwide (Sung et al., 2021). These numbers reflect the harsh reality that the treatment of these malignancies still is, and has always been, considered one of the most arduous challenges faced by the scientific community (Sapio & Naviglio, 2022) despite long decades of investigation.

Cancer is one of the pathologies with the highest rates of drug attrition. This can be partially attributed to the lack of reliable preclinical models with acceptable translational potential (Liu, Delavan, Roberts, & Tong, 2017). Cultured cells are often utilized as in vitro models, although the most widely used method, 2D culture, does not accurately represent a real tumor (Costa et al., 2016; Pampaloni, Reynaud, & Stelzer, 2007). Additionally, the utilization of animal models has several shortcomings because these cannot fully mimic the complex processes of carcinogenesis, progression, and human immune system interaction with cancer. While the utilization of humanized animal models represents a significant advancement in preclinical cancer research, these models retain limitations (Walsh et al., 2017). Notably, their generation can be laborious, their scalability for high-throughput drug screening is limited and they have associated ethical concerns (Robinson et al., 2019).

3D cell culture is a powerful alternative that more closely resembles the biophysical and biochemical signaling of the native tumor microenvironment.

One of the first reports on this culture method was published in 1971 (Sutherland, McCredie, & Inch, 1971). Accumulating evidence suggests that cells in this environment can replicate cancerous behavior better (Birgersdotter, Sandberg, & Ernberg, 2005; Bissell, Rizki, & Mian, 2003; Ghosh et al., 2005; Kilian, Bugarija, Lahn, & Mrksich, 2010; Kita et al., 1993; Li et al., 2006; Mazzoleni, Di Lorenzo, & Steimberg, 2009; Pampaloni et al., 2007; Yamada & Cukierman, 2007). Nonetheless, their use in high-throughput preclinical studies is dependent on the development and optimization of robust scalable platforms for production.

Concerning anticancer agents, extracellular vesicles (EVs) (including exosomes), which are spherical phospholipid structures that mediate cell-to-cell communication, have been recently proposed as promising drug delivery systems (DDS) (Elsharkasy et al., 2020). The steep rise in publications in this field is indicative of a growing interest in their promising features (Lin, Yang, Wang, Xing, & Lin, 2022). For example, EVs present high biocompatibility, efficient biological barrier crossing (Chen et al., 2016), and amenability to be engineered (Conlan, Pisano, Oliveira, Ferrari, & Mendes Pinto, 2017) towards more selective cancer therapies. However, the fact that studies generally evaluate the anticancer effect of EVs in 2D cultured tumor cells is a challenge for their clinical translation.

Omics technologies are promising for further advancing the use of EV-based anticancer agents (Chitoiu, Dobranici, Gherghiceanu, Dinescu, & Costache, 2020; Shaba et al., 2022) and 3D cancer models (Pieters, Co, Wu, & McGuigan, 2021) for more effective therapies. This can be achieved by the in-depth characterization of models, possibly allowing the extraction of important cancer signatures associated with biomarkers. Moreover, the molecular responses of cancer cells to EV therapeutic products can be more comprehensively determined using and integrating data generated from high-throughput technologies, such as genomics, transcriptomics, proteomics or glycomics, amongst others. The potential of this approach is further enhanced by single-cell and spatially resolved omics (Chen, Wang, & Ko, 2023; Shaba et al., 2022; Simão, Gomes, Alves, & Brito, 2022).

Here, we highlight the importance of spheroids and organoids as promising cancer models and scalable approaches for their manufacturing are enumerated. We also explore the potential of EVs as DDS for cancer therapy and their possible synergy with 3D cancer models. Finally, we discuss how omics profiling can further advance the study of 3D cancer models and EV-based therapeutics.

2. Beyond 2D: exploring the potential of 3D models in cancer research

The lack of widespread implementation of more reliable tumor models is one of the reasons for elevated cancer drug attrition rates (Heem Wong, Wei Siah, & Lo, 2019; Liu et al., 2017), which can culminate in the stagnation of treatment modalities for cancer patients (Forsythe et al., 2022). Thus, it is necessary to develop pre-clinical models that support better clinical outcomes by aiding drug testing and enhancing comprehension of cancer biology (Medle et al., 2022).

Two-dimensional (2D) cell culture techniques are traditionally used for in vitro drug testing (Breslin & O'driscoll, 2012; Jensen & Teng, 2020; Kolenda et al., 2016). Despite their low cost, reproducibility, wide establishment, and simplicity, these cannot accurately mimic cancer complexity (Jensen & Teng, 2020; Kolenda et al., 2016). The fact that cultured cells grow attached to a high-stiffness plastic or glass surface leads to cytoskeleton remodeling (Solon, Levental, Sengupta, Georges, & Janmey, 2007) and altered proliferation (Jubelin et al., 2022; Nemir & West, 2010). Additionally, it leads to improper cell signaling, cell-to-cell, and cell-to-ECM interactions (Bissell et al., 2003; Breslin & O'driscoll, 2012), tissue organization, and cell polarity (Mseka, Bamburg, & Cramer, 2007) and provides unlimited access to nutrients in the culture medium (Jubelin et al., 2022), which is often not the case for in situ solid tumors. Moreover, 2D culture is usually established with cancer cell lines (rather than primary cells), that when maintained for a high number of passages may lose tumor heterogeneity (Drost & Clevers, 2018; Jubelin et al., 2022), a key biological feature involved in drug resistance (Jubelin et al., 2022). As a result, flat biology, that is 2D culture, can lead to altered cell behavior and is inadequate for accurate drug response prediction.

Three-dimensional (3D) models are an alternative that more closely resemble solid tumors (Jubelin et al., 2022; Manduca, Maccafeo, De Maria, Sistigu, & Musella, 2023; Rodrigues, Heinrich, Teixeira, & Prakash, 2021). These models can offer a more appropriate cell-to-cell and cell-to-extracellular environment interactions (Baker & Chen, 2012; LaBarbera, Reid, & Yoo, 2012). Different layers are present owing to the diffusion gradients of nutrients, oxygen, and metabolic waste (Gunti, Hoke, Vu, & London, 2021). As a result, several cell features, such as morphology, gene and protein expression, metabolism, proliferation, viability, and the epithelial to mesenchymal transition status, resemble more closely the tumors in a real clinical setting (Baker & Chen, 2012; Friedrich, Seidel, Ebner, & Kunz-Schughart, 2009; Lovitt, Shelper, & Avery, 2013; Luca et al., 2013; Myungjin Lee et al., 2013; Song et al., 2018; Timmins, Maguire, Grimmond, & Nielsen, 2004; Weiswald et al., 2013). Thus, it is expected that the response to drugs is also modulated to more accurately emulate actual solid tumors (Ekert et al., 2014; Song et al., 2018).

2.1 Leveraging spheroids as reliable preclinical cancer models

The most investigated 3D cancer models are spheroids (Hirschhaeuser et al., 2010; Sant & Johnston, 2017), which are self-assembled cancer cell clusters, with multiple cell layers, that initiate from a single-cell suspension (Han, Kwon, & Kim, 2021). Despite its simplicity, a relevant tumor-mimicking capacity is provided.

When growing as 3D spheroids, the cells in the central core of the spheroid may have a limited access to nutrients and oxygen and accumulate more metabolic waste, which leads to the presence of necrotic cells when a size threshold is reached (approximately 200–500 µm in diameter) (Däster et al., 2017; Hirschhaeuser et al., 2010; Pinto, Henriques, Silva, & Bousbaa, 2020). In contrary, an intermediate layer of quiescent cells, and an outer spheroid layer active proliferative cells are present due to an easy access to nutrients and oxygen (Ferreira, Martins, & Vilela, 2003; Gebhard, Gabriel, & Walter, 2016; Khanna, Chauhan, Bhatt, & Dwarakanath, 2020).

These different cell layers may be relevant for the ability of cancer spheroids to predict drug responses more accurately (Lovitt, Shelper, & Avery, 2015; Nunes, Barros, Costa, Moreira, & Correia, 2019). It has been demonstrated that the hypoxic core can modulate the effects of reactive oxygen species (ROS)-dependent chemotherapeutic agents such as doxorubicin and 5-fluorouracil (5-FU) (Al-Akra, Bae, Leck, Richardson, & Jansson, 2019; Däster et al., 2017; Deo et al., 2016). In addition, cells in less proliferative layers can also be more resistant to antiproliferative drugs such as cisplatin, doxorubicin, and paclitaxel (Imamura et al., 2015; Reynolds et al., 2017), which contributes to higher drug resistance profiles in cancer spheroids in comparison to 2D models (Han et al., 2021).

The cells used for tumor spheroid formation can come from different sources. The most utilized are established cancer cell lines (Hofmann, Cohen-Harazi, Maizels, & Koman, 2022), because of their easy access, reproducibility, and low cost. However, certain cancer cell lines do not spontaneously form adequate spheroids owing to loose compaction. This is especially the case for cancerous cells with metastatic potential, because these have weaker cell-to-cell adhesion and can break off from neighboring cells (Kenny et al., 2007), for example, the breast cancer cell line MDA-MB-231 (Froehlich et al., 2016). The exogenous inclusion of ECM-based materials, like Matrigel and collagen, among others, can facilitate the formation of spheroids from these cells (Froehlich et al., 2016; Jubelin et al., 2022).

There is a wide range of techniques for generating spheroids. Depending on the method, the cells are kept under static culture conditions or in a dynamic environment. One protocol to produce spheroids is the hanging drop method in which small droplets of a cell suspension are applied to a planar surface. Then, this surface is inverted and the cellular aggregate assembles owing to gravity (Foty, 2011; Jubelin et al., 2022; Timmins & Nielsen, 2007). Droplet attachment is made possible by the surface tension and the spheroid size can be defined by controlling the volume and cell seeding density. However, this method is laborious, and it can be difficult to test drugs if the spheroid is not collected from the drop.

Another method is the liquid overlay, in which cells are seeded onto a non-adhesive hydrophilic substrate and as a result aggregate to each other. Ultralow attachment (ULA) plates are widely utilized for this strategy and usually, the plate is centrifuged to aid the cell aggregation (Ivascu & Kubbies, 2006; Malhão, Macedo, Ramos, & Rocha, 2022). Moreover, micromolding, for instance with agarose, is also an option to prevent cell adhesion to the surface to promote the spheroid formation (Esa et al., 2023; Fobian et al., 2022; Martins et al., 2023).

Other option for the formation of spheroids is magnetic levitation, in which the cells assimilate biocompatible magnetic particles and aggregate due to the application of a magnetic field (Jaganathan et al., 2014; Kim et al., 2013; Libring, Enríquez, Lee, & Solorio, 2021), or the pellet culture in which cells are centrifugated to gather at the bottom of a tube (Zanoni et al., 2016).

Besides tumor cells, other types of cells, such as cancer-associated fibroblasts, endothelial cells or immune cells, among others, are commonly found in the tumor microenvironment (Anderson & Simon, 2020). These cells can influence various aspects in the behavior of cancer cells, like drug resistance (Salemme et al., 2023; Son et al., 2017). Thus, to enhance the modeling capacity and acquire a deeper understanding of the impact of specific tumor microenvironment cells on drug sensitivity, it is vital to incorporate those into cancer spheroids. For example, the presence of macrophages and fibroblasts in a spheroid model of non-small cell lung cancer was shown to impact the response to paclitaxel leading to decreased apoptosis in comparison to monocultured spheroids, emphasizing that both cell types impact the response to this drug (Rebelo et al., 2018). In the same study, the authors showed that the macrophages in these spheroids were repolarized to the M1 anti-cancer phenotype upon treatment with the CSF1R (Colony Stimulating Factor 1 Receptor) inhibitor BLZ954 (Rebelo et al., 2018), resembling the response obtained in relevant models of the human tumor microenvironment (Pyonteck et al., 2013; Zhu et al., 2014). However, this comparison is limited because these models were based on mice, which may limit drawing conclusions about the responses to anti-cancer drugs in humans (Pyonteck et al., 2013; Zhu et al., 2014). Nonetheless, this study suggests that heterotypic spheroids, i.e., spheroids with various types of cells, offer the opportunity to evaluate therapeutic agents that aim to treat cancer by targeting immune cells (Rebelo et al., 2018).

However, developing heterotypic models can be challenging. In fact, the ratio of different cells in spheroids needs to be carefully optimized (Bauleth-Ramos et al., 2020; Carvalho et al., 2023; Domingues, Leite Pereira, Sarmento, & Castro, 2023; Franchi-Mendes, Lopes, & Brito, 2021), and incorporating these models into high-throughput workflows may be even more challenging. Notably, several groups developed complex heterotypic spheroid models with up to four different cell types (Bauleth-Ramos et al., 2020; Carvalho et al., 2023; Domingues et al., 2023; Franchi-Mendes et al., 2021), highlighting the advances already made in this field.

Additionally, patient-derived cells can also be utilized for more relevant cancer modeling of individual tumors (Halfter et al., 2015, 2016; Hofmann et al., 2022; Idrisova, Simon, & Gomzikova, 2023). In this case, some patient-specific cell features such as cancer cell heterogeneity can be replicated to a certain degree (Halfter et al., 2016), which is not the case for established cancer cell lines (Halfter et al., 2016).

2.2 Scalable strategies for tumor spheroid production: bioreactor systems for robust preclinical modeling

Even though static platforms for the formation of spheroids are more conventional, these methods can be labor-intensive, time-consuming and represent a challenge to implement spheroid mass production (Han et al., 2021). When a scalable production of spheroids is required it may be helpful to employ dynamic culture conditions, with perfusion of the culture medium, to encourage the diffusion of oxygen and the transfer of nutrients into spheroids (Jensen & Teng, 2020; Nath & Devi, 2016). Agitation-based culture systems, including bioreactors, are scalable options that possess these qualities. And since spheroid mass production is paramount for sustaining high-throughput preclinical studies (Hickman et al., 2014; Santo et al., 2016) optimization of bioreactors for the culture of cancer spheroids holds significant value (Van Kessel, Zuiverloon, Alberts, Boormans, & Zwarthoff, 2015).

In these culture systems, generally the stirring prevents the adherence of cells to the vessel surface and promotes cell collisions thereby forming spheroids (Park, Avera, & Kim, 2022). To further avoid the adhesion of cells to the vessel's surface, bioreactor walls can be coated for example with siliconizing reagents for example (Panchalingam et al., 2010; Youn, Sen, Behie, Girgis-Gabardo, & Hassell, 2006).

The most well-established dynamic systems for cancer spheroid production are stirred vessels, where the culture medium is agitated by an impeller which offers a homogenous culture environment. These can be computer-assisted (Fig. 1), allowing on-line monitoring and automated control over culture conditions—stirred tank reactor (Hickman et al., 2014).

The spinner flask (Fig. 1, where the spinner flask is to the right of the stirred tank vessels and the stirred tank reactor is on the left), which is a less costly and a simpler option, is constituted of a three-neck vessel where the medium is agitated through a magnetic impeller. Its capacity is within the range of a few hundred milliliters, which is the size that most studies adopt (Table 1). Among others, this methodology was already employed for the formation of breast (Batalha, Gomes, & Brito, 2023; Santo et al., 2016),

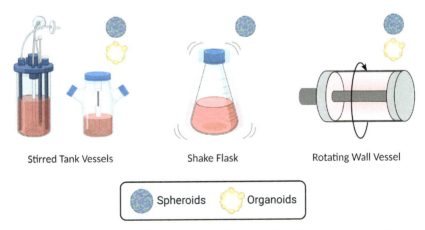

Fig. 1 Types of dynamic culture systems already utilized for the culture of cancer spheroids and organoids. *Created with Biorender.com.*

Table 1 Selected studies employing bioreactors for culture of cancer spheroids.

Culture vessel	Source of cells in the spheroid	Added matrix	Bioreactor culture time	Agitation	Main outcomes	References
Stirred Tank Reactor (125 mL)	Breast cancer (MCF-7) Co-cultured with fibroblasts	Alginate hydrogel	15 days	80 rpm	- Recapitulation of tumor-stroma crosstalk, for example *de novo* synthesis and accumulation of ECM and cytokine accumulation. - Enhanced angiogenic potential in co-cultures	Estrada et al. (2016)
Stirred Tank Reactor	Cancer stem cell lines derived from the lung and colorectal	None	8 days	Not disclosed	- Serum-free medium promoted CSC enrichment in both 3D aggregate and microcarriers - Donor dependent cell proliferation. - Only lung cancer cells formed aggregates.	Serra et al. (2019)
Spinner Flaks (125 mL or 250 mL) Stirred Tank (500 mL)	Breast tumors resected from mouse mammary tumor virus Her2	None	5–6 days	50 rpm, spinner flask 52 rpm, stirred tank	- Maximum cell quantity of 1.2×10^6 cells/mL (20-fold increase) - Successfully scaled up spheroid culture to a 500 mL vessel	Youn et al. (2006)

(continued)

Table 1 Selected studies employing bioreactors for culture of cancer spheroids. (*cont'd*)

Culture vessel	Source of cells in the spheroid	Added matrix	Bioreactor culture time	Agitation	Main outcomes	References
Spinner Flask	Head and neck cancer (FaDu) Co-cultured with peripheral blood mononuclear cells	None	11 days	130 rpm	- The model reflected pathophysiological characteristics, heterogeneity and the complexity of the in vivo tumor tissue - Model was adequate for testing the efficacy immunotherapeutic agents, such as catumaxomab	Hirschhaeuser, Leidig, Rodday, Lindemann, and Mueller-Klieser (2009)
Spinner Flask (125 or 500 mL)	Breast cancer (MCF-7 and BT474, HCC1954 and HCC1806) (breast cancer) Lung cancer (A549, H460, H1650 and H157) Colorectal cancer (HT29) Co-cultured with fibroblasts	Alginate hydrogel	17 days	60, 80 and 100 rpm	- Spheroids were obtained in large numbers (1000–1500 spheroids/mL), displaying characteristics of real tumors such as morphology, proliferation, and hypoxia gradients.	Santo et al. (2016)

Spinner Flask (125 mL)	Brain cancer (G144)	None	24 days, batch 32 days, fed–batch	100 rpm	- The culture system maintained a high percentage of CD133 + (putative neural stem cell marker) and cells showed minimal changes in gene expression throughout the culture.	Panchalingam et al. (2010)
Spinner Flask (125 mL)	Lung cancer (NCI-H157) Coculture with lung derived cancer associated fibroblasts and THP-1 (monocytic cell line) or peripheral blood-derived monocytes	Alginate hydrogel	21 days when using THP-1 10 days when using peripheral blood-derived monocytes	80–100 rpm	- The model accumulated cytokines and ECM components. - Monocytes infiltrated the model and polarized to the M2 phenotype, resembling TAM. This phenotype was modulated by immunotherapy treatment - Heterotypic spheroids showed increased resistance to paclitaxel in comparison to a homotypic model.	Rebelo et al. (2018)

(continued)

Table 1 Selected studies employing bioreactors for culture of cancer spheroids. (*cont'd*)

Culture vessel	Source of cells in the spheroid	Added matrix	Bioreactor culture time	Agitation	Main outcomes	References
Spinner Flask (125 mL, agregation) Shake Flask (125 mL, cell culture and antibody assays)	Breast cancer (SKBR3 and HCC1954) Co-cultured with human peripheral blood mononuclear derived NK cells	Alginate hydrogel	7 days	100 rpm	- HC1954 formed cohesive cell aggregates, while SKBR3 formed loose and irregular aggregates. - Preservation of NK cell compartment in the spheroids - Models displayed cell-line dependent response to trastuzumab and pertuzumab - Recapitulation of immunotherapy responses (NK cell activation and decline in immunosuppressive PD-L1$^+$ cells)	Batalha et al. (2023)
Spinner Flask (125 mL, cell agregation) Shake Flask (125 mL, cell culture)	Breast cancer (HCC1954) Co-cultured with peripheral blood derived monocytes	Alginate hydrogel	Not disclosed	80–100 rpm	- Monocytes contributed to tumor progression by being integrated in the tumor neo-vasculature	Lopes-Coelho et al. (2020)

Shake Flask	Breast cancer (MCF-7, BT474, HCC1954, HCC1806, and MDA-MB-231) Co-cultured with human fibroblasts and EC	None	1 month	100 rpm	- HCC1954 was the most promising for heterotypic long-term culture. - Maintenance of EC in the spheroid is cell line dependent and agitation based dependent for some cells. - Cancer cells:fibroblasts: EC ratio of 1:3:10 led to larger EC area in comparison to other ratios	Franchi-Mendes et al. (2021)
Rotating Wall Vessel (50 mL)	Lung cancer (A549)	None	25 days	12 rpm, with increasing speeds as the spheroids enlarged	- Several spheroid morphology parameters impacted the response to treatment. As such it is advised to pre-select spheroids of similar volume and shape to reduce result variability of cytotoxic tests	Zanoni et al. (2016a)
Rotating Wall Vessel	Lung cancer (A549)	None	3 days	10 rpm	The innovative design reduced the impact of air bubbles on cell aggregate formation leading to increased aggregate sizes	Phelan et al. (2019)

(continued)

Table 1 Selected studies employing bioreactors for culture of cancer spheroids. (*cont'd*)

Culture vessel	Source of cells in the spheroid	Added matrix	Bioreactor culture time	Agitation	Main outcomes	References
Novel bioreactor design	Lung cancer (Calu-3)	None	5 days	5 mL/min recirculation	Ultralow shear stress provided in the novel bioreactor design was beneficial for cancer spheroid culturing. It preserved the spheroid morphological features, promoted intercellular connection, increased size (2.4-fold), and the number of cycling cells (1.58-fold) compared to static suspension culture.	Massai et al. (2016)

ECM, Extracellular matrix; CSC, cancer stem cells; TME, tumor microenvironment; TAM, tumor associated macrophages; EC, endothelial cells; NK cells, natural killer cells.

lung (Rebelo et al., 2018), and brain (Panchalingam et al., 2010) cancer spheroids in a homotypic and heterotypic fashion.

Moreover, through the selection of the cell seeding, agitation speed, and an appropriate impeller design it is possible to roughly control spheroid density, size, and shape. For instance, a trapezoid-shaped impeller in comparison to a pitched 4-blade impeller led to the formation of aggregates with an increased spherical shape and approximately 2-fold higher final cell concentration (a 21-fold increase after 8 days) (Serra et al., 2019). However, precisely fine-tuning and homogenizing the spheroid size and shape is still challenging with this approach (Park, Hong, & Lee, 2022).

In Table 1, a summary of selected studies which employ bioreactor technology to study large-scale production of cancer spheroids is presented.

One of the disadvantages of these methodologies is the substantial shear stress applied to the cells through impeller-based agitation (Jubelin et al., 2022). Hence, certain shear-sensitive cell types may not be compatible with this approach. Spheroid microencapsulation in scaffolds can be utilized as a protective measure to stop this stress from negatively affecting the cells. For example, when aiming to establish stirred tank cultures of breast and lung cancer cell lines in heterotypic spheroids also constituted by fibroblasts, Santo *et al.* microencapsulated the cells in an alginate hydrogel demonstrating the suitability of this approach to culture heterotypic spheroids. The alginate scaffold changed the cell distribution in comparison to non-encapsulated cells, possibly because this hydrogel acts as an ECM that cells interact with. In this study, the spheroids were also challenged with the chemotherapeutic drug docetaxel, whereby the drug efficiency was not compromised by the microencapsulation (Santo et al., 2016).

This approach has also been used to microencapsulate non-small lung cancer cells, fibroblasts, and monocytes to form triple heterotypic tumor spheroids. This allowed for a long-term culture of up to 21 days. The macrophages showed the expected M2 pro-tumor-like phenotype observed in an actual cancer. Also, the inclusion of different cells into the spheroid modulated the drug response to paclitaxel, suggesting that tumoral and non-tumoral cell interaction can modulate therapeutic response and that the model is suitable for studying how the tumor microenvironment impacts drug responses (Rebelo et al., 2018). Once again alginate microencapsulation was demonstrated to be successful for developing heterotypic cancer spheroids (Rebelo et al., 2018).

Another scalable option for cancer spheroid production is the shake flask system (Fig. 1), which is not as frequently reported in the literature. In this system, agitation is achieved by setting the vessel in motion, for example, using an orbital shaker (Table 1). This strategy requires that spheroids are formed before inoculation, something that is not necessary for the reactors mentioned above, which allow inoculation with a single cell suspension.

The rotating wall vessel (RWV) comprises another option (Fig. 1). It is a bioreactor developed by NASA to culture cells and tissues under a microgravity environment (Schwarz, Goodwin, & Wolf, 1992) and allowing to culture cells in suspension or attached to microcarriers in a low-shear stress environment (Goodwin, Prewett, Wolf, & Spaulding, 1993; Muhitch et al., 2000). This design is comprised by a rotating cylindrical solid body filled with culture medium and a rubber membrane for oxygenation. The spinning of the container transfers its motion to the cell culture medium (Nath & Devi, 2016) and consequently, the fluid inside spins. In comparison to impeller-based agitation, employed in the stirred tank bioreactor, the RWV applies lower shear stresses while also promoting the mass transfer of nutrients and diffusion of oxygen (Schwarz et al., 1992).

Zanoni et al. used a 50 mL RWV successfully to culture lung cancer spheroids for 15 days. In this study, it was shown that pre-selection of spheroids of similar sizes and shapes is indispensable to reduce variability in the results obtained in cytotoxicity assays (Zanoni et al., 2016).

Nonetheless, this configuration has several disadvantages. One is the generation of air bubbles in the culture medium, because of the gas exchange system (Phelan, Gianforcaro, Gerstenhaber, & Lelkes, 2019; Yang et al., 2023). Air bubbles increase the shear stress transmitted to cells in certain regions of the vessel, impacting spheroid and organoid formation, cell aggregation, and viability. A RWV design in which air bubbles were removed has already been validated using human lung adenocarcinoma cells (A549) encapsulated in alginate beads. This evolved configuration improved cell aggregation, allowing more spheroid growth (Phelan et al., 2019). Moreover, another limitation may be the microgravity generated which can lead to altered spheroid and organoid characteristics in comparison to the actual cancer aimed to be modeled.

Overall, bioreactors are compatible with long-term culture, which is valuable for longitudinal anticancer drug assays. For example, Panchalingam et al. were able to culture glioblastoma spheroids for 32 days in a spinner flask operated under a fed-batch mode, maintaining the cell viability above 85% (Panchalingam et al., 2010). In another study, Franchi-Mendes and colleagues cultured heterotypic breast cancer spheroids, with fibroblasts and

Advancing cancer therapeutics

endothelial cells in a shake flask for a month (Franchi-Mendes et al., 2021). This demonstrates that heterotypic spheroids are also amenable to long-term culturing. Interestingly, in this study, not all cell types, like MDA-MB-231 and HCC1806, could be maintained in long-term culture, which, as previously stated, may be related to the poor capacity of these cells to form spheroids. Finally, the RWV also allowed the culture of A549 lung cancer spheroids for 15 days (Zanoni et al., 2016).

In sum, spheroids are simple yet effective models for the recapitulation of cancer cell features being an improvement over 2D cultures. The fact that large quantities of spheroids [e.g., 7000 spheroids were used to produce high quality dose-dependent drug responses for 25 chemotherapeutic drugs and three different cancer cells lines (Thakuri, Ham, Luker, & Tavana, 2016)] may be needed for their meaningful application in drug testing renders optimization of their culture in scalable bioreactors important. However, despite the advances, challenges such as the lack of standardization (Peirsman et al., 2021) as well as the high variability of morphologies and sizes associated to the different methods of spheroid formation still need to be overcome for the potential of cancer spheroids to be fully realized.

2.3 Advances and challenges in cancer organoid versus spheroid culture

Cancer organoids are miniaturized organs that are formed through the self-organization of neoplastic cells in vitro. They are an upgrade to cancer spheroid models, which have simplified complexity, for instance, in cell composition, impacting their clinical relevance. Organoids offer a more intricate model for cancer simulation, demonstrating enhanced fidelity to the tumor architecture (Luo et al., 2021), and closely resembling the genetic and phenotypic profile of in situ cancer (Verduin, Hoeben, De Ruysscher, & Vooijs, 2021). Therefore, they can be utilized in cancer biology, e.g., to discover cancer-driving genes, and for predicting cancer drug responses and safety at a personalized patient level (Verduin et al., 2021). Moreover, along with spheroids they aim to overcome the limitations of 2D cell culture as well as of animal models in cancer research (Bhandari, Patil, Pingale, & Amrutkar, 2021).

Organoids depict more accurately cancer heterogeneity, which is important for predicting the effectiveness of treatments. For instance, Hubert et al. already demonstrated that glioblastoma organoids could recapitulate the stem cell tumor heterogeneity found in vivo (Hubert et al., 2016). This heterogeneity is displayed by the presence of different cell subpopulations with distinct genotypes, phenotypes, and biological behavior, and can be originated within the primary tumor or metastatic site or between tumors of the same histopathological subtype (Dagogo-Jack & Shaw, 2018). Different molecular signatures between cells translate into different drug suscept-ibilities, explaining the wide range of therapeutic responses (Dagogo-Jack & Shaw, 2018) and the development of drug resistance in cancer. This emphasizes how important customized treatments can be (Londoño-Berrio, Castro, Cañas, Ortiz, & Osorio, 2022) and how organoids can be a valuable tool in this context.

Cancer organoids are often established from resected tumor tissue, although previously established cancer cell lines can also be used (Hall et al., 2019; Park et al., 2022). Forsythe et al. recently developed sarcoma organoids from freshly surgically resected tumor specimens, demonstrating this direct approach to generate trustworthy models that failed to respond to chemotherapy and immunotherapy as observed clinically (Forsythe et al., 2022). Other studies reported the genetic stability of long-term organoid culture suggesting that, in spite of prolonged culture times, organoids can be adequate for drug screening (Liu et al., 2020). However, it is also described that the genetic composition of tumor-derived organoids can change over time (Fujii et al., 2016). Thus, although these phenomena could be more prevalent in some tumors, further efforts are needed to mimic the proper clonal dynamics of cancer.

Conventional methods for organoid formation involve tissue processing through mechanical or enzymatic methods. Then the cells are encapsulated in an ECM-mimicking material and finally cultured in cell culture media of variable composition, depending on the origin of cells utilized (LeSavage, Suhar, Broguiere, Lutolf, & Heilshorn, 2022). These can be cultured submerged in the cell culture medium (Gunti et al., 2021), or in the air-liquid interface to increase the oxygen availability (Gunti et al., 2021).

More advanced and dynamic organoid culturing methods are also available. For instance, microfluidic-based systems have also been widely used for spheroid culture (Moshksayan et al., 2018; Wu, Zhou, Qin, & Liu, 2021), being a cost-effective and high throughput approach for cancer organoid culture, offering extensive control over the culture conditions (Licata, Schwab, Har-El, El Gerstenhaber, & Lelkes, 2023). Microfluidic devices can have several culture compartments separated by porous membranes where cells arranged in three-dimensional structures are (co-)cultured. This can allow simulating multi-organ interactions which is useful for drug toxicity assessment (Imparato, Urciuolo, & Netti, 2022). Until now, such devices were already able to mimic key features of tumor growth and to more accurately predict responses to anticancer drugs (Shang, Soon, Lim, Khoo, & Han, 2019).

Similarly to tumor spheroids, the scalability of organoid cultures is needed, since drug testing may require large quantities of organoids for a parallel testing of multiple drugs (Licata et al., 2023). This becomes even more important when envisaging drug screenings tailored to each patient. Therefore, the fast manufacturing of cancer organoids that closely emulate in vivo tumors in a scalable and cost-effective manner is paramount for their widespread adoption in translational research and clinical settings.

Bioreactors may serve as valuable tools for organoid culture by providing enhanced control of the growth and organoid maturation compared to static systems, thereby improving cell organization and function (Fatehullah, Tan, & Barker, 2016; Lancaster & Knoblich, 2014; Ovando-Roche et al., 2018). Additionally, bioreactors improve mass transport of nutrients and diffusion of oxygen (Licata et al., 2023; Santo et al., 2016), and can allow real-time monitoring of culture conditions, including oxygen tension, pH, shear stress, and temperature. To date, several bioreactor systems have been employed for cancer organoid culture including the stirred-tank and the rotating wall vessel (Fig. 1 and Table 2).

While the stirred-tank bioreactor has been previously employed for the culture of healthy organoids (Harrison et al., 2023; Lancaster et al., 2013; Ovando-Roche et al., 2018; Saglam-Metiner et al., 2023), its application in cancer organoid culture is relatively infrequent. An illustrative example of such approach was presented by Park et al., where matrix-free glioblastoma organoids assembled from patient-derived cancer cell lines were success-fully generated (Park et al., 2022). Bioreactor conditions were optimized to culture the glioblastoma organoids which attained diameters over one millimeter within a timeframe of four to five weeks. Furthermore, these organoids exhibited pronounced stemness and a strong cell-to-cell inter-action. Through the employment of single-cell transcriptomics, it was also elucidated that glioblastoma cells underwent transdifferentiation to peri-cytes, endothelial cells, and astrocytes within the organoid, also displaying a hierarchically organized and heterogenous tumor microenvironment as observed in actual glioblastomas (Park et al., 2022). Overall, this was an efficient approach, that yielded glioblastoma organoids of consistent sizes and with clinically relevant cancerous in vivo-like characteristics (Park et al., 2022), showing potential to be applied for drug screening.

Table 2 Selected studies employing bioreactor systems for culture of cancer organoids.

Bioreactor	Cell source	Objective	Main outcomes	References
Rotating Wall Vessel	Heterotypic organoids with HepG2 hepatocyte like cells and HCT–116 metastatic colon cancer	Investigate the difference between tumor cells in 2D and inside organoids	- 3D presented mesenchymal mobile and metastatic phenotype that is more similar to colon carcinoma metastasis in vivo in comparison to the 2D model evaluated. - The Wnt pathway, with higher activation in 3D increased resistance to 5-FU	Skardal et al. (2015)
Rotating Wall Vessel	Heteroypic organoids with bone marrow derived mesenhcymal stem cells (MSC), primary human hepatocytes and colon carcinoma HCT116 cells	Evaluate organoid potential for studying tumor progression and drug response modeling	- MSC presence in organoids induced a rapid formation of large tumor foci and the formation of stoma-like tissue surrounding the tumor foci and hepatocytes. - Mature organoids were less sensitive to the chemotherapeutic drug 5-FU	Devarasetty, Wang, Soker, and Skardal (2017)
Rotating Wall Vessel	SH-SY5Y, Neuroblastoma cells (wild type, and transfected with the tyrosine kinases TrkA and TrkB)	Evaluate whether the clinical heterogeneity of neuroblastomas impacted aggregation kintetics and organoid morphology	- TrKA-expressing cells agregated more slowly and formed less morphological complex organoids in comparison to Trk-null and TrkB-expressing cells. Suggesting that the behavior of neuroblastoma	Redden, Iyer, Brodeur, and Doolin (2014)

| Spinner Flask | U251, malignant glioblastoma JX6, patient derived xenograft glioblastoma cell line | Develop a simple scalable approach for creating glioblastoma organoids that closely mimic the in vivo tumor microenvironment. | - Tumor microenvironment similar to glioblastoma tumors with transdifferentiation of glioblatoma cells into perycites, endothelial cells and astrocytes with the organoids, demonstrating hierarchical and heterogeneous organization. - Glioblastoma organoids also exhibited high stemness, strong cell-to-cell interactions, spatial gradients of hypoxia-inducible factor 1α, and heterogeneous spatial expression of NOTCH and its ligands | Park et al. (2022) |
| | | | *cells* in vitro may correlate with clinical behavior and tumorigenicity, where slower aggregation and less complexity may indicate a less aggressive phenotype | |

The rotating wall vessel (RWV) (Fig. 1) was also previously utilized for cancer organoid cultures, such as in the production of HepG2 liver organoids inoculated with metastatic colon carcinoma cells HCT-116. Under the culture conditions provided by the RWV, the organoids developed tumor foci with cells showing a mesenchymal and metastatic phenotype which better mimics colon carcinoma metastasis in comparison to the two-dimensional model also evaluated in this study. The higher activation of the Wnt pathway in 3D may be the reason for this difference, leading to increased drug resistance to the chemotherapeutic 5-FU in organoids (Skardal, Devarasetty, Rodman, Atala, & Soker, 2015).

Despite the progress in the field, organoid culture in the context of cancer research and beyond still has several limitations. Firstly, tissue accessibility can pose challenges to the establishment of organoid cultures. Secondly, the lack of standardization of culture conditions is still a major obstacle to the broad application of organoids in fundamental and translational research (Londoño-Berrio et al., 2022). This limitation affects not only the cell collection and tissue processing protocols, but also the culture medium, which is usually composed of several costly elements, like purified growth factors and others supplements, such as animal serums e.g., fetal bovine serum (FBS). Moreover, the ill definition of some of these components prevents the fine-tuning of formulations to the specificities of the study or the cancer patient characteristics. Such adjustments could represent a significant advance in what concerns drug response prediction since models could be more closely matched to the actual tumor growth conditions. Moreover, some of the ECM options available (e.g. Matrigel) are also ill-defined, posing similar challenges to the standardization and prediction accuracy of cancer organoids (LeSavage et al., 2022).

Overall, spheroid and organoid models have inherent variabilities concerning morphology, function, and formation efficiency. The use of fully controlled bioreactors featuring culture automation can decrease such inconsistencies (Hofer & Lutolf, 2021).

To summarize, cancer organoids offer the possibility to match their parental tissues more closely in comparison to spheroids. The prospect of large-scale and automated cultivation holds significant promise for their widespread adoption. However, meticulous fine-tuning is imperative to maintain accurate cell phenotypes and overall characteristics, which distinguish organoids from simpler models. Such accuracy is critical for fully harnessing their enhanced predictive potential.

3. Leveraging omics technologies for robust validation and application of 3D cancer models

The study of the molecular profile of 3D models such as spheroids and organoids can provide valuable insights into their mimicking capacity of native tumors. Such analysis contributes to the ongoing effort to more robustly validate the claim that these models recapitulate more faithfully human tissues (Simão et al., 2022). Overall omics technologies represent valuable tools for achieving these objectives (Simão et al., 2022). It can also enable the discovery of molecular signatures indicative of cancer behavior and response to therapies (Boghaert et al., 2017), holding the potential for identifying cancer biomarkers, for instance, those of treatment resistance (Wensink et al., 2021).

Omics profiling was previously employed to validate the robustness of 3D models in comparison to 2D culture. For example, genomic analysis of glioblastoma cells derived from primary samples showed that after just two weeks of 2D culture, the cells deviated from their parental tumors. In contrast, spheroids maintained characteristic genetic abnormalities associated with glioblastoma for at least 12 weeks, providing evidence that these 3D cultures offer a more representative and stable cancer model (De Witt Hamer et al., 2008).

From a transcriptomic perspective, RNA sequencing stands out as the gold standard, already previously providing valuable insights into gene expression patterns in various 3D cultures. For example, Boghaert et al. comprehensively characterized the transcriptomic profiles of 2D and 3D cultures, as well as xenografts of lung adenocarcinoma and squamous cell lung carcinoma. Their findings challenged the notion that the mode of culture per se dictates the predictiveness of cancer behavior, suggesting that factors such as the progression of the culture, such as the models' size, rather play a crucial role. Transcriptomic analysis revealed that beyond a critical size of 0.8 mm^3, 3D models recapitulated several aspects of cancer, including hypoxia, angiogenesis, and differentiation signatures, which were not observed in 2D culture (Boghaert et al., 2017).

Similarly, in a study investigating the gene expression profiles of 2D and 3D colorectal cancer cultures and xenografts through Next-Generation Sequencing (NGS) technology, Däster et al. revealed that spheroids exhibited hypoxia and necrosis only when the size exceeded 500 µm. This critical size threshold endowed the spheroids with a gene expression profile alike to that observed in murine xenografts. Furthermore, this study demonstrated the significance of spheroid size in obtaining a realistic drug resistance profile, as evidenced by challenging the models with the chemotherapeutic drug 5-FU (Däster et al., 2017). These conclusions underscore how omics technologies can inform the development of 3D models that more faithfully replicate real tumors.

RNA-seq-based methods can also be used to support drug screening. Previously Norkin et al. developed a high-content, high throughput, and cost-effective organoid sequencing drug screening platform. This system, TORNADO-seq, was validated in normal and cancerous intestinal organoids allowing the quantification of cell types and differentiation states as well as the discovery that differentiation-inducing drugs could induce the loss of cancer stem cells in organoids which holds therapeutic potential (Norkin, Ordóñez-Morán, & Huelsken, 2021).

Furthermore, the emergence of single-cell RNA sequencing represents a significant advance over bulk sequencing, offering improved capabilities to unravel intratumoral heterogeneity within models (Kim et al., 2011; Lee et al., 2014; Savage et al., 2017; Simão et al., 2022; Tirosh et al., 2016). Being a critical factor influencing cancer responses to treatment, it is thus imperative to closely replicate tumor heterogeneity in in vitro 3D models (Zhang, Miao, Sun, & Deng, 2022). Usman et al. already attempted to do so. The authors used single-cell whole genome sequencing to observe genomic heterogeneity in chromosome copy number in two independent lines of pancreatic ductal adenocarcinoma organoids. Transcriptomic profiling revealed that the gene copy number was positively correlated with the gene expression regulation. Overall, with this study it was demonstrated that the genetic heterogeneity of the organoids changes over prolonged culture times, with some of the clones having growth advantages over others. Besides the genomic instability of the pancreatic ductal adenocarcinoma cells, the culture conditions employed in this study, involving matrigel, may impact the clonal shift observed. As such, studies evaluating how the culture conditions modulate the heterogeneity observed in organoids are necessary to allow accurate data interpretation (Usman et al., 2022).

Additionally, single-cell omics technologies generally prove to be a distinguished tool, especially for heterotypic 3D model characterization, offering a unique capability to discern between different cell types, an advantage not shared by bulk analysis. This improved level of detail is critical for comprehending the different roles of distinct cell populations in various cancer-associated processes, including drug resistance and disease progression (Guo & Deng, 2018; Ni et al., 2021). As such, the increased resolution afforded by single-cell analysis allows working with more complex models and more thoroughly interpreting the results which is advantageous for drug testing.

Moreover, spatially resolved analysis is a great progression in omics technologies. This can be of great importance since the spatial distribution of tumors affects important cancer features such as drug resistance, which suggests the potential for the discovery of cancer biomarkers (Kulasinghe et al., 2022; Ma, Black, & Qian, 2022; Schmelz et al., 2021). Additionally, the spatial distribution of the different cell types in the tumor micro-environment can shed light on potential therapeutic targets (Lee, Kim, Lee, Lee, & Kwon, 2024).

Besides the transcriptome and genome, other omics dimensions that target different types of molecules are important to be included in 3D model analysis. For instance, metabolomics is also important for unraveling critical features of 3D in vitro cultures. This is exemplified in a study by Mendes et al. with patient-derived ovarian cancer explants. Through metabolomic profiling, distinct molecular signatures within the cultured tissues were identified, delineating variations in drug sensitivities, particularly to carboplatin and paclitaxel. The integration of machine learning methodologies played a crucial role in elucidating these metabolic makeups, demonstrating the potential of integrating omics and machine learning to comprehensively characterize in vitro cancer models (Mendes et al., 2022).

Furthermore, proteomic profiling adds another dimension to the evaluation of the cell state as post-translational information is also insightful into the regulatory processes of cells (Codrich et al., 2021). Other less frequently analyzed omics data, such as glycomics, which targets carbohydrate structures, and proteoglycomics that analyses protein–glycan sites, also hold the potential to unravel important cell signatures that can be applied in the clinical practice (Pinho & Reis, 2015; Radhakrishnan et al., 2014).

Conventional characterization methods for in vitro 3D models, such as imaging and evaluation of the growth kinetics (Costa et al., 2016), should be complemented with omics technologies for a thorough characterization. When compared to single-omics, multi-omic approaches integrating proteomics, glycomics, transcriptomics, metabolomics, genomics, and proteoglycomics, increase the detail of molecular profiles obtained, which can lead to a more comprehensive analysis and association of biomolecules with pathological phenotypes (Li, Li, You, Wei, & Xu, 2023).

Shi et al. demonstrated the potential of multi-omics for drug screening in organoids. The authors combined whole genome sequencing, RNA sequencing, and the transposase accessible chromatin assay using sequencing as well as drug sensitivity screening to reveal gene regulatory networks of different pancreatic cancer subtypes and chromatin accessibility profiles related to drug sensibility (Shi et al., 2022).

In conclusion, the combination of advanced 3D models with omics is a valuable strategy to better understand cancer and create ever more effective drugs and models to assist in quality patient treatment, an endeavor that already showed to benefit from the support of machine learning and artificial intelligence tools to dissect the data generated (Bai et al., 2024; Kong et al., 2020; Li et al., 2023; Serra et al., 2019). Scalable manufacturing using bioreactor technology can also support this effort since it will enable feeding large numbers of spheroids and organoids into omics analysis, further enhancing the identification of meaningful molecular features within these models, for instance, linked to treatment susceptibility.

4. Beyond cells: extracellular vesicles as drug delivery systems for targeted cancer therapy

The integration of spheroids and organoids in nanomedicine screening can further enhance this therapeutic strategy by promoting more relevant findings. In particular, the translation of EVs (including exosomes) as anticancer agents can substantially benefit from this alliance.

In recent years, EVs emerged as promising candidates for drug delivery systems (DDS) due to their ability to induce phenotypical and functional responses in recipient cells. Importantly, as EVs naturally occur in the organism, these present high biocompatibility and do not show significant cytotoxic and immunogenic effects (Smyth et al., 2015). These systems are also explored for their high bioavailability and organotropic properties which can render them especially effective in cell targeting (Jayasinghe et al., 2021). Also, EVs demonstrate increased stability in circulation, attributed to their negatively charged surface and the exposition of the CD47 surface protein, enabling them to avoid the mononuclear phagocytic system (Kamerkar et al., 2017). Additionally, EVs can also efficiently cross biological barriers, namely the blood-brain barrier (Alvarez-Erviti et al., 2011; EL Andaloussi, Mäger, Breakefield, & Wood, 2013). With these characteristics, EVs come up as promising options to be applied as DDS for cancer therapy.

Overall, drug delivery vehicles play a crucial role in enhancing the targeting abilities of free drugs and reducing off-target effects, which are prevalent in cancer therapies (Herrmann, 2020; Schirrmacher, 2019). By improving parameters such as bioavailability, fluctuations in plasma quantity, and sustained release, DDS contribute to a more favorable pharmacokinetic and pharmacodynamic profile, enhancing overall efficacy while minimizing toxicity (Adepu & Ramakrishna, 2021).

For instance, EV-based DDS hold promise for improving RNA and DNA-based therapies, addressing challenges such as susceptibility to blood RNases and DNases, instability in circulation, and inefficient membrane penetration due to negative charges (Damase et al., 2021; Kulkarni et al., 2021). The encapsulation of these molecules within EVs presents a valuable solution, facilitating penetration into cells while protecting circulation. These cargoes can be used alone or in combination with conventional drugs in the EV lumen to create multifunctional anticancer therapeutic agents.

Even though EVs are recognized for their stability in circulation, some studies have already shown that the biodistribution profile of EVs is similar to that of liposomes upon intravenous injection (Van Der Koog, Gandek, & Nagelkerke, 2022) where most of the EVs end up in the liver and spleen within hours or minutes (Kooijmans et al., 2016; Smyth et al., 2015). The utilization of different cellular sources, and downstream processing methods, which impact the functionality and biophysical properties of EVs may explain the contrasting findings. Another factor that can impact the biodistribution of EVs is the protein and lipid profiles displayed on their surface. Additionally, even though the expression patterns of surface molecules such as tetraspanins (Rana, Yue, Stadel, & Zöller, 2012) and integrins (Hoshino et al., 2015) on EVs' surface endow these with organotropic ability upon systemic administration, naive EVs rarely achieve relevant clinical numbers in the targeted tissue. A better knowledge about the biodistribution profile of EVs is imperative, as well as how bioprocessing parameters affect it, to develop safe and effective EV-based DDS.

To improve EVs' targeting capabilities, surface engineering can be used to anchor targeting moieties into the EV's surface. This has the potential to solve the off-target accumulation, which could consequently lead to insufficient EV accumulation in desired sites. In addition, surface modifications, such as the addition of polyethylene glycol (PEG) or the inclusion of CD47 in their membrane, confer stealth properties to EVs, avoiding phagocytic uptake and improving accumulation in targeted sites (Richter, Vader, & Fuhrmann, 2021). Cargo anchoring on the EV surface is another possibility particularly useful for large and polar molecules, though it is more susceptible to damage (Richter et al., 2021). Fluorescent labeling based on EV membrane properties serves as a valuable tool for using EVs in theranostic applications and studying uptake and trafficking routes.

Strategies for EV surface modifications can vary based on the intended therapeutic applications. These are broadly categorized as before-isolation and post-isolation approaches. Before isolation involves the engineering of producer cells through genetic, metabolic, and plasma membrane alterations (Jayasinghe et al., 2021). Post-isolation techniques focus on EV surface modification employing physical methods like fusion with liposomes, lipid post-insertion, chemical modifications, and surface adsorption (Gao et al., 2018; Jayasinghe et al., 2021). The molecules amenable to be used for each technique differ. In general, antibodies, fragments of antibodies, antibody mimetics, proteins, aptamers, glycans (Jayasinghe et al., 2021) and polysaccharides can be used (Tamura, Uemoto, & Tabata, 2017). EVs can also be loaded with several therapeutic cargoes, namely nucleic acids (e.g., RNA molecules, such as coding RNAs, like mRNA, and non-coding RNAs, like miRNA and siRNA), proteins, small molecules (e.g., curcumin a naturally occurring substance, and paclitaxel or doxorubicin, two chemotherapeutic drugs), showing its versatility.

Parental cell selection is a crucial step in DDS development, impacting the qualities of isolated EVs, including biological activity, cargo, tissue homing abilities, immunogenicity, and carcinogenicity. The selected cell type must also be evaluated for usability in large-scale production. Immortalized cell lines, such as MDA-MB-231 and HeLa, offer high EV production quantities due to almost indefinite culture time but may present oncogenic potential (György, Hung, Breakefield, & Leonard, 2015). Primary cells, while offering extensive characterization, produce lower amounts of EVs, and the culture duration is limited by the number of passages until reaching senescence (György et al., 2015).

Studies have utilized various cell sources, including mesenchymal stromal cells (MSC), dendritic cells, macrophages, established cancer cell lines, and primary tumor cells. Cancer cells, with advantages like ease of acquisition, number of EVs released, and high abundance, produce EVs with tumor-homing abilities, which may be especially effective in traveling back to their parental cells (Qiao et al., 2020). If this is the case, these can be utilized as a

Trojan horse allowing the encapsulation of chemotherapeutic drugs and delivery to their parental cells. Another advantage worth noting is that cancer cell-derived EVs loaded with chemotherapeutics can overcome the drug resistance usually faced in cancer stem cells upon free drug delivery (Ma et al., 2016), which constitutes a major mechanism of cancer cell resistance to chemotherapy. However, safety concerns arise as cancer-derived EVs could contain signaling messengers associated with tumor development, progression, and metastasis (Qiao et al., 2020). Moreover, cancer cell-derived EVs' interaction with the immune system is also not well understood yet, with their role as immune system inhibitors or activators not being clear, nonetheless, it is reported that such characteristics may vary according to the type of tumor considered, as well as its progression stage (Chaput & Théry, 2011).

Allogeneic MSC-derived EVs featuring low immunogenicity (Walker et al., 2019) have emerged as promising candidates for off-the-shelf DDS products. Clinical-grade MSC, manufactured under conditions able to comply with good manufacturing practices (GMP) for cell therapy applications, facilitate their use as EV producers. MSC-EVs offer high modification flexibility and tumor-homing properties (Kalimuthu et al., 2016), crucial for targeted cancer therapies. Pre-clinical studies demonstrate the anti-tumor effects of MSC-EVs, loaded with therapeutic components such as non-coding RNAs and chemotherapeutic drugs. For example, adipose tissue-derived MSC-EVs loaded with miR424–5p exhibited anti-tumor cytotoxicity in triple-negative breast cancer cells, both in vitro and in vivo, by down-regulating the PD-L1 pathway (Zhou et al., 2021). Another study showed that exosomes derived from MSC transfected with miR-584 reduce malignant glioma progression in vivo (Kim et al., 2018). Importantly, these studies employed animal models, which lack several relevant features related to cancer progression and treatment response and thus could benefit from the integration of human cancer spheroids or organoids (Zhou et al., 2021).

In conclusion, EVs stand out as promising candidates for drug delivery in cancer therapy, offering unique advantages. The extensive toolbox available for working with EVs includes various engineering strategies, drug-loading techniques, and upstream/downstream processing methodologies (de Almeida Fuzeta et al., 2022). Careful consideration of parental cell selection is crucial, with options ranging from immortalized cell lines to primary cells. While challenges such as reliable scalable EV production still exist, ongoing research continues to unveil the potential of EVs in revolutionizing drug delivery in cancer therapy.

4.1 From production to therapeutic application: omics in EV research

The suitability of EVs to be applied as effective DDS for cancer therapy depends on their molecular profile. For example, while integrins and tetraspanins can provide EVs with specific organotropic profiles (Hoshino et al., 2015; Rana et al., 2012), the presence of major histocompatibility complex (MHC) molecules impacts their immunogenicity (Robbins & Morelli, 2014; Synowsky et al., 2017), both of which are crucial for effective drug delivery. Since the process of EV production, which includes upstream production and downstream purification steps (Fig. 2), can modulate the characteristics of EVs (Rekker et al., 2014), for instance by isolating a specific subpopulation, it is crucial to fine-tune and monitor the manufacturing process of EVs to have a product with the desired properties. Besides being suitable for investigating molecular pathways involved in cell behavior, omics technologies can also be employed to study EVs (Chitoiu et al., 2020; Kugeratski et al., 2021; Lam et al., 2021; Rai, Fang, Claridge, Simpson, & Greening, 2021; Shaba et al., 2022), contributing to a better understanding on how the production protocols affect EVs, EV biology and the mechanism of action of EVs on target cells.

In particular, omics can be used to examine how changes to the cell culture environment affect the composition of EVs (Cha et al., 2018; Palviainen et al., 2019; Rayamajhi, Sulthana, Ferrel, Shrestha, & Aryal, 2023). For instance, by proteomic profiling, it was previously shown that EVs

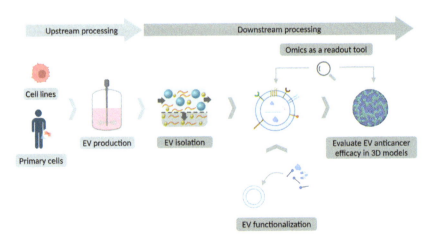

Fig. 2 Possible pipeline to further develop extracellular vesicle-based drug delivery systems integrating 3D cancer models and omics. *Created with Biorender.com.*

produced under hypoxic conditions (1% O_2 for example (Xue et al., 2018)) have modified content and function (Bister et al., 2020; Kucharzewska et al., 2013; Xue et al., 2018) compared to normoxia. To date, ultrasound waves, electrical stimuli, and shear stress, among other stimuli, were also shown to modulate EV production (Erwin, Serafim, & He, 2023), for example towards increased yields.

Other cell culture conditions can also impact the EV cargo. How different culture medium formulations impact EVs has already been evaluated through omics. For example, FBS starvation of melanoma cells, in comparison to culture in small EV-depleted FBS, led to the secretion of EVs with substantially distinct proteomes. The isolated EVs showed differences in protein metabolism, cell adhesion, and membrane raft assembly on target cells (Böröczky et al., 2023). In another study, the EV proteome was also altered when comparing OptiMEM and DMEM supplemented with FBS depleted of EVs by ultracentrifugation. Small GTPases, mRNA processing proteins, and splicing factors were the most considerably altered, and, interestingly, when cultured in serum-free OptiMEM cells increased the production of EVs likely due to the higher expression of ARF6 and other small GTPases which may have led to increased multivesicular body docking and release of EVs harboring these proteins. Moreover, the lack of serum may also have stimulated cellular stress which induced an increase in the release of EVs containing stress related proteins (Li et al., 2015).

The culture setup used for EV production can also impact the EV omic profile and thus EV function. In a work by Palviainen et al. the use of the Celline 1000 AD model bioreactor (Integra-Biosciences) induced significant differences on the EV metabolome in comparison to culture on conventional tissue flasks (Palviainen et al., 2019). Importantly, large-scale omic analysis of EV composition can be favored by large-scale EV production (Grangier et al., 2021; Jeppesen et al., 2014).

In what concerns downstream processing, the isolation technique's impact on EV properties can also be dissected through omics. For instance, the miRNA EV content was altered when comparing ultracentrifugation with ExoQuick precipitation solution (Rekker et al., 2014). In another study, the comparison of ultracentrifugation with density gradient and immunoaffinity also revealed differences in EV content. Analysis of proteins associated with EV biogenesis, release, and trafficking routes and enrichment in EV markers such as CD81, CD9, and TSG101 rendered the EpCAM immunoaffinity the best for EV isolation among the three methods tested (Tauro et al., 2012). Furthermore, different EV isolation techniques were also compared in other works through omics profiling (Benedikter et al., 2017; Helwa et al., 2017; Kalra et al., 2013; Tang et al., 2017; Van Deun et al., 2014). Overall, the fine-tuning of EV-based DDS may benefit from such comparisons.

Currently, EV-related databases Exocarta (Keerthikumar et al., 2016; Mathivanan & Simpson, 2009; Mathivanan, Fahner, Reid, & Simpson, 2012; Simpson, Kalra, & Mathivanan, 2012), EVpedia (Choi, Kim, Kim, & Gho, 2015; Kim et al., 2013, 2015), and vesiclepedia (Kim et al., 2015) contain several EV-related transcriptomic, proteomic, lipidomic, and metabolomic information. This can help researchers in designing new studies and implementing a more comprehensive analysis. For that, it is also important to follow the guidelines provided by competent sources, e.g., mass spectroscopy EV proteomics recommendations by the International Society of Extracellular Vesicles MISEV2023 (Welsh et al., 2024), in order to assure transparency and reproducibility. The establishment of collaborations between EV researchers, bioinformaticians, and data analysts is also invaluable to promote harnessing the full potential of omics in the development of EVs as DDS (Simão et al., 2022). Presently, machine learning and artificial intelligence developments are rendering the integration of these tools a valuable resource for further supporting this objective (Greenberg, Graim, & He, 2023; Kulkarni et al., 2024).

Furthermore, although regulatory agencies as FDA do not require knowledge of the mechanism of action of drugs for their clinical application ("Mechanism Matters," 2010), this understanding can be crucial for improving the design of EV-based therapeutics. As already mentioned, advanced 3D tumor models could take the relevance of such screenings to the next level (Bordanaba-Florit et al., 2021) (Fig. 2) by providing a more accurate cancer mimicry, which translates into more authentic drug responses in comparison to 2D cultures. 3D in vitro models can also minimize the number of animals used in in vivo studies, which, by their complexity, can hinder conclusions specially at early stages of investigation. Moreover, animal models also carry ethical concerns related with animal welfare. With the recent progress in omics, including single-cell and spatially resolved approaches, more detailed data, for example concerning EV biodistribution and intercellular interactions, can be acquired (Zhou et al., 2023). In sum, the integration of 3D advanced cancer cell models and omics can allow further advances in nanomedicine, including for EVs.

5. Conclusions and future perspectives

Due to their enhanced tumor-mimicking ability, 3D tumor models hold great potential to be the future of in vitro cancer research. However, gaps still exist concerning the wide application of these models. The efficient scalable production is already under investigation, however, better fine-tuning of the sizes and shapes of the models is important to achieve conclusive findings. Moreover, until now, their production has been generally limited to culture systems of a few hundred milliliters and scaling-up to higher volumes may present challenges.

In addition, these models are still not commonly analyzed concerning different omics dimensions, which is especially important for developing matched patient models. The application of multi-omic profiling could solve this research gap. Advances in single-cell and spatially resolved methodologies further add to this potential. Nonetheless, high costs and informatic analysis can still be challenging.

We envision that EVs have the potential to be one of the most promising next-generation DDS for cancer therapy, overcoming several limitations of their synthetic counterparts and free drugs. However, EV production and an in-depth understanding of EV biology still pose significant challenges to achieve this goal. Greater integration of 3D models and omics technologies in EV research is a promising solution for some of these gaps and for screening their therapeutic potential to fully harness EVs as drug delivery tools.

Acknowledgments

The authors acknowledge Fundação para a Ciência e a Tecnologia for the projects UIDB/04565/2020 and UIDP/04565/2020 of the Research Unit Institute for Bioengineering and Biosciences—iBB and the project LA/P/0140/2020 of the Associate Laboratory Institute for Health and Bioeconomy—i4HB. PPG acknowledges FCT for the PhD fellowship (2023.00546. BD).

References

Adepu, S., & Ramakrishna, S. (2021). Controlled drug delivery systems: Current status and future directions. *Molecules (Basel, Switzerland), 26*(19), https://doi.org/10.3390/MOLECULES26195905.

Al-Akra, L., Bae, D. H., Leck, L. Y. W., Richardson, D. R., & Jansson, P. J. (2019). The biochemical and molecular mechanisms involved in the role of tumor micro-environment stress in development of drug resistance. *Biochimica et Biophysica Acta. General Subjects, 1863*(9), 1390–1397. https://doi.org/10.1016/J.BBAGEN.2019.06.007.

Alvarez-Erviti, L., Seow, Y., Yin, H., Betts, C., Lakhal, S., & Wood, M. J. A. (2011). Delivery of siRNA to the mouse brain by systemic injection of targeted exosomes. *Nature Biotechnology, 29*(4), 341–345. https://doi.org/10.1038/NBT.1807.

Anderson, N. M., & Simon, M. C. (2020). Tumor microenvironment. *Current Biology: CB, 30*(16), R921. https://doi.org/10.1016/J.CUB.2020.06.081.

Bai, L., Wu, Y., Li, G., Zhang, W., Zhang, H., & Su, J. (2024). AI-enabled organoids: Construction, analysis, and application. *Bioactive Materials, 31*, 525–548. https://doi.org/10.1016/J.BIOACTMAT.2023.09.005.

Baker, B. M., & Chen, C. S. (2012). Deconstructing the third dimension—How 3D culture microenvironments alter cellular cues. *Journal of Cell Science, 125*(13), 3015. https://doi.org/10.1242/JCS.079509.

Batalha, S., Gomes, C. M., & Brito, C. (2023). Immune microenvironment dynamics of HER2 overexpressing breast cancer under dual anti-HER2 blockade. *Frontiers in Immunology, 14*, 1267621. https://doi.org/10.3389/FIMMU.2023.1267621/BIBTEX.

Bauleth-Ramos, T., Feijão, T., Gonçalves, A., Shahbazi, M.-A., Liu, Z., Barrias, C., ... Sarmento, B. (2020). Colorectal cancer triple co-culture spheroid model to assess the biocompatibility and anticancer properties of polymeric nanoparticles. *Journal of Controlled Release, 323*, 398–411. https://doi.org/10.1016/j.jconrel.2020.04.025.

Benedikter, B. J., Bouwman, F. G., Vajen, T., Heinzmann, A. C. A., Grauls, G., Mariman, E. C., ... Stassen, F. R. M. (2017). Ultrafiltration combined with size exclusion chromatography efficiently isolates extracellular vesicles from cell culture media for compositional and functional studies. *Scientific Reports, 7*(1), 1–13. https://doi.org/10.1038/s41598-017-15717-7.

Bhandari, K., Patil, R., Pingale, P., & Amrutkar, S. (2021). Emphasis on organoids in cancer research. *Cancer and Oncology Research (CEASE PUBLICATION), 7*(2), 11–22. https://doi.org/10.13189/COR.2021.070201.

Birgersdotter, A., Sandberg, R., & Ernberg, I. (2005). Gene expression perturbation in vitro—A growing case for three-dimensional (3D) culture systems. *Seminars in Cancer Biology, 15*(5), 405–412. https://doi.org/10.1016/J.SEMCANCER.2005.06.009.

Bissell, M. J., Rizki, A., & Mian, I. S. (2003). Tissue architecture: The ultimate regulator of breast epithelial function. *Current Opinion in Cell Biology, 15*(6), 753. https://doi.org/10.1016/J.CEB.2003.10.016.

Bister, N., Pistono, C., Huremagic, B., Jolkkonen, J., Giugno, R., & Malm, T. (2020). Hypoxia and extracellular vesicles: A review on methods, vesicular cargo and functions. *Journal of Extracellular Vesicles, 10*(1), https://doi.org/10.1002/JEV2.12002.

Boghaert, E. R., Lu, X., Hessler, P. E., McGonigal, T. P., Oleksijew, A., Mitten, M. J., ... Vaidya, K. S. (2017). The volume of three-dimensional cultures of cancer cells in vitro influences transcriptional profile differences and similarities with monolayer cultures and xenografted tumors. *Neoplasia (New York, N. Y.), 19*(9), 695–706. https://doi.org/10.1016/J.NEO.2017.06.004.

Bordanaba-Florit, G., Madarieta, I., Olalde, B., Falcón-P é rez, J. M., & Royo, F. (2021). 3D cell cultures as prospective models to study extracellular vesicles in cancerspan&gt. *Cancers, 13*(2), 307. https://doi.org/10.3390/CANCERS13020307.

Böröczky, T., Dobra, G., Bukva, M., Gyukity-Sebestyén, E., Hunyadi-Gulyás, É., Darula, Z., ... Harmati, M. (2023). Impact of experimental conditions on extracellular vesicles' proteome: A comparative study. *Life (Chicago, Ill.: 1978), 13*(1), https://doi.org/10.3390/LIFE13010206/S1.

Breslin, S., & O'driscoll, L. (2012). Three-dimensional cell culture: The missing link in drug discovery. *Drug Discovery Today, 00*. https://doi.org/10.1016/j.drudis.2012.10.003.

Carvalho, S., Silveira, M. J., Domingues, M., Ferreira, B., Pereira, C. L., Gremião, M. P., & Sarmento, B. (2023). Multicellular quadruple colorectal cancer spheroids as an in vitro tool for antiangiogenic potential evaluation of nanoparticles. *Advanced Therapeutics*, 2200282. https://doi.org/10.1002/ADTP.202200282.

Cha, J. M., Shin, E. K., Sung, J. H., Moon, G. J., Kim, E. H., Cho, Y. H., ... Bang, O. Y. (2018). Efficient scalable production of therapeutic microvesicles derived from human mesenchymal stem cells. *Scientific Reports, 8*(1), 1–16. https://doi.org/10.1038/s41598-018-19211-6.

Chaput, N., & Théry, C. (2011). Exosomes: Immune properties and potential clinical implementations. *Seminars in Immunopathology, 33*(5), 419–440. https://doi.org/10.1007/S00281-010-0233-9.

Chen, C. C., Liu, L., Ma, F., Wong, C. W., Guo, X. E., Chacko, J. V., ... Zhao, W. (2016). Elucidation of exosome migration across the blood-brain barrier model in vitro. *Cellular and Molecular Bioengineering, 9*(4), 509–529. https://doi.org/10.1007/S12195-016-0458-3.

Chen, J., Wang, Y., & Ko, J. (2023). Single-cell and spatially resolved omics: Advances and limitations. *Journal of Pharmaceutical Analysis, 13*(8), 833. https://doi.org/10.1016/J.JPHA.2023.07.002.

Chitoiu, L., Dobranici, A., Gherghiceanu, M., Dinescu, S., & Costache, M. (2020). Multi-omics data integration in extracellular vesicle biology—Utopia or future reality? *International Journal of Molecular Sciences, 21*(22), 8550. https://doi.org/10.3390/IJMS21228550.

Choi, D. S., Kim, D. K., Kim, Y. K., & Gho, Y. S. (2015). Proteomics of extracellular vesicles: Exosomes and ectosomes. *Mass Spectrometry Reviews, 34*(4), 474–490. https://doi.org/10.1002/MAS.21420.

Codrich, M., Dalla, E., Mio, C., Antoniali, G., Malfatti, M. C., Marzinotto, S., ... Tell, G. (2021). Integrated multi-omics analyses on patient-derived CRC organoids highlight altered molecular pathways in colorectal cancer progression involving PTEN. *Journal of Experimental & Clinical Cancer Research: CR, 40*(1), https://doi.org/10.1186/S13046-021-01986-8.

Conlan, R. S., Pisano, S., Oliveira, M. I., Ferrari, M., & Mendes Pinto, I. (2017). Exosomes as reconfigurable therapeutic systems. *Trends in Molecular Medicine, 23*(7), 636–650. https://doi.org/10.1016/J.MOLMED.2017.05.003.

Costa, E. C., Moreira, A. F., de Melo-Diogo, D., Gaspar, V. M., Carvalho, M. P., & Correia, I. J. (2016). 3D tumor spheroids: An overview on the tools and techniques used for their analysis. *Biotechnology Advances, 34*(8), 1427–1441. https://doi.org/10.1016/J.BIOTECHADV.2016.11.002.

Dagogo-Jack, I., & Shaw, A. T. (2018). Tumour heterogeneity and resistance to cancer therapies. *Nature Reviews Clinical Oncology, 15*(2), 81–94. https://doi.org/10.1038/nrclinonc.2017.166.

Damase, T. R., Sukhovershin, R., Boada, C., Taraballi, F., Pettigrew, R. I., & Cooke, J. P. (2021). The limitless future of RNA therapeutics. *Frontiers in Bioengineering and Biotechnology, 9*, 161. ⟨https://doi.org/10.3389/FBIOE.2021.628137/BIBTEX⟩.

Däster, S., Amatruda, N., Calabrese, D., Ivanek, R., Turrini, E., Droeser, R. A., ... Muraro, M. G. (2017). Induction of hypoxia and necrosis in multicellular tumor spheroids is associated with resistance to chemotherapy treatment. *Oncotarget, 8*(1), 1725 https://doi.org/10.18632/ONCOTARGET.13857.

De Almeida Fuzeta, M., Gonçalves, P. P., Fernandes-Platzgummer, A., Cabral, J. M. S., Bernardes, N., & da Silva, C. L. (2022). From promise to reality: Bioengineering strategies to enhance the therapeutic potential of extracellular vesicles. *Bioengineering, 9*(11), 675. https://doi.org/10.3390/bioengineering9110675.

De Witt Hamer, P. C., Van Tilborg, A. A. G., Eijk, P. P., Sminia, P., Troost, D., Van Noorden, C. J. F., ... Leenstra, S. (2008). The genomic profile of human malignant glioma is altered early in primary cell culture and preserved in spheroids. *Oncogene, 27*(14), 2091–2096. https://doi.org/10.1038/SJ.ONC.1210850.

Deo, S. K., Moschou, E. A., Peteu, S. F., Bachas, L. G., Daunert, S., Eisenhardt, P. E., & Madou, M. J. (2016). ROS-responsive drug delivery systems. *Bioengineering & Translational Medicine, 1*(3), 239–251. ⟨https://doi.org/10.1002/BTM2.10014⟩.

Devarasetty, M., Wang, E., Soker, S., & Skardal, A. (2017). Mesenchymal stem cells support growth and organization of host-liver colorectal-tumor organoids and possibly resistance to chemotherapy. *Biofabrication, 9*(2), 021002. https://doi.org/10.1088/1758-5090/AA7484.

Domingues, M., Leite Pereira, C., Sarmento, B., & Castro, F. (2023). Mimicking 3D breast tumor-stromal interactions to screen novel cancer therapeutics. *European Journal of Pharmaceutical Sciences, 190*, 106560. https://doi.org/10.1016/j.ejps.2023.106560.

Drost, J., & Clevers, H. (2018). Organoids in cancer research. *Nature Reviews. Cancer, 18*(7), 407–418. https://doi.org/10.1038/S41568-018-0007-6.

Ekert, J. E., Johnson, K., Strake, B., Pardinas, J., Jarantow, S., Perkinson, R., & Colter, D. C. (2014). Three-dimensional lung tumor microenvironment modulates therapeutic compound responsiveness in vitro—Implication for drug development. *PLoS One, 9*(3), 92248. https://doi.org/10.1371/JOURNAL.PONE.0092248.

EL Andaloussi, S., Mäger, I., Breakefield, X. O., & Wood, M. J. A. (2013). Extracellular vesicles: Biology and emerging therapeutic opportunities. *Nature Reviews. Drug Discovery, 12*(5), 347–357. https://doi.org/10.1038/nrd3978.

Elsharkasy, O. M., Nordin, J. Z., Hagey, D. W., de Jong, O. G., Schiffelers, R. M., Andaloussi, S. E. L., & Vader, P. (2020). Extracellular vesicles as drug delivery systems: Why and how? *Advanced Drug Delivery Reviews, 159*, 332–343. https://doi.org/10.1016/j.addr.2020.04.004.

Erwin, N., Serafim, M. F., & He, M. (2023). Enhancing the cellular production of extracellular vesicles for developing therapeutic applications. *Pharmaceutical Research, 40*(4), 833. https://doi.org/10.1007/S11095-022-03420-W.

Esa, R., Steinberg, E., Dagan, A., Yekhtin, Z., Tischenko, K., & Benny, O. (2023). Newly synthesized methionine aminopeptidase 2 inhibitor hinders tumor growth. *Drug Delivery and Translational Research, 13*(5), 1170–1182. ⟨https://doi.org/10.1007/S13346-022-01187-6/FIGURES/6⟩.

Estrada, M. F., Rebelo, S. P., Davies, E. J., Pinto, M. T., Pereira, H., Santo, V. E., ... Brito, C. (2016). Modelling the tumour microenvironment in long-term microencapsulated 3D co-cultures recapitulates phenotypic features of disease progression. *Biomaterials, 78*, 50–61. https://doi.org/10.1016/J.BIOMATERIALS.2015.11.030.

Fatehullah, A., Tan, S. H., & Barker, N. (2016). Organoids as an in vitro model of human development and disease. *Nature Cell Biology, 18*(3), 246–254. https://doi.org/10.1038/ncb3312.

Ferreira, S. C., Martins, M. L., & Vilela, M. J. (2003). Morphology transitions induced by chemotherapy in carcinomas in situ. *Physical Review E—Statistical Physics, Plasmas, Fluids, and Related Interdisciplinary Topics, 67*(5), 9. https://doi.org/10.1103/PHYSREVE.67.051914.

Fobian, S. F., Petzer, M., Vetten, M., Steenkamp, V., Gulumian, M., & Cordier, W. (2022). Mechanisms facilitating the uptake of carboxyl–polythene glycol-functionalized gold nanoparticles into multicellular spheroids. *Journal of Pharmacy and Pharmacology, 74*(9), 1282–1295. https://doi.org/10.1093/JPP/RGAC017.

Forsythe, S. D., Sivakumar, H., Erali, R. A., Wajih, N., Li, W., Shen, P., ... Votanopoulos, K. I. (2022). Patient-specific sarcoma organoids for personalized translational research: Unification of the operating room with rare cancer research and clinical implications. *Annals of Surgical Oncology, 29*(12), 7354–7367. https://doi.org/10.1245/s10434-022-12086-y.

Foty, R. (2011). A simple hanging drop cell culture protocol for generation of 3D spheroids. *Journal of Visualized Experiments: JoVE, 51*(51), https://doi.org/10.3791/2720.

Advancing cancer therapeutics — 173

Franchi-Mendes, T., Lopes, N., & Brito, C. (2021). Heterotypic tumor spheroids in agitation-based cultures: A scaffold-free cell model that sustains long-term survival of endothelial cells. *Frontiers in Bioengineering and Biotechnology, 9*. https://doi.org/10.3389/FBIOE.2021.649949/PDF.

Friedrich, J., Seidel, C., Ebner, R., & Kunz-Schughart, L. A. (2009). Spheroid-based drug screen: Considerations and practical approach. *Nature Protocols, 4*(3), 309–324. https://doi.org/10.1038/NPROT.2008.226.

Froehlich, K., Haeger, J. D., Heger, J., Pastuschek, J., Photini, S. M., Yan, Y., ... Schmidt, A. (2016). Generation of multicellular breast cancer tumor spheroids: Comparison of different protocols. *Journal of Mammary Gland Biology and Neoplasia, 21*(3–4), 89–98. https://doi.org/10.1007/S10911-016-9359-2/FIGURES/4.

Fujii, M., Shimokawa, M., Date, S., Takano, A., Matano, M., Nanki, K., ... Sato, T. (2016). A colorectal tumor organoid library demonstrates progressive loss of niche factor requirements during tumorigenesis. *Cell Stem Cell, 18*(6), 827–838. https://doi.org/10.1016/J.STEM.2016.04.003.

Gao, X., Ran, N., Dong, X., Zuo, B., Yang, R., Zhou, Q., ... Yin, H. (2018). Anchor peptide captures, targets, and loads exosomes of diverse origins for diagnostics and therapy. *Science Translational Medicine, 10*(444), eaat0195. https://doi.org/10.1126/scitranslmed.aat0195.

Gebhard, C., Gabriel, C., & Walter, I. (2016). Morphological and immunohistochemical characterization of canine osteosarcoma spheroid cell cultures. *Anatomia, Histologia, Embryologia, 45*(3), 219. https://doi.org/10.1111/AHE.12190.

Ghosh, S., Spagnoli, G. C., Martin, I., Ploegert, S., Demougin, P., Heberer, M., & Reschner, A. (2005). Three-dimensional culture of melanoma cells profoundly affects gene expression profile: A high density oligonucleotide array study. *Journal of Cellular Physiology, 204*(2), 522–531. https://doi.org/10.1002/JCP.20320.

Goodwin, T. J., Prewett, T. L., Wolf, D. A., & Spaulding, G. F. (1993). Reduced shear stress: A major component in the ability of mammalian tissues to form three-dimensional assemblies in simulated microgravity. *Journal of Cellular Biochemistry, 51*(3), 301–311. https://doi.org/10.1002/JCB.240510309.

Grangier, A., Branchu, J., Volatron, J., Piffoux, M., Gazeau, F., Wilhelm, C., & Silva, A. K. A. (2021). Technological advances towards extracellular vesicles mass production. *Advanced Drug Delivery Reviews, 176*. https://doi.org/10.1016/J.ADDR.2021.113843.

Greenberg, Z. F., Graim, K. S., & He, M. (2023). Towards artificial intelligence-enabled extracellular vesicle precision drug delivery. *Advanced Drug Delivery Reviews, 199*, 114974. https://doi.org/10.1016/J.ADDR.2023.114974.

Gunti, S., Hoke, A. T. K., Vu, K. P., & London, N. R. (2021). Organoid and spheroid tumor models: Techniques and applications. *Cancers, 13*(4), 1–18. https://doi.org/10.3390/CANCERS13040874.

Guo, S., & Deng, C. X. (2018). Effect of stromal cells in tumor microenvironment on metastasis initiation. *International Journal of Biological Sciences, 14*(14), 2083. https://doi.org/10.7150/IJBS.25720.

György, B., Hung, M. E., Breakefield, X. O., & Leonard, J. N. (2015). Therapeutic applications of extracellular vesicles: Clinical promise and open questions. *Annual Review of Pharmacology and Toxicology, 55*, 439–464. https://doi.org/10.1146/annurev-pharmtox-010814-124630.

Halfter, K., Ditsch, N., Kolberg, H. C., Fischer, H., Hauzenberger, T., von Koch, F. E., ... Mayer, B. (2015). Prospective cohort study using the breast cancer spheroid model as a predictor for response to neoadjuvant therapy—The SpheroNEO study. *BMC Cancer, 15*(1), https://doi.org/10.1186/S12885-015-1491-7.

Halfter, K., Hoffmann, O., Ditsch, N., Ahne, M., Arnold, F., Paepke, S., ... Mayer, B. (2016). Testing chemotherapy efficacy in HER2 negative breast cancer using patient-derived spheroids. *Journal of Translational Medicine, 14*(1), 112. https://doi.org/10.1186/S12967-016-0855-3.

Hall, B. R., Cannon, A., Thompson, C., Santhamma, B., Chavez-Riveros, A., Bhatia, R., ... Kumar, S. (2019). Utilizing cell line-derived organoids to evaluate the efficacy of a novel LIFR-inhibitor, EC359 in targeting pancreatic tumor stroma. *Genes & Cancer, 10*(1–2), 1. https://doi.org/10.18632/GENESANDCANCER.184.

Han, S. J., Kwon, S., & Kim, K. S. (2021). Challenges of applying multicellular tumor spheroids in preclinical phase. *Cancer Cell International, 21*(1), 152. https://doi.org/10.1186/s12935-021-01853-8.

Harrison, S. P., Siller, R., Tanaka, Y., Chollet, M. E., de la Morena-Barrio, M. E., Xiang, Y., ... Sullivan, G. J. (2023). Scalable production of tissue-like vascularized liver organoids from human PSCs. *Experimental & Molecular Medicine, 55*(9), 2005–2024. https://doi.org/10.1038/s12276-023-01074-1.

Heem Wong, C., Wei Siah, K., & Lo, A. W. (2019). Estimation of clinical trial success rates and related parameters. *Biostatistics (Oxford, England), 20*, 273–286. https://doi.org/10.1093/biostatistics/kxx069.

Helwa, I., Cai, J., Drewry, M. D., Zimmerman, A., Dinkins, M. B., Khaled, M. L., ... Liu, Y. (2017). A comparative study of serum exosome isolation using differential ultra-centrifugation and three commercial reagents. *PLoS One, 12*(1), e0170628. https://doi.org/10.1371/JOURNAL.PONE.0170628.

Herrmann, J. (2020). Adverse cardiac effects of cancer therapies: Cardiotoxicity and arrhythmia. *Nature Reviews Cardiology, 17*(8), 474–502. https://doi.org/10.1038/s41569-020-0348-1.

Hickman, J. A., Graeser, R., de Hoogt, R., Vidic, S., Brito, C., Gutekunst, M., & van der Kuip, H. Imi Predect consortium. (2014). Three-dimensional models of cancer for pharmacology and cancer cell biology: Capturing tumor complexity in vitro/ex vivo. *Biotechnology Journal, 9*(9), 1115–1128. https://doi.org/10.1002/BIOT.201300492.

Hirschhaeuser, F., Leidig, T., Rodday, B., Lindemann, C., & Mueller-Klieser, W. (2009). Test system for trifunctional antibodies in 3D MCTS culture. *Journal of Biomolecular Screening, 14*(8), 980–990. https://doi.org/10.1177/1087057109341766.

Hirschhaeuser, F., Menne, H., Dittfeld, C., West, J., Mueller-Klieser, W., & Kunz-Schughart, L. A. (2010). Multicellular tumor spheroids: An underestimated tool is catching up again. *Journal of Biotechnology, 148*(1), 3–15. https://doi.org/10.1016/J.JBIOTEC.2010.01.012.

Hofer, M., & Lutolf, M. P. (2021). Engineering organoids. *Nature Reviews Materials, 6*(5), 402–420. https://doi.org/10.1038/s41578-021-00279-y.

Hofmann, S., Cohen-Harazi, R., Maizels, Y., & Koman, I. (2022). Patient-derived tumor spheroid cultures as a promising tool to assist personalized therapeutic decisions in breast cancer. *Translational Cancer Research, 11*(1), 134–147. https://doi.org/10.21037/TCR-21-1577/COIF.

Hoshino, A., Costa-Silva, B., Shen, T. L., Rodrigues, G., Hashimoto, A., Tesic Mark, M., ... Lyden, D. (2015). Tumour exosome integrins determine organotropic metastasis. *Nature, 527*(7578), 329–335. https://doi.org/10.1038/NATURE15756.

Hubert, C. G., Rivera, M., Spangler, L. C., Wu, Q., Mack, S. C., Prager, B. C., ... Rich, J. N. (2016). A three-dimensional organoid culture system derived from human glioblastomas recapitulates the hypoxic gradients and cancer stem cell heterogeneity of tumors found in vivo. *Cancer Research, 76*(8), 2465–2477. https://doi.org/10.1158/0008-5472.CAN-15-2402.

Idrisova, K. F., Simon, H. U., & Gomzikova, M. O. (2023). Role of patient-derived models of cancer in translational oncology. *Cancers, 15*(1), https://doi.org/10.3390/CANCERS15010139.

Imamura, Y., Mukohara, T., Shimono, Y., Funakoshi, Y., Chayahara, N., Toyoda, M., ... Minami, H. (2015). Comparison of 2D- and 3D-culture models as drug-testing platforms in breast cancer. *Oncology Reports, 33*(4), 1837–1843. https://doi.org/10.3892/OR.2015.3767/DOWNLOAD.

Imparato, G., Urciuolo, F., & Netti, P. A. (2022). Organ on chip technology to model cancer growth and metastasis. *Bioengineering (Basel, Switzerland), 9*(1), https://doi.org/10.3390/BIOENGINEERING9010028.

Ivascu, A., & Kubbies, M. (2006). Rapid generation of single-tumor spheroids for high-throughput cell function and toxicity analysis. *Journal of Biomolecular Screening, 11*(8), 922–932. https://doi.org/10.1177/1087057106292763.

Jaganathan, H., Gage, J., Leonard, F., Srinivasan, S., Souza, G. R., Dave, B., & Godin, B. (2014). Three-dimensional in vitro co-culture model of breast tumor using magnetic levitation. *Scientific Reports, 4*(1), 1–9. https://doi.org/10.1038/srep06468.

Jayasinghe, M. K., Tan, M., Peng, B., Yang, Y., Sethi, G., Pirisinu, M., & Le, M. T. N. (2021). New approaches in extracellular vesicle engineering for improving the efficacy of anti-cancer therapies. *Seminars in Cancer Biology, 74*, 62–78. https://doi.org/10.1016/j.semcancer.2021.02.010.

Jensen, C., & Teng, Y. (2020). Is it time to start transitioning from 2D to 3D cell culture? *Frontiers in Molecular Biosciences, 7*, 513823. https://doi.org/10.3389/FMOLB.2020.00033/BIBTEX.

Jeppesen, D. K., Nawrocki, A., Jensen, S. G., Thorsen, K., Whitehead, B., Howard, K. A., ... Ostenfeld, M. S. (2014). Quantitative proteomics of fractionated membrane and lumen exosome proteins from isogenic metastatic and nonmetastatic bladder cancer cells reveal differential expression of EMT factors. *Proteomics, 14*(6), 699–712. https://doi.org/10.1002/PMIC.201300452.

Jubelin, C., Muñoz-Garcia, J., Griscom, L., Cochonneau, D., Ollivier, E., Heymann, M. F., ... Heymann, D. (2022). Three-dimensional in vitro culture models in oncology research. *Cell & Bioscience, 12*(1), 1–28. https://doi.org/10.1186/S13578-022-00887-3.

Kalimuthu, S., Gangadaran, P., Li, X. J., Oh, J. M., Lee, H. W., Jeong, S. Y., ... Ahn, B. C. (2016). In Vivo therapeutic potential of mesenchymal stem cell-derived extracellular vesicles with optical imaging reporter in tumor mice model. *Scientific Reports, 6*. https://doi.org/10.1038/SREP30418.

Kalra, H., Adda, C. G., Liem, M., Ang, C. S., Mechler, A., Simpson, R. J., ... Mathivanan, S. (2013). Comparative proteomics evaluation of plasma exosome isolation techniques and assessment of the stability of exosomes in normal human blood plasma. *Proteomics, 13*(22), 3354–3364. https://doi.org/10.1002/PMIC.201300282.

Kamerkar, S., Lebleu, V. S., Sugimoto, H., Yang, S., Ruivo, C. F., Melo, S. A., ... Kalluri, R. (2017). Exosomes facilitate therapeutic targeting of oncogenic KRAS in pancreatic cancer. *Nature, 546*(7659), 498–503. https://doi.org/10.1038/nature22341.

Keerthikumar, S., Chisanga, D., Ariyaratne, D., Al Saffar, H., Anand, S., Zhao, K., ... Mathivanan, S. (2016). ExoCarta: A web-based compendium of exosomal cargo. *Journal of Molecular Biology, 428*(4), 688–692. https://doi.org/10.1016/J.JMB.2015.09.019.

Kenny, P. A., Lee, G. Y., Myers, C. A., Neve, R. M., Semeiks, J. R., Spellman, P. T., ... Bissell, M. J. (2007). The morphologies of breast cancer cell lines in three-dimensional assays correlate with their profiles of gene expression. *Molecular Oncology, 1*(1), 84. https://doi.org/10.1016/J.MOLONC.2007.02.004.

Khanna, S., Chauhan, A., Bhatt, A. N., & Dwarakanath, B. S. R. (2020). Multicellular tumor spheroids as in vitro models for studying tumor responses to anticancer therapies. *Animal Biotechnology: Models in Discovery and Translation*, 251–268. https://doi.org/10.1016/B978-0-12-811710-1.00011-2.

Kilian, K. A., Bugarija, B., Lahn, B. T., & Mrksich, M. (2010). Geometric cues for directing the differentiation of mesenchymal stem cells. *Proceedings of the National Academy of Sciences of the United States of America, 107*(11), 4872–4877. https://doi.org/10.1073/PNAS.0903269107/SUPPL_FILE/PNAS.0903269107_SI.PDF.

Kim, D. K., Kang, B., Kim, O. Y., Choi, D. S., Lee, J., Kim, S. R., ... Gho, Y. S. (2013). EVpedia: An integrated database of high-throughput data for systemic analyses of extracellular vesicles. *Journal of Extracellular Vesicles, 2*(1), https://doi.org/10.3402/JEV.V2I0.20384.

Kim, D. K., Lee, J., Kim, S. R., Choi, D. S., Yoon, Y. J., Kim, J. H., ... Gho, Y. S. (2015). EVpedia: A community web portal for extracellular vesicles research. *Bioinformatics (Oxford, England), 31*(6), 933–939. https://doi.org/10.1093/BIOINFORMATICS/BTU741.

Kim, J. A., Choi, J. H., Kim, M., Rhee, W. J., Son, B., Jung, H. K., & Park, T. H. (2013). High-throughput generation of spheroids using magnetic nanoparticles for three-dimensional cell culture. *Biomaterials, 34*(34), 8555–8563. https://doi.org/10.1016/J.BIOMATERIALS.2013.07.056.

Kim, K. -T., Lee, H. W., Lee, H. -O., Kim, S. C., Jee Seo, Y., Chung, W., & Park, W.-Y. (2011). *Single-cell mRNA sequencing identifies subclonal heterogeneity in anti-cancer drug responses of lung adenocarcinoma cells.* https://doi.org/10.1186/s13059-015-0692-3.

Kim, R., Lee, S., Lee, J., Kim, M., Kim, W. J., Lee, H. W., ... Chang, W. (2018). Exosomes derived from microRNA-584 transfected mesenchymal stem cells: Novel alternative therapeutic vehicles for cancer therapy. *BMB Reports, 51*(8), 406. https://doi.org/10.5483/BMBREP.2018.51.8.105.

Kita, K., Oya, H., Gennis, R. B., Ackrell, B. A. C., Kasa-Hara, M., Lombardo, A., ... Scheffler, I. E. (1993). Interaction with basement membrane serves to rapidly distinguish growth and differentiation pattern of normal and malignant human breast epithelial cells. *Proceedings of the National Academy of Sciences of the United States of America, 90*(6), 2556. https://www.ncbi.nlm.nih.gov/pmc/articles/PMC1083027/.

Kolenda, T., Kapałczyńska, M., Przybyła, W., Zajączkowska, M., Teresiak, A., Filas, V., ... Lamperska, K. (2016). 2D and 3D cell cultures–a comparison of different types of cancer cell cultures. *Termedia. PlM Kapałczyńska, T Kolenda, W Przybyła, M Zajączkowska, A Teresiak, V Filas, M IbbsArchives of Medical Science, 2018•termedia. Pl, 14*(4), 910–919. https://doi.org/10.5114/aoms.2016.63743.

Kong, J. H., Lee, H., Kim, D., Han, S. K., Ha, D., Shin, K., & Kim, S. (2020). Network-based machine learning in colorectal and bladder organoid models predicts anti-cancer drug efficacy in patients. *Nature Communications, 11*(1), 1–13. https://doi.org/10.1038/s41467-020-19313-8.

Kooijmans, S. A. A., Fliervoet, L. A. L., Van Der Meel, R., Fens, M. H. A. M., Heijnen, H. F. G., Van Bergen En Henegouwen, P. M. P., ... Schiffelers, R. M. (2016). PEGylated and targeted extracellular vesicles display enhanced cell specificity and circulation time. *Journal of Controlled Release: Official Journal of the Controlled Release Society, 224*, 77–85. https://doi.org/10.1016/J.JCONREL.2016.01.009.

Kucharzewska, P., Christianson, H. C., Welch, J. E., Svensson, K. J., Fredlund, E., Ringnér, M., ... Belting, M. (2013). Exosomes reflect the hypoxic status of glioma cells and mediate hypoxia-dependent activation of vascular cells during tumor development. *Proceedings of the National Academy of Sciences of the United States of America, 110*(18), 7312–7317. https://doi.org/10.1073/PNAS.1220998110/-/DCSUPPLEMENTAL/PNAS.201220998SI.PDF.

Kugeratski, F. G., Hodge, K., Lilla, S., McAndrews, K. M., Zhou, X., Hwang, R. F., ... Kalluri, R. (2021). Quantitative proteomics identifies the core proteome of exosomes with syntenin-1 as the highest abundant protein and a putative universal biomarker. *Nature Cell Biology, 23*(6), 631–641. https://doi.org/10.1038/S41556-021-00693-Y.

Kulasinghe, A., Monkman, J., Shah, E. T., Matigian, N., Adams, M. N., & O'Byrne, K. (2022). Spatial profiling identifies prognostic features of response to adjuvant therapy in triple negative breast cancer (TNBC). *Frontiers in Oncology, 11*. https://doi.org/10.3389/FONC.2021.798296/FULL.

Kulkarni, J. A., Witzigmann, D., Thomson, S. B., Chen, S., Leavitt, B. R., Cullis, P. R., & van der Meel, R. (2021). The current landscape of nucleic acid therapeutics. *Nature Nanotechnology, 16*(6), 630–643. https://doi.org/10.1038/s41565-021-00898-0.

Kulkarni, M., Kar, R., Ghosh, S., Sonar, S., Mirgh, D., Sivakumar, I., ... Muthusamy, R. (2024). Clinical impact of multi-omics profiling of extracellular vesicles in cancer liquid biopsy. *The Journal of Liquid Biopsy, 3*, 100138. https://doi.org/10.1016/J.JLB.2024.100138.

LaBarbera, D. V., Reid, B. G., & Yoo, B. H. (2012). The multicellular tumor spheroid model for high-throughput cancer drug discovery. *Expert Opinion on Drug Discovery, 7*(9), 819–830. https://doi.org/10.1517/17460441.2012.708334.

Lam, S. M., Zhang, C., Wang, Z., Ni, Z., Zhang, S., Yang, S., ... Shui, G. (2021). A multi-omics investigation of the composition and function of extracellular vesicles along the temporal trajectory of COVID-19. *Nature Metabolism, 3*(7), 909–922. https://doi.org/10.1038/s42255-021-00425-4.

Lancaster, M. A., & Knoblich, J. A. (2014). Organogenesis in a dish: Modeling development and disease using organoid technologies. *Science (New York, N. Y.), 345*(6194), https://doi.org/10.1126/SCIENCE.1247125.

Lancaster, M. A., Renner, M., Martin, C. A., Wenzel, D., Bicknell, L. S., Hurles, M. E., ... Knoblich, J. A. (2013). Cerebral organoids model human brain development and microcephaly. *Nature, 501*(7467), 373–379. https://doi.org/10.1038/nature12517.

Lee, M.-C. W., Lopez-Diaz, F. J., Khan, S. Y., Tariq, M. A., Dayn, Y., Vaske, C. J., ... Pourmand, N. (2014). Single-cell analyses of transcriptional heterogeneity during drug tolerance transition in cancer cells by RNA sequencing. *Proceedings of the National Academy of Sciences, 111*(44), https://doi.org/10.1073/pnas.1404656111.

Lee, S., Kim, G., Lee, J. Y., Lee, A. C., & Kwon, S. (2024). Mapping cancer biology in space: Applications and perspectives on spatial omics for oncology. *Molecular Cancer, 23*(1), 1–27. https://doi.org/10.1186/S12943-024-01941-Z.

LeSavage, B. L., Suhar, R. A., Broguiere, N., Lutolf, M. P., & Heilshorn, S. C. (2022). Next-generation cancer organoids. *Nature Materials, 21*(2), 143–159. https://doi.org/10.1038/s41563-021-01057-5.

Li, C., Kato, M., Shiue, L., Shively, J. E., Ares, M., & Lin, R. J. (2006). Cell type and culture condition-dependent alternative splicing in human breast cancer cells revealed by splicing-sensitive microarrays. *Cancer Research, 66*(4), 1990–1999. https://doi.org/10.1158/0008-5472.CAN-05-2593.

Li, J., Lee, Y., Johansson, H. J., Mäger, I., Vader, P., Nordin, J. Z., ... El Andaloussi, S. (2015). Serum-free culture alters the quantity and protein composition of neuroblastoma-derived extracellular vesicles. *Journal of Extracellular Vesicles, 4*(2015), 1–12. https://doi.org/10.3402/JEV.V4.26883.

Li, J., Li, L., You, P., Wei, Y., & Xu, B. (2023). Towards artificial intelligence to multi-omics characterization of tumor heterogeneity in esophageal cancer. *Seminars in Cancer Biology, 91*, 35–49. https://doi.org/10.1016/J.SEMCANCER.2023.02.009.

Libring, S., Enríquez, Á., Lee, H., & Solorio, L. (2021). In vitro magnetic techniques for investigating cancer progression. *Cancers, 13*(17), https://doi.org/10.3390/CANCERS13174440.

Licata, J. P., Schwab, K. H., Har-El, Y., El Gerstenhaber, J. A., & Lelkes, P. I. (2023). Bioreactor technologies for enhanced organoid culture. *International Journal of Molecular Sciences, 24*(14), https://doi.org/10.3390/IJMS241411427.

Lin, J., Yang, Z., Wang, L., Xing, D., & Lin, J. (2022). Global research trends in extracellular vesicles based on stem cells from 1991 to 2021: A bibliometric and visualized study. *Frontiers in Bioengineering and Biotechnology, 10*, 956058. https://doi.org/10.3389/FBIOE.2022.956058/BIBTEX.

Liu, C., Qin, T., Huang, Y., Li, Y., Chen, G., & Sun, C. (2020). Drug screening model meets cancer organoid technology. *Translational Oncology, 13*(11), https://doi.org/10.1016/J.TRANON.2020.100840.

Liu, Z., Delavan, B., Roberts, R., & Tong, W. (2017). Lessons learned from two decades of anticancer drugs. *Trends in Pharmacological Sciences, 38*(10), 852–872. https://doi.org/10.1016/j.tips.2017.06.005.

Londoño-Berrio, M., Castro, C., Cañas, A., Ortiz, I., & Osorio, M. (2022). Advances in tumor organoids for the evaluation of drugs: A bibliographic review. MDPI *Pharmaceutics, 14*(12), https://doi.org/10.3390/pharmaceutics14122709.

Lopes-Coelho, F., Silva, F., Gouveia-Fernandes, S., Martins, C., Lopes, N., Domingues, G., ... Serpa, J. (2020). Monocytes as endothelial progenitor cells (EPCs), another brick in the wall to disentangle tumor angiogenesis. *Cells, 9*(1), https://doi.org/10.3390/CELLS9010107.

Lovitt, C. J., Shelper, T. B., & Avery, V. M. (2013). Miniaturized three-dimensional cancer model for drug evaluation. *Assay and Drug Development Technologies, 11*(7), 435–448. https://doi.org/10.1089/ADT.2012.483.

Lovitt, C. J., Shelper, T. B., & Avery, V. M. (2015). Evaluation of chemotherapeutics in a three-dimensional breast cancer model. *Journal of Cancer Research and Clinical Oncology, 141*(5), 951–959. https://doi.org/10.1007/S00432-015-1950-1.

Luca, A. C., Mersch, S., Deenen, R., Schmidt, S., Messner, I., Schäfer, K. L., ... Stoecklein, N. H. (2013). Impact of the 3D microenvironment on phenotype, gene expression, and EGFR inhibition of colorectal cancer cell lines. *PLoS One, 8*(3), e59689. https://doi.org/10.1371/JOURNAL.PONE.0059689.

Luo, Z., Zhou, X., Mandal, K., He, N., Wennerberg, W., Qu, M., ... Khademhosseini, A. (2021). Reconstructing the tumor architecture into organoids. *Advanced Drug Delivery Reviews, 176*, 113839. https://doi.org/10.1016/J.ADDR.2021.113839.

Ma, J., Zhang, Y., Tang, K., Zhang, H., Yin, X., Li, Y., ... Huang, B. (2016). Reversing drug resistance of soft tumor-repopulating cells by tumor cell-derived chemotherapeutic microparticles. *Cell Research, 26*(6), 713–727. https://doi.org/10.1038/cr.2016.53.

Ma, R. Y., Black, A., & Qian, B. Z. (2022). Macrophage diversity in cancer revisited in the era of single-cell omics. *Trends in Immunology, 43*(7), 546–563. https://doi.org/10.1016/J.IT.2022.04.008.

Malhão, F., Macedo, A. C., Ramos, A. A., & Rocha, E. (2022). Morphometrical, morphological, and immunocytochemical characterization of a tool for cytotoxicity research: 3D cultures of breast cell lines grown in ultra-low attachment plates. *Toxics, 10*(8), 415. https://doi.org/10.3390/TOXICS10080415/S1.

Manduca, N., Maccafeo, E., De Maria, R., Sistigu, A., & Musella, M. (2023). 3D cancer models: One step closer to in vitro human studies. *Frontiers in Immunology, 14*, 1175503. https://doi.org/10.3389/FIMMU.2023.1175503/BIBTEX.

Martins, C., Pacheco, C., Moreira-Barbosa, C., Marques-Magalhães, Â., Dias, S., Araújo, M., ... Sarmento, B. (2023). Glioblastoma immuno-endothelial multicellular microtissue as a 3D in vitro evaluation tool of anti-cancer nano-therapeutics. *Journal of Controlled Release, 353*, 77–95. https://doi.org/10.1016/J.JCONREL.2022.11.024.

Massai, D., Isu, G., Madeddu, D., Cerino, G., Falco, A., Frati, C., ... Morbiducci, U. (2016). A versatile bioreactor for dynamic suspension cell culture. Application to the culture of cancer cell spheroids. *PLoS One, 11*(5), e0154610. https://doi.org/10.1371/journal.pone.0154610.

Mathivanan, S., Fahner, C. J., Reid, G. E., & Simpson, R. J. (2012). ExoCarta 2012: Database of exosomal proteins, RNA and lipids. *Nucleic Acids Research, 40*(Database issue), https://doi.org/10.1093/NAR/GKR828.

Mathivanan, S., & Simpson, R. J. (2009). ExoCarta: A compendium of exosomal proteins and RNA. *Proteomics, 9*(21), 4997–5000. https://doi.org/10.1002/PMIC.200900351.

Mazzoleni, G., Di Lorenzo, D., & Steimberg, N. (2009). Modelling tissues in 3D: The next future of pharmaco-toxicology and food research? *Genes & Nutrition, 4*(1), 13. https://doi.org/10.1007/S12263-008-0107-0.

Mechanism matters. (2010). *Nature Medicine, 16*(4), 347. https://doi.org/10.1038/nm0410-347.

Medle, B., Sjödahl, G., Eriksson, P., Liedberg, F., Höglund, M., & Bernardo, C. (2022). Patient-derived bladder cancer organoid models in tumor biology and drug testing: A systematic review. *Cancers.* https://doi.org/10.3390/cancers14092062.

Mendes, R., Graça, G., Silva, F., Guerreiro, A. C. L., Gomes-Alves, P., Serpa, J., ... Isidro, I. A. (2022). Exploring metabolic signatures of ex vivo tumor tissue cultures for prediction of chemosensitivity in ovarian cancer. *Cancers, 14*(18), https://doi.org/10.3390/CANCERS14184460.

Moshksayan, K., Kashaninejad, N., Warkiani, M. E., Lock, J. G., Moghadas, H., Firoozabadi, B., ... Nguyen, N. T. (2018). Spheroids-on-a-chip: Recent advances and design considerations in microfluidic platforms for spheroid formation and culture. *Sensors and Actuators B: Chemical, 263*, 151–176. https://doi.org/10.1016/J.SNB.2018.01.223.

Mseka, T., Bamburg, J. R., & Cramer, L. P. (2007). ADF/cofilin family proteins control formation of oriented actin-filament bundles in the cell body to trigger fibroblast polarization. *Journal of Cell Science, 120*(Pt 24), 4332–4344. https://doi.org/10.1242/JCS.017640.

Muhitch, J. W., O'Connor, K. C., Blake, D. A., Lacks, D. J., Rosenzweig, N., & Spaulding, G. F. (2000). Characterization of aggregation and protein expression of bovine corneal endothelial cells as microcarrier cultures in a rotating-wall vessel. *Cytotechnology, 32*(3), 253. https://doi.org/10.1023/A:1008117410827.

Myungjin Lee, J., Mhawech-Fauceglia, P., Lee, N., Cristina Parsanian, L., Gail Lin, Y., Andrew Gayther, S., & Lawrenson, K. (2013). A three-dimensional microenvironment alters protein expression and chemosensitivity of epithelial ovarian cancer cells in vitro. *Laboratory Investigation; A Journal of Technical Methods and Pathology, 93*(5), 528–542. https://doi.org/10.1038/LABINVEST.2013.41.

Nath, S., & Devi, G. R. (2016). Three-dimensional culture systems in cancer research: Focus on tumor spheroid model. *Pharmacology & Therapeutics, 163*, 94–108. https://doi.org/10.1016/J.PHARMTHERA.2016.03.013.

Nemir, S., & West, J. L. (2010). Synthetic materials in the study of cell response to substrate rigidity. *Annals of Biomedical Engineering, 38*(1), 2–20. https://doi.org/10.1007/S10439-009-9811-1.

Ni, Y., Zhou, X., Yang, J., Shi, H., Li, H., Zhao, X., & Ma, X. (2021). The role of tumor-stroma interactions in drug resistance within tumor microenvironment. *Frontiers in Cell and Developmental Biology, 9.* https://doi.org/10.3389/FCELL.2021.637675.

Norkin, M., Ordóñez-Morán, P., & Huelsken, J. (2021). High-content, targeted RNA-seq screening in organoids for drug discovery in colorectal cancer. *Graphical Abstract.* https://doi.org/10.1016/j.celrep.2021.109026.

Nunes, A. S., Barros, A. S., Costa, E. C., Moreira, A. F., & Correia, I. J. (2019). 3D tumor spheroids as in vitro models to mimic in vivo human solid tumors resistance to therapeutic drugs. *Biotechnology and Bioengineering, 116*(1), 206–226. https://doi.org/10.1002/BIT.26845.

Ovando-Roche, P., West, E. L., Branch, M. J., Sampson, R. D., Fernando, M., Munro, P., ... Ali, R. R. (2018). Use of bioreactors for culturing human retinal organoids improves photoreceptor yields. *Stem Cell Research and Therapy, 9*(1), 1–14. https://doi.org/10.1186/S13287-018-0907-0/FIGURES/7.

Palviainen, M., Saari, H., Kärkkäinen, O., Pekkinen, J., Auriola, S., Yliperttula, M., ... Siljander, P. R. M. (2019). Metabolic signature of extracellular vesicles depends on the cell culture conditions. *Journal of Extracellular Vesicles, 8*(1), 1596669. https://doi.org/10.1080/20013078.2019.1596669.

Pampaloni, F., Reynaud, E. G., & Stelzer, E. H. K. (2007). The third dimension bridges the gap between cell culture and live tissue. *Nature Reviews. Molecular Cell Biology, 8*(10), 839–845. https://doi.org/10.1038/nrm2236.

Panchalingam, K. M., Paramchuk, W. J., Chiang, C. Y. K., Shah, N., Madan, A., Hood, L., ... Behie, L. A. (2010). Bioprocessing of human glioblastoma brain cancer tissue. *Tissue Engineering. Part A, 16*(4), 1169–1177. https://doi.org/10.1089/TEN.TEA.2009.0490.

Park, S., Avera, A. D., & Kim, Y. (2022). Biomanufacturing of glioblastoma organoids exhibiting hierarchical and spatially organized tumor microenvironment via transdifferentiation. *Biotechnology and Bioengineering, 119*(11), 3252–3274. https://doi.org/10.1002/bit.28191.

Park, S. Y., Hong, H. J., & Lee, H. J. (2022). Fabrication of cell spheroids for 3D cell culture and biomedical applications. *BioChip Journal, 17*(1), 24–43. https://doi.org/10.1007/S13206-022-00086-9.

Peirsman, A., Blondeel, E., Ahmed, T., Anckaert, J., Audenaert, D., Boterberg, T., ... De Wever, O. (2021). MISpheroID: A knowledgebase and transparency tool for minimum information in spheroid identity. *Nature Methods, 18*(11), 1294–1303. https://doi.org/10.1038/s41592-021-01291-4.

Phelan, M. A., Gianforcaro, A. L., Gerstenhaber, J. A., & Lelkes, P. I. (2019). An air bubble-isolating rotating wall vessel bioreactor for improved spheroid/organoid formation. *Tissue Engineering Part C: Methods, 25*(8), 479–488. https://doi.org/10.1089/ten.tec.2019.0088.

Pieters, V. M., Co, I. L., Wu, N. C., & McGuigan, A. P. (2021). Applications of omics technologies for three-dimensional in vitro disease models. *Tissue Engineering. Part C, Methods, 27*(3), 183–199. https://doi.org/10.1089/TEN.TEC.2020.0300.

Pinho, S. S., & Reis, C. A. (2015). Glycosylation in cancer: Mechanisms and clinical implications. *Nature Reviews. Cancer, 15*(9), 540–555. https://doi.org/10.1038/NRC3982.

Pinto, B., Henriques, A. C., Silva, P. M. A., & Bousbaa, H. (2020). Three-dimensional spheroids as in vitro preclinical models for cancer research. *Pharmaceutics, 12*(12), 1–38. https://doi.org/10.3390/PHARMACEUTICS12121186.

Pyonteck, S. M., Akkari, L., Schuhmacher, A. J., Bowman, R. L., Sevenich, L., Quail, D. F., ... Joyce, J. A. (2013). CSF-1R inhibition alters macrophage polarization and blocks glioma progression. *Nature Medicine, 19*(10), 1264–1272. https://doi.org/10.1038/nm.3337.

Qiao, L., Hu, S., Huang, K., Su, T., Li, Z., Vandergriff, A., ... Cheng, K. (2020). Tumor cell-derived exosomes home to their cells of origin and can be used as Trojan horses to deliver cancer drugs. *Theranostics, 10*(8), 3474–3487. https://doi.org/10.7150/thno.39434.

Radhakrishnan, P., Dabelsteen, S., Madsen, F. B., Francavilla, C., Kopp, K. L., Steentoft, C., ... Wandall, H. H. (2014). Immature truncated O-glycophenotype of cancer directly induces oncogenic features. *Proceedings of the National Academy of Sciences of the United States of America, 111*(39), E4066–E4075. https://doi.org/10.1073/PNAS.1406619111/SUPPL_FILE/PNAS.1406619111.SAPP.PDF.

Rai, A., Fang, H., Claridge, B., Simpson, R. J., & Greening, D. W. (2021). Proteomic dissection of large extracellular vesicle surfaceome unravels interactive surface platform. *Journal of Extracellular Vesicles, 10*(13), https://doi.org/10.1002/JEV2.12164.

Rana, S., Yue, S., Stadel, D., & Zöller, M. (2012). Toward tailored exosomes: The exosomal tetraspanin web contributes to target cell selection. *The International Journal of Biochemistry & Cell Biology, 44*(9), 1574–1584. https://doi.org/10.1016/J.BIOCEL.2012.06.018.

Rayamajhi, S., Sulthana, S., Ferrel, C., Shrestha, T. B., & Aryal, S. (2023). Extracellular vesicles production and proteomic cargo varies with incubation time and temperature. *Experimental Cell Research, 422*(2), 113454. https://doi.org/10.1016/J.YEXCR.2022. 113454.

Rebelo, S. P., Pinto, C., Martins, T. R., Harrer, N., Estrada, M. F., Loza-Alvarez, P., ... Brito, C. (2018). 3D-3-culture: A tool to unveil macrophage plasticity in the tumour microenvironment. *Biomaterials, 163*, 185–197. https://doi.org/10.1016/J.BIOMATERIALS.2018.02.030.

Redden, R. A., Iyer, R., Brodeur, G. M., & Doolin, E. J. (2014). Rotary bioreactor culture can discern specific behavior phenotypes in Trk-null and Trk-expressing neuroblastoma cell lines. *In Vitro Cellular & Developmental Biology. Animal, 50*(3), 188–193. https://doi.org/10.1007/S11626-013-9716-Z.

Rekker, K., Saare, M., Roost, A. M., Kubo, A. L., Zarovni, N., Chiesi, A., ... Peters, M. (2014). Comparison of serum exosome isolation methods for microRNA profiling. *Clinical Biochemistry, 47*(1–2), 135–138. https://doi.org/10.1016/J.CLINBIOCHEM. 2013.10.020.

Reynolds, D. S., Tevis, K. M., Blessing, W. A., Colson, Y. L., Zaman, M. H., & Grinstaff, M. W. (2017). Breast cancer spheroids reveal a differential cancer stem cell response to chemotherapeutic treatment. *Scientific Reports, 7*(1), https://doi.org/10.1038/S41598-017-10863-4.

Richter, M., Vader, P., & Fuhrmann, G. (2021). Approaches to surface engineering of extracellular vesicles. *Advanced Drug Delivery Reviews, 173*, 416–426. https://doi.org/10. 1016/j.addr.2021.03.020.

Robbins, P. D., & Morelli, A. E. (2014). Regulation of immune responses by extracellular vesicles. *Nature Reviews. Immunology, 14*(3), 195. https://doi.org/10.1038/NRI3622.

Robinson, N. B., Krieger, K., Khan, F., Huffman, W., Chang, M., Naik, A., ... Gaudino, M. (2019). The current state of animal models in research: A review. *International Journal of Surgery, 72*, 9–13. https://doi.org/10.1016/J.IJSU.2019.10.015.

Rodrigues, J., Heinrich, M. A., Teixeira, L. M., & Prakash, J. (2021). 3D in vitro model (r) evolution: Unveiling tumor–stroma interactions. *Trends in Cancer, 7*(3), 249–264. https://doi.org/10.1016/j.trecan.2020.10.009.

Saglam-Metiner, P., Devamoglu, U., Filiz, Y., Akbari, S., Beceren, G., Goker, B., ... Yesil-Celiktas, O. (2023). Spatio-temporal dynamics enhance cellular diversity, neuronal function and further maturation of human cerebral organoids. *Communications Biology, 6*(1), 1–18. https://doi.org/10.1038/s42003-023-04547-1.

Salemme, V., Centonze, G., Avalle, L., Natalini, D., Piccolantonio, A., Arina, P., ... Defilippi, P. (2023). The role of tumor microenvironment in drug resistance: Emerging technologies to unravel breast cancer heterogeneity. *Frontiers in Oncology, 13*, 1170264. https://doi.org/10.3389/FONC.2023.1170264/BIBTEX.

Sant, S., & Johnston, P. A. (2017). The production of 3D tumor spheroids for cancer drug discovery. *Drug Discovery Today: Technologies, 23*, 27–36. https://doi.org/10.1016/J. DDTEC.2017.03.002.

Santo, V. E., Estrada, M. F., Rebelo, S. P., Abreu, S., Silva, I., Pinto, C., ... Brito, C. (2016). Adaptable stirred-tank culture strategies for large scale production of multicellular spheroid-based tumor cell models. *Journal of Biotechnology, 221*, 118–129. https://doi.org/10.1016/j.jbiotec.2016.01.031.

Sapio, L., & Naviglio, S. (2022). Innovation through tradition: The current challenges in cancer treatment. *International Journal of Molecular Sciences, 23*(10), https://doi.org/10. 3390/IJMS23105296.

Savage, P., Blanchet-Cohen, A., Revil, T., Badescu, D., Saleh, S. M. I., Wang, Y. C., ... Ragoussis, J. (2017). A targetable EGFR-dependent tumor-initiating program in breast cancer. *Cell Reports, 21*(5), 1140–1149. https://doi.org/10.1016/J.CELREP.2017.10.015.

Schirrmacher, V. (2019). From chemotherapy to biological therapy: A review of novel concepts to reduce the side effects of systemic cancer treatment (review). *International Journal of Oncology, 54*(2), 407. https://doi.org/10.3892/IJO.2018.4661.

Schmelz, K., Toedling, J., Huska, M., Cwikla, M. C., Kruetzfeldt, L. M., Proba, J., ... Eggert, A. (2021). Spatial and temporal intratumour heterogeneity has potential consequences for single biopsy-based neuroblastoma treatment decisions. *Nature Communications, 12*(1), 1–13. https://doi.org/10.1038/s41467-021-26870-z.

Schwarz, R. P., Goodwin, T. J., & Wolf, D. A. (1992). Cell culture for three-dimensional modeling in rotating-wall vessels: An application of simulated microgravity. *Journal of Tissue Culture Methods: Tissue Culture Association Manual of Cell, Tissue, and Organ Culture Procedures, 14*(2), 51–57. https://doi.org/10.1007/BF01404744.

Serra, A. T., Serra, M., Silva, A. C., Brckalo, T., Seshire, A., Brito, C., ... Alves, P. M. (2019). Scalable culture strategies for the expansion of patient-derived cancer stem cell lines. *Stem Cells International, 2019*. https://doi.org/10.1155/2019/8347595.

Shaba, E., Vantaggiato, L., Governini, L., Haxhiu, A., Sebastiani, G., Fignani, D., ... Landi, C. (2022). Multi-omics integrative approach of extracellular vesicles: A future challenging milestone. *Proteomes, 10*(2), https://doi.org/10.3390/PROTEOMES10020012.

Shang, M., Soon, R. H., Lim, C. T., Khoo, B. L., & Han, J. (2019). Microfluidic modelling of the tumor microenvironment for anti-cancer drug development. *Lab on a Chip, 19*(3), 369–386. https://doi.org/10.1039/C8LC00970H.

Shi, X., Li, Y., Yuan, Q., Tang, S., Guo, S., Zhang, Y., ... Gao, D. (2022). Integrated profiling of human pancreatic cancer organoids reveals chromatin accessibility features associated with drug sensitivity. *Nature Communications, 13*(1), 1–16. https://doi.org/10.1038/s41467-022-29857-6.

Simão, D., Gomes, C. M., Alves, P. M., & Brito, C. (2022). Capturing the third dimension in drug discovery: Spatially-resolved tools for interrogation of complex 3D cell models. *Biotechnology Advances, 55*, 107883. https://doi.org/10.1016/J.BIOTECHADV.2021.107883.

Simpson, R. J., Kalra, H., & Mathivanan, S. (2012). ExoCarta as a resource for exosomal research. *Journal of Extracellular Vesicles, 1*(1), https://doi.org/10.3402/JEV.V1I0.18374.

Skardal, A., Devarasetty, M., Rodman, C., Atala, A., & Soker, S. (2015). Liver-tumor hybrid organoids for modeling tumor growth and drug response in vitro. *Annals of Biomedical Engineering, 43*(10), 2361. https://doi.org/10.1007/S10439-015-1298-3.

Smyth, T., Kullberg, M., Malik, N., Smith-Jones, P., Graner, M. W., & Anchordoquy, T. J. (2015). Biodistribution and delivery efficiency of unmodified tumor-derived exosomes. *Journal of Controlled Release, 199*, 145–155. https://doi.org/10.1016/J.JCONREL.2014.12.013.

Solon, J., Levental, I., Sengupta, K., Georges, P. C., & Janmey, P. A. (2007). Fibroblast adaptation and stiffness matching to soft elastic substrates. *Biophysical Journal, 93*(12), 4453–4461. https://doi.org/10.1529/BIOPHYSJ.106.101386.

Son, B., Lee, S., Youn, H. S., Kim, E. G., Kim, W., & Youn, B. H. (2017). The role of tumor microenvironment in therapeutic resistance. *Oncotarget, 8*(3), 3933. https://doi.org/10.18632/ONCOTARGET.13907.

Song, Y., Kim, J. S., Kim, S. H., Park, Y. K., Yu, E., Kim, K. H., ... Seo, H. R. (2018). Patient-derived multicellular tumor spheroids towards optimized treatment for patients with hepatocellular carcinoma. *Journal of Experimental and Clinical Cancer Research, 37*(1), 1–13. https://doi.org/10.1186/S13046-018-0752-0/TABLES/3.

Sung, H., Ferlay, J., Siegel, R. L., Laversanne, M., Soerjomataram, I., Jemal, A., & Bray, F. (2021). Global cancer statistics 2020: GLOBOCAN estimates of incidence and mortality worldwide for 36 cancers in 185 countries. *CA: A Cancer Journal for Clinicians, 71*(3), 209–249. https://doi.org/10.3322/CAAC.21660.

Sutherland, R. M., McCredie, J. A., & Inch, W. R. (1971). Growth of multicell spheroids in tissue culture as a model of nodular carcinomas. *JNCI: Journal of the National Cancer Institute.* https://doi.org/10.1093/JNCI/46.1.113.

Synowsky, S. A., Shirran, S. L., Cooke, F. G. M., Antoniou, A. N., Botting, C. H., & Powis, S. J. (2017). The major histocompatibility complex class I immunopeptidome of extracellular vesicles. *The Journal of Biological Chemistry, 292*(41), 17084. https://doi.org/10.1074/JBC.M117.805895.

Tamura, R., Uemoto, S., & Tabata, Y. (2017). Augmented liver targeting of exosomes by surface modification with cationized pullulan. *Acta Biomaterialia, 57,* 274–284. https://doi.org/10.1016/J.ACTBIO.2017.05.013.

Tang, Y.-T., Huang, Y.-Y., Zheng, L., Qin, S.-H., Xu, X.-P., An, T.-X., ... Wang, Q. (2017). Comparison of isolation methods of exosomes and exosomal RNA from cell culture medium and serum. *International Journal of Molecular Medicine, 40*(3), 834–844. https://doi.org/10.3892/ijmm.2017.3080.

Tauro, B. J., Greening, D. W., Mathias, R. A., Ji, H., Mathivanan, S., Scott, A. M., & Simpson, R. J. (2012). Comparison of ultracentrifugation, density gradient separation, and immunoaffinity capture methods for isolating human colon cancer cell line LIM1863-derived exosomes. *Methods (San Diego, Calif.), 56*(2), 293–304. https://doi.org/10.1016/J.YMETH.2012.01.002.

Thakuri, P. S., Ham, S. L., Luker, G. D., & Tavana, H. (2016). Multiparametric analysis of oncology drug screening with aqueous two-phase tumor spheroids. *Molecular Pharmaceutics, 13*(11), 3724–3735. https://doi.org/10.1021/ACS.MOLPHARMACEUT.6B00527/SUPPL_FILE/MP6B00527_SI_001.PDF.

Timmins, N. E., Maguire, T. L., Grimmond, S. M., & Nielsen, L. K. (2004). Identification of three gene candidates for multicellular resistance in colon carcinoma. *Cytotechnology, 46*(1), 9–18. https://doi.org/10.1007/s10616-005-1476-5.

Timmins, N. E., & Nielsen, L. K. (2007). Generation of multicellular tumor spheroids by the hanging-drop method. *Methods in Molecular Medicine, 140,* 141–151. https://doi.org/10.1007/978-1-59745-443-8_8.

Tirosh, I., Venteicher, A. S., Hebert, C., Escalante, L. E., Patel, A. P., Yizhak, K., ... Suvà, M. L. (2016). Single-cell RNA-seq supports a developmental hierarchy in human oligodendroglioma. *Nature, 539*(7628), 309–313. https://doi.org/10.1038/nature20123.

Usman, O. H., Zhang, L., Xie, G., Kocher, H. M., Hwang, C. il, Wang, Y. J., ... Irianto, J. (2022). Genomic heterogeneity in pancreatic cancer organoids and its stability with culture. *Npj Genomic Medicine, 7*(1), 1–10. https://doi.org/10.1038/s41525-022-00342-9.

Van Der Koog, L., Gandek, T. B., & Nagelkerke, A. (2022). Liposomes and extracellular vesicles as drug delivery systems: A comparison of composition, pharmacokinetics, and functionalization. *Advanced Healthcare Materials, 11*(5), 2100639. https://doi.org/10.1002/ADHM.202100639.

Van Deun, J., Mestdagh, P., Sormunen, R., Cocquyt, V., Vermaelen, K., Vandesompele, J., ... Hendrix, A. (2014). The impact of disparate isolation methods for extracellular vesicles on downstream RNA profiling. *Journal of Extracellular Vesicles, 3*(1), https://doi.org/10.3402/JEV.V3.24858.

Van Kessel, K. E. M., Zuiverloon, T. C. M., Alberts, A. R., Boormans, J. L., & Zwarthoff, E. C. (2015). Targeted therapies in bladder cancer: An overview of in vivo research. *Nature Reviews Urology, 12*(12), 681–694. https://doi.org/10.1038/nrurol.2015.231.

Verduin, M., Hoeben, A., De Ruysscher, D., & Vooijs, M. (2021). Patient-derived cancer organoids as predictors of treatment response. *Frontiers in Oncology, 11.* https://doi.org/10.3389/FONC.2021.641980.

Walker, S., Busatto, S., Pham, A., Tian, M., Suh, A., Carson, K., ... Wolfram, J. (2019). Extracellular vesicle-based drug delivery systems for cancer treatment. *Theranostics, 9*(26), 8001–8017. https://doi.org/10.7150/thno.37097.

Walsh, N. C., Kenney, L. L., Jangalwe, S., Aryee, K. E., Greiner, D. L., Brehm, M. A., & Shultz, L. D. (2017). Humanized mouse models of clinical disease. *Annual Review of Pathology, 12*, 187. https://doi.org/10.1146/ANNUREV-PATHOL-052016-100332.

Weiswald, L. B., Richon, S., Massonnet, G., Guinebretière, J. M., Vacher, S., Laurendeau, I., ... Dangles-Marie, V. (2013). A short-term colorectal cancer sphere culture as a relevant tool for human cancer biology investigation. *British Journal of Cancer, 108*(8), 1720–1731. https://doi.org/10.1038/bjc.2013.132.

Welsh, J. A., Goberdhan, D. C. I., O'Driscoll, L., Buzas, E. I., Blenkiron, C., Bussolati, B., ... Witwer, K. W. (2024). Minimal information for studies of extracellular vesicles (MISEV2023): From basic to advanced approaches. *Journal of Extracellular Vesicles, 13*(2), e12404. https://doi.org/10.1002/JEV2.12404.

Wensink, G. E., Elias, S. G., Mullenders, J., Koopman, M., Boj, S. F., Kranenburg, O. W., & Roodhart, J. M. L. (2021). Patient-derived organoids as a predictive biomarker for treatment response in cancer patients. *Npj Precision Oncology, 5*(1), 1–13. https://doi.org/10.1038/s41698-021-00168-1.

World Health Organization. (2024). *Global cancer burden growing, amidst mounting need for services.* https://www.who.int/news/item/01-02-2024-global-cancer-burden-growing-amidst-mounting-need-for-services.

Wu, Y., Zhou, Y., Qin, X., & Liu, Y. (2021). From cell spheroids to vascularized cancer organoids: Microfluidic tumor-on-a-chip models for preclinical drug evaluations. *Biomicrofluidics, 15*(6), 61503. https://doi.org/10.1063/5.0062697/13900226/061503_1_ACCEPTED_MANUSCRIPT.PDF.

Xue, C., Shen, Y., Li, X., Li, B., Zhao, S., Gu, J., ... Zhao, R. C. (2018). Exosomes derived from hypoxia-treated human adipose mesenchymal stem cells enhance angiogenesis through the PKA signaling pathway. *Stem Cells and Development, 27*(7), 456–465. https://doi.org/10.1089/SCD.2017.0296.

Yamada, K. M., & Cukierman, E. (2007). Modeling tissue morphogenesis and cancer in 3D. *Cell, 130*(4), 601–610. https://doi.org/10.1016/J.CELL.2007.08.006.

Yang, Q., Li, M., Yang, X., Xiao, Z., Tong, X., Tuerdi, A., ... Lei, L. (2023). Flourishing tumor organoids: History, emerging technology, and application. *Bioengineering & Translational Medicine, 8*(5), e10559. https://doi.org/10.1002/BTM2.10559.

Youn, B. S., Sen, A., Behie, L. A., Girgis-Gabardo, A., & Hassell, J. A. (2006). Scale-up of breast cancer stem cell aggregate cultures to suspension bioreactors. *Biotechnology Progress, 22*(3), 801–810. https://doi.org/10.1021/BP050430Z.

Zanoni, M., Piccinini, F., Arienti, C., Zamagni, A., Santi, S., Polico, R., ... Tesei, A. (2016). 3D tumor spheroid models for in vitro therapeutic screening: A systematic approach to enhance the biological relevance of data obtained. *Scientific Reports, 6*(1), 1–11. https://doi.org/10.1038/srep19103.

Zhang, A., Miao, K., Sun, H., & Deng, C. X. (2022). Tumor heterogeneity reshapes the tumor microenvironment to influence drug resistance. *International Journal of Biological Sciences, 18*(7), 3019. https://doi.org/10.7150/IJBS.72534.

Zhou, Y., Jiang, X., Wang, X., Huang, J., Li, T., Jin, H., & He, J. (2023). Promise of spatially resolved omics for tumor research. *Journal of Pharmaceutical Analysis, 13*(8), 851–861. https://doi.org/10.1016/J.JPHA.2023.07.003.

Zhou, Y., Yamamoto, Y., Takeshita, F., Yamamoto, T., Xiao, Z., & Ochiya, T. (2021). Delivery of miR-424-5p via extracellular vesicles promotes the apoptosis of MDA-MB-231 TNBC cells in the tumor microenvironment. *International Journal of Molecular Sciences, 22*(2), 844. https://doi.org/10.3390/IJMS22020844.

Zhu, Y., Knolhoff, B. L., Meyer, M. A., Nywening, T. M., West, B. L., Luo, J., ... De Nardo, D. G. (2014). CSF1/CSF1R blockade reprograms tumor-infiltrating macrophages and improves response to T-cell checkpoint immunotherapy in pancreatic cancer models. *Cancer Research, 74*(18), 5057–5069. https://doi.org/10.1158/0008-5472.CAN-13-3723/657718/AM/CSF1-CSF1R-BLOCKADE-REPROGRAMS-TUMOR-INFILTRATING.

Further reading

Abdollahi, S., Sara Abdollahi, C., & Broadway, N. (2021). Extracellular vesicles from organoids and 3D culture systems. *Biotechnology and Bioengineering, 118*(3), 1029–1049. https://doi.org/10.1002/BIT.27606.

Jurj, A., Pasca, S., Braicu, C., Rusu, I., Korban, S. S., & Berindan-Neagoe, I. (2022). Focus on organoids: cooperation and interconnection with extracellular vesicles—Is this the future of in vitro modeling? *Seminars in Cancer Biology, 86*, 367–381. https://doi.org/10.1016/J.SEMCANCER.2021.12.002.

Szvicsek, Z., Oszvald, Á., Szabó, L., Sándor, G. O., Kelemen, A., Soós, A.Á., Pálóczi, K., Harsányi, L., Tölgyes, T., Dede, K., Bursics, A., Buzás, E. I., Zeöld, A., & Wiener, Z. (2019). Extracellular vesicle release from intestinal organoids is modulated by Apc mutation and other colorectal cancer progression factors. *Cellular and Molecular Life Sciences, 76*(12), 2463–2476. https://doi.org/10.1007/S00018-019-03052-1/FIGURES/5.

Uddin, M. J., Mohite, P., Munde, S., Ade, N., Oladosu, T. A., Chidrawar, V. R., Patel, R., Bhattacharya, S., Paliwal, H., & Singh, S. (2024). Extracellular vesicles: The future of therapeutics and drug delivery systems. *Intelligent Pharmacy.* https://doi.org/10.1016/J.IPHA.2024.02.004.

Zhang, H., Zhang, N., Zhou, Y., Maria Jó, T., Maria Erdmá Nska, P., Stachowicz-Karpí Nska, A., ... Jó, W. (2024). Exosomes—Promising carriers for regulatory therapy in oncology. *Cancers, 16*(5), 923. https://doi.org/10.3390/CANCERS16050923.

CHAPTER SIX

Crosstalk between tumor and microenvironment: Insights from spatial transcriptomics

Malvika Sudhakar, Harie Vignesh, and Kedar Nath Natarajan*
DTU Bioengineering, Technical University of Denmark, Kongens Lyngby, Denmark
*Corresponding author. e-mail address: kenana@dtu.dk

Contents

1.	Introduction	188
2.	Cancer heterogeneity and landscape	189
	2.1 Unique cancer genomes	189
	2.2 Intertumor heterogeneity	191
	2.3 Intratumor heterogeneity	192
3.	Spatial transcriptomics and insights	192
	3.1 Spatial technologies	192
	3.2 Analysis of spatial data	196
4.	Evolution of tumor	201
	4.1 Clonal architecture	201
	4.2 Tumor evolution types	205
	4.3 Insights from spatial transcriptomics	206
	4.4 Metastasis	207
5.	Tumor microenvironment (TME)	208
	5.1 Dynamics of cancer associated fibroblasts in cancer progression	209
	5.2 Insights about the tumor microenvironment from single cell and ST studies	210
6.	Integrating ST and scRNA-seq	212
7.	Challenges and future directions	213
	Acknowledgments	214
	Conflict of interest statement	214
	References	214

Abstract

Cancer is a dynamic disease, and clonal heterogeneity plays a fundamental role in tumor development, progression, and resistance to therapies. Single-cell and spatial multimodal technologies can provide a high-resolution molecular map of underlying genomic, epigenomic, and transcriptomic alterations involved in inter- and intra-tumor heterogeneity and interactions with the microenvironment. In this review, we

Advances in Cancer Research, Volume 163
ISSN 0065-230X, https://doi.org/10.1016/bs.acr.2024.06.009
Copyright © 2024 Elsevier Inc. All rights are reserved, including those for text and data mining, AI training, and similar technologies.

187

provide a perspective on factors driving cancer heterogeneity, tumor evolution, and clonal states. We briefly describe spatial transcriptomic technologies and summarize recent literature that sheds light on the dynamical interactions between tumor states, cell-to-cell communication, and remodeling local microenvironment.

1. Introduction

Cancer, with its intricate web of genetic mutations, signaling pathways, and microenvironmental interactions, presents a formidable challenge in both diagnosis and therapy. The cellular heterogeneity underscores the complexity of this disease, with tumors exhibiting diverse genomic, epigenomics, transcriptomic, and proteomic landscapes (Stratton, Campbell, & Futreal, 2009; Vogelstein et al., 2013). Such variability not only poses significant challenges for precision medicine but also emphasizes on the urgent need to decipher the underlying mechanisms driving tumor evolution.

Understanding the intricacies of tumor evolution is paramount for designing better cancer therapeutics and in addressing the vexing issue of treatment resistance (Schoenfeld et al., 2020; Stewart et al., 2020; Zhao et al., 2021). As tumors evolve, clones undergo dynamic changes, acquiring additional mutations and adaptive phenotypes that confer resistance to conventional therapies (Aas et al., 1996; Bozic & Nowak, 2014; Misale et al., 2012; Robinson et al., 2013; Yun et al., 2008). There is an increasing need and rapid adoption of state-of-the-art spatial multimodal profiling technologies, which capture the intricate architecture of tumors and their surrounding microenvironment alongside individual molecular readouts (transcriptome, epigenome, proteome, etc.). Such spatial methods can provide novel insights into the distribution of resistant clones, different clonotypes within sampled tissue, the dynamics of tumor cell migration, and cell-to-cell communication (both within tumors and with microenvironment). The spatially resolved datasets also offer a window into the selective pressures exerted by the tumor microenvironment (TME), revealing how cancer cells exploit spatial niches to evade immune surveillance and therapeutic interventions. By deciphering these spatial dynamics, researchers can identify novel vulnerabilities and therapeutic targets, paving the way for the development of more effective treatment strategies.

In essence, the application of spatial data to study cancers not only expands our understanding of tumor heterogeneity, clonal architecture and evolution; but can provide a deeper appreciation of the spatially localized niches, their adaptive responses to therapeutic interventions and ultimately guide the development of more personalized and effective cancer therapies.

In this review, we delve into the intricate landscape of cancer genomes and its profound impact on shaping clonal architecture. We provide an overview of imaging-based and sequence-based spatial transcriptomics (ST) methods and applications to unravel mechanisms behind tumor progression and evolution. We highlight studies with scRNA-seq, spatial and integrative computational analysis that provide new insights into the factors driving the dynamic processes underlying tumor states, interplay between tumor cells and their microenvironment and role of cell-to-cell communication in tumor growth, progression and metastasis.

2. Cancer heterogeneity and landscape
2.1 Unique cancer genomes

Cancer genomes exhibit remarkable diversity, each presenting a distinctive landscape shaped by genomic instability. This instability drives the accumulation of genomic changes, including mutations, gene fusions, and chromothripsis. The complexity is expounded by a feedback loop of such alterations, heritable epigenomic changes drive varied transcriptional readouts within individual cells and clones thereby contributing to tumor evolution and adaptation (Fig. 1).

2.1.1 Mutational landscapes

Cancer genomic landscapes are intricate tapestries of genetic alterations, revealing a multitude of mutations that drive tumorigenesis and shape tumor behavior. These mutations span a wide spectrum, including point mutations, insertions, deletions, and span one or more nucleotides (Fig. 1 (**Mutational**)). Within this landscape, a subset of mutations called driver mutations (Stratton et al., 2009; Vogelstein et al., 2013) stand out as pivotal alterations that confer selective growth advantages to cancer cells, promoting their clonal expansion and driving tumor progression. In contrast, passenger mutations are incidental alterations that do not directly contribute to tumor growth but accompany the expansion of cancerous clones. By mapping these mutational landscapes, researchers uncover key driver mutations (Dees et al., 2012; Tokheim & Karchin, 2019) in driver genes (tumor suppressive (TSG) or oncogenes (OG)) (Dinstag & Shamir, 2020;

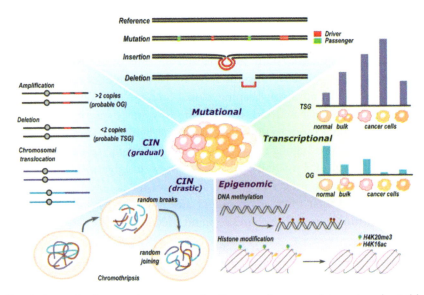

Fig. 1 **Genomic changes in cancer cells.** Cancer cells accumulate various heritable and non-heritable genetic and epigenetic changes, including mutations, copy number variations, gene fusions and chromothripsis. These heritable changes can drive tumor evolution and can be carried forward as passenger events within the individual subclones. Profiling of cellular populations by bulk methods (DNA-seq, RNA-seq, ATAC-seq, etc.) masks underlying cellular heterogeneity. Single-cell and spatial multimodal technologies can help capture heterogeneity, cell types and states but also distinguish cancer subclones and interactions with other cell types.

Lawrence et al., 2013; Sudhakar, Rengaswamy, & Raman, 2022a, 2022b) responsible for initiating and promoting cancer growth (Fig. 1).

The identification of recurrent mutations across different cancer types provides insights into common underlying pathways (Takeuchi, Doi, Ikari, Yamamoto, & Furukawa, 2018; Wooster et al., 1995; Zabransky et al., 2015) and potential therapeutic targets (Hsu et al., 2016; Paez et al., 2004). Spatial studies in lung adenocarcinomas (LUADs) (Dietz et al., 2017) and breast cancer (Yates et al., 2015) reveal driver gene mutations show wide heterogeneity between samples and within the tumor (described in §4.1.1). The selective advantage conferred by driver mutations is often minuscule, necessitating the occurrence of multiple driver events for significant impact (Bozic et al., 2010). Tumors typically accumulate a series of these genetic alterations to drive their growth and survival (Tomasetti, Marchionni, Nowak, Parmigiani, & Vogelstein, 2015), underlining the complex and multifactorial nature of cancer evolution.

2.1.2 Fusion genes

Fusion genes, arising from chromosomal rearrangements, represent another hallmark of cancer genomic landscapes. These chimeric genes result from the fusion of two previously separate genes, often leading to the aberrant activation of OG or the disruption of tumor suppressor genes. An example of fusion genes is the translocation event between chromosomes 22 and 9 in chronic myelogenous leukemia (CML) patients, resulting in the formation of the BCL-ABL2 fusion gene (Rowley, 1973). Similarly, the EML4-ALK fusion gene was identified in Asian populations with non-small cell lung cancer (NSCLC) (Soda et al., 2007). The rate of fusion events in different cancer types, for a panel of 6 fusion genes, is found to vary between 12.9% and 85% (Yu et al., 2019). Fusion genes play a significant role in driving tumorigenesis by promoting cell proliferation, inhibiting apoptosis, or enhancing metastatic potential. Furthermore, the detection of fusion genes has important implications for diagnosis, prognosis, and targeted therapy in various cancer types (Kohno et al., 2013; Ross et al., 2016; Umemoto & Sunakawa, 2021).

2.1.3 Chromosomal instability

Chromosomal instability (CIN) and chromothripsis are intertwined processes driving genomic complexity in cancer (Fig. 1 **(CIN)**). CIN is marked by increased chromosomal alterations and aneuploidy, which stems from faulty DNA repair and mitotic mechanisms. CIN leads to chromosomal translocation, causing deletion and amplification of genetic segments and, consequently, genes. Copy number variations are conserved for some genes such as FAM58A across cancer types, while other genes are specific to cancer types (Drews et al., 2022; Yan et al., 2023). Chromothripsis, a dramatic event involving localized shattering and rearrangement of chromosomal segments, contributes to extensive genomic rearrangements (Meyerson & Pellman, 2011) (Fig. 1). While individual tumors (stratified by aggressiveness and grades) and their evolution have been equated to gradual Darwinian evolution, tumors are also interspersed with sudden drastic events like chromothripsis (Vendramin, Litchfield, & Swanton, 2021). Nonetheless, both phenomena fuel tumor heterogeneity and clonal evolution.

2.2 Intertumor heterogeneity

Intertumor heterogeneity refers to the variability observed between tumors of the same cancer type across different individuals. Despite belonging to the same cancer subtype, only a small fraction of patients may share specific driver mutations, leading to diverse molecular profiles and clinical outcomes. In breast cancer, the presence or absence of hormone receptors such as HER2 (HER2-positive or HER2-negative) can significantly impact treatment strategies and prognosis (Meric-Bernstam & Hung, 2006). Similarly, in CML, while the majority of patients harbor the characteristic BCR-ABL fusion gene, variations in disease progression and response to therapy highlight the underlying intertumor heterogeneity (An et al., 2010). Intertumor heterogeneity is the major reason for the need for personalized medicine.

2.3 Intratumor heterogeneity

Intratumor heterogeneity (ITH) underscores the diverse landscape within tumors, encompassing genetic variability transmitted to future cell generations and non-genetic variability through epigenetic changes. In breast cancer patients with minimal hereditary mutations (<10%), targeted treatments initially show promise but often result in tumor recurrence, dominated by resistant clones (Bai, Ni, Beretov, Graham, & Li, 2018). The presence of self-renewing cancer stem cells further exacerbates cellular heterogeneity, while their strong differentiation potential contributes to multiple tumor malignancies, metastasis, tumor evolution, and multi-drug resistance (Yang et al., 2020). Tumor clones interacting with the diverse normal stromal cells affect each other to remodel the local microenvironmental, contributing further to ITH (discussed in detail in §5).

3. Spatial transcriptomics and insights
3.1 Spatial technologies

ST methods represent a powerful approach to quantify RNA expression while preserving spatial information within tissues or organisms. The field of spatial technologies has seen an incredible development of highly-sensitive methods capable of profiling multiple targets (transcripts, proteins, etc.) at multi-, single-cell and sub-cellular resolution (Fig. 2). These methods can be broadly categorized into image-based and sequencing-based approaches, and further stratified on input tissue material, resolution and throughput (Fig. 2). As a prerequisite, ST approaches require tissue slices of relevant thickness and are stained with hematoxylin and eosin on matched or adjacent sections. Here we briefly introduced the methods and highlight some of the differences between them (Fig. 2). These methods have their own merits and

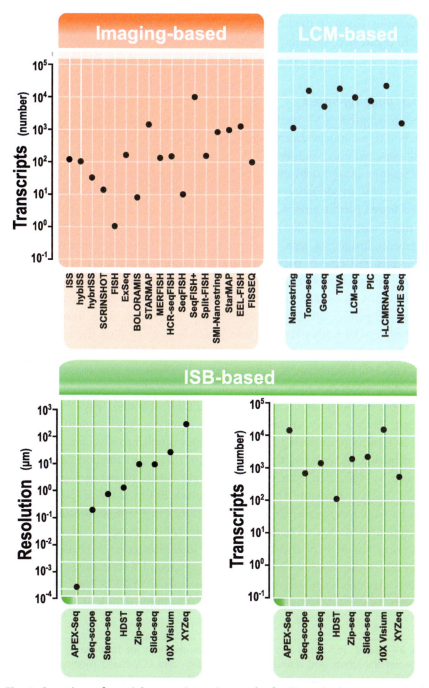

Fig. 2 Overview of spatial transcriptomics methods. ST methods are categorized into imaging-based (including in situ hybridization) and sequencing-based approaches

(Continued)

limitations, based on the underlying principle, and we refer readers to recent method-oriented reviews (Moffitt, Lundberg, & Heyn, 2022; Moses & Pachter, 2022; Vandereyken, Sifrim, Thienpont, & Voet, 2023).

3.1.1 Imaging-based

The imaging-based ST methods capture signals generated upon RNA hybridization to a complementary fluorescent tagged oligonucleotide (20–30 mers), and measured using high-resolution microscopy. While the number of targets (throughput of approach) and specific labeling approaches may vary, the general steps include hybridization of single- or sequential oligonucleotides to fixed, fixed-frozen paraffin embedded (FFPE) tissue sections, labeling of general DNA or membrane markers (for segmentation), followed by washing out unbound fragments, imaging to capture signals from either single- or branched-amplified molecules, real-time or post-processing of images and data storage. Depending on the approach, optical barcodes are employed to distinguish between different RNA molecules and high-resolution microscopy allows for deconvolution of precise cellular and sub-cellular localization. The imaging-based methods can be further classified into in situ sequencing (ISS) (Ke et al., 2013) and fluorescence in situ hybridization (FISH)-based methods.

The FISH-based methods label target RNA using fluorescently labeled oligonucleotides that hybridize to complementary sequences, conjugated to a fluorophore. Single-molecule FISH (smFISH) (Femino, Fay, Fogarty, & Singer, 1998) is a popular method that quantifies gene expression in fixed cells but is limited to analyzing handful of genes simultaneously, due to limited combination of available fluoroprobes. smFISH-based methods are further limited by diffraction limit toward resolving closely placed molecules. SeqFISH (Lubeck, Coskun, Zhiyentayev, Ahmad, & Cai, 2014) improves upon this by employing sequential rounds of smFISH in a semi-automated custom microfluidics device to generate a unique color barcode combination for detecting hundreds of transcripts within individual cells. A challenge with SeqFISH is that measurement errors can compound with combinatorial barcodes. Another approach multiplex error-robust FISH (Chen, Boettiger, Moffitt, Wang, & Zhuang, 2015) also applies multiple rounds of hybridization

Fig. 2—Cont'd (ISB-based and LCM-based). The methods are stratified by number of detected transcripts and resolution. The imaging- and LCM-based methods have an almost single-cell resolution but vary in number of transcripts detected. The sequencing-based methods have variable resolution (spot size, μm) and number of transcripts detected.

using a probe set with modified Hamming distance-4 scheme for error correction, followed by primary and secondary branch amplification to quantification of several hundreds, thousands of transcripts, and error correction.

The in situ sequencing methods, such as ISS (Ke et al., 2013) and BOLORAMIS (Liu et al., 2021), utilize padlock probes that hybridize to the RNA molecule of interest. These probes contain single-stranded oligonucleotides with 3′ and 5′ ends that hybridize to pre-selected target genes, followed by amplification by rolling circle amplification (Baner, Nilsson, Mendel-Hartvig, & Landegren, 1998). While these methods can offer specific and bright signals, they have a relatively lower throughput, i.e., the number of captured, quantified targets, and typically detect around 100 RNAs. STARmap (Wang et al., 2018) improves upon this by bypassing the cDNA synthesis step using SNAIL probes, allowing for the quantification of hundreds to thousands of genes simultaneously. Similarly, Gap-filled ISS incorporates RNA sequences into the barcode (instead of padlock probes) thereby expanding its throughout and broader applicability to complex tissues (Ke et al., 2013). There are also non-targeted based methods (without transcript-specific probes) such as FIS-SEQ (Lee et al., 2014) and ExSeq (Alon et al., 2021), which can quantify all transcripts in a sequence-agnostic manner (also including next-generation sequencing). These non-target sequence-based methods allow the detection of different targets, splice junctions and any single-nucleotide variations.

3.1.2 Sequencing-based

Intertumor and ITH are key components of cancer progression and the lack of response for treatments in patients. Tumor heterogeneity is a phenomenon where a tumor is populated by different sub-populations of cells. This creates a dire need to explore the spatial distribution of different sub-population of cells within and across tumors. The advent of sequencing-based spatially resolved transcriptomics has helped in gaining crucial insights into the progression of cancer. Recent technological advancements have advanced our understanding of the physical localization of the components of tumor heterogeneity (Yu, Jiang, & Wu, 2022).

Sequencing-based ST approaches typically include an ISB (in situ barcoding) step containing either a targeted probe-set or array of barcoded OligodT probes. The ST approaches generally utilize a two-step process, first where cells/nuclei are permeabilized to release and captured RNA in situ with spatial barcodes, and thereafter the cDNA extraction, amplification and counting is performed ex situ either using proprietary devices or next generation sequencing (NGS). The in situ part involves tagging RNA molecules with spatially barcoded oligonucleotides directly within the tissue sample, allowing for the identification of the spatial origin of each transcript and integration during downstream sequencing analysis. Based on the barcoding methods, ISBs can be further subdivided. The first group includes Digital Spatial Profiling and ZipSeq which involves the spatially resolved analysis of gene expression within tissue samples using uniquely barcoded oligonucleotide probes. It allows for the simultaneous detection of multiple transcripts (and proteins) within specific regions of the tissue. Here the DNA barcodes are extracted from a specific tissue location, which makes it suitable for transcriptome-wide comparison across tissues. The second group includes NGS-based ISB techniques like APEX-seq (Fazal et al., 2019), Seq-Scope (Cho et al., 2021), Stereo-Seq (Chen et al., 2022), PIXEL-Seq (Fu et al., 2021) and HDST (Vickovic et al., 2019) that can provide multi-cellular, single-cell and even subcellular resolution. Some of the ISB techniques include the Solid-Based Capture Method for in situ indexing, containing a whitelist of possible barcodes for error correction and minimizing spurious spatial assignment.

LCM (laser capture microdissection) is an important method for pre-selecting tissue regions from rare samples and can be coupled to ST approaches. In LCM, a lazer beam is employed to cut out a tissue region of interest of desired thickness under a microscope. LCM-based methods possess a range of applicability and methods like the TIVA seq and NICHE seq can be performed on live cells. One of the major limitations of the LCM-based methods is the labor intensive workflows and lower throughput to sequencing-based ST approaches. Several new ST methods such as are constantly developed to overcome the challenges and limitations of existing methods, including improved probe architecture, signal generation, error correction and applicability to wide variety of tissues such as FFPE.

3.2 Analysis of spatial data

The dataset from an ST experiment typically comprises of a map of spatial coordinates (i.e. X–Y positions with barcodes sequences) and a corresponding de-multiplexed gene expression matrix. This high-dimensional dataset consisting of counts or unique molecular identifiers individual detected features (genes/transcripts) corresponding to a spatial position, which maps to multi-cellular, single-cell or sub-cellular resolution. For ST methods with multi-cellular resolution, discrimination of homotypic (several cells of same cell type in a spot) versus heterotypic (several cells of different types in a sport) is a critical challenge. The ultimate aim of ST data analysis is to amalgamate the spatial distribution of gene-expression profiles across a tissue from the experimentally captured data. The ST data analysis and interpretation is considered as a de novo process, serving the dual purpose of hypothesis testing as well as data-driven exploratory analysis for hypothesis generation.

The ST analysis can be categorized as a two-stage process, (i) pre-processing and (ii) downstream analysis. We sumamarise the different ST analysis steps alongside tools in Table 1. For both imaging and sequencing-based methods, the tissue image has to be processed to deconvolute expression into specific resolution based on the respective methods (multicellular, single-cell or sub-cellular). The experimental ST steps include staining with hematoxylin and eosin, ssDNA, DAPI or nuclear/membrane dye that enables identification of cell/nuclei borders (segmentation). This segmentation corresponds to the specific resolution of the ST methods and is critical for mapping the expression values (pixels in imaging; reads from sequencing) onto spatial coordinates, resulting in a cell/spot-level gene expression matrix. This pre-processing step also enables elimination of low-quality data from spatial barcoded regions as well as filtering low-quality transcripts and technical errors. Often smoothing algorithms are applied to increase sensitivity by averaging expression over pre-defined smoothing window between adjacent coordinates, minimizing technical expression noise. The pre-processing serves as an initial quality control step for assessment of quality of the ST experiment. Depending upon the quality, the initial biological attributes of the data can be appreciated. For instance, lower number of transcripts per spot in the pre-processed gene expression matrix may denote experimental limitations, uneven library complexity which can be driven by a combination of technical variation and true biological factors.

The downstream analysis of spatial data generally includes assessment of a combination of technical features and biological variation in the dataset. A key primary step in ST analysis is coupling of spatial neighborhood data with expression data, followed by scaling data to uniform units (TPM: transcripts per million, same mean and variance across spots), regression (removing mitochondrial, ribosomal, cell cycle and unwanted variance in datasets) and normalization (non-negative matrix factorization among others). The biological interpretation of ST data is performed initially by reducing the high-dimensional dataset to lower interpretative dimensions for visualization and further analysis. Approaches such as principal component analysis (PCA), t-distributed stochastic neighbor

Table 1 Spatial transcriptomics methods used for different stages of data analysis and interpretation.

Analysis steps	Tools	References
Pre-processing	Seurat, DeepBlink, BarDensr, graph-ISS	Hao et al. (2021), Eichenberger, Zhan, Rempfler, Giorgetti, and Chao (2021), Chen et al. (2021), and Partel et al. (2020)
Cell segmentation	DeepCell, Sparcle, Spage2vec, JSTA	Prabhakaran (2022), Partel and Wählby (2021), and Littman et al. (2021)
Batch effect correction	Seurat, harmony	Hao et al. (2021) and Korsunsky et al. (2019)
Dimensionality reduction and clustering	Seurat, PCA, t-SNE, weighted PCA	Hao et al. (2021), Maćkiewicz and Ratajczak (1993), van der Maaten and Hinton (2008), and Hong, Yang, Fessler, and Balzano (2023)
Spatial cell type annotation	Seurat, Cell2location, CellTrek	Hao et al. (2021), Kleshchevnikov et al. (2022), and Wei et al. (2022)
Spatially variable genes	SpatialIDE, Seurat	Svensson, Teichmann, and Stegle (2018) and Hao et al. (2021)
Identification of gene patterns	CellTrek, MERINGUE	Wei et al. (2022) and Miller, Bambah-Mukku, Dulac, Zhuang, and Fan (2021)
Mapping of spatial regions	RESEPT, CellTrek	Chang et al. (2022) and Wei et al. (2022)
Cell–cell interactions	SpaOTsc, SVCA	Cang and Nie (2020) and Arnol et al. (2019)
Gene–gene interactions	scHOT	Ghazanfar et al. (2020)
Spatial trajectory analysis	SPATA, StLearn	Kueckelhaus et al. (2020) and Pham et al. (2023)

embedding (t-SNE) and uniform manifold approximation and projection (UMAP) provide both linear and non-linear approach to extract meaningful variance within ST datasets UMAP (McInnes, Healy, & Melville, 2018). To reveal the patterns within the data, the expression values are used to cluster spots (Louvain, Leiden, k-means or hierarchical clustering) to capture cell types, states or distinct spatial regions. The annotation of these spatial features is performed using a combination of data-driven analysis (enriched marker genes) and prior biological knowledge.

For ST methods with sub-cellular resolution, cell type estimation is usually performed by a maximum likelihood-estimate (smoothing, normalized expression, local density) that links expression of neighboring spots over segmented borders. The challenge is often annotating transcripts localization to sub-cellular organelles and compartments. In ST methods with multicellular resolution, a key challenge is to infer the cell-type composition of each spot (homo- vs heterotypic spots), which is crucial for biological interpretation. This can be overcome either by performing matched scRNA-seq or integrating with available scRNA-seq data and applying computational methods (linear regression, graph-based neighborhood estimation, local density mapping and deep-learning) for robust multiple cell-type estimates within a spot (Cable et al., 2022; Kleshchevnikov et al., 2022). A variety of further downstream analyses can further provide information of cell states, types or regional specification, which include non-negative matrix factor analysis, cell-to-cell correlation, gene ontology, gene set enrichment analysis and newer advanced methods which can sensitively prioritize features for pattern extraction at sub-cellular resolution (Kleshchevnikov et al., 2022; Zhao et al., 2021).

3.2.1 Cell-to-cell communication

Within complex tissues, cellular communication within and across niches ensures a coordinated responses as well as for maintaining homeostasis. The cell-to-cell communication enables two way communication (i.e. *operator and receiver cell*), for biochemical, signaling cues within a population of cells for efficient cell type function and phenotype. The molecules that mediate both long (paracrine, endocrine) and short-range communication (autocrine, juxtacrine) include signaling factors, ligands, receptors, cell-secreted proteins, structural proteins, ions and metabolites (Armingol, Officer, Harismendy, & Lewis, 2021). In healthy cells, this two-way communication ensures tightly regulated and coordinated response for cell differentiation, development (Camp et al., 2017), apoptosis, elimination of unfit cells as well as robust activation and response to fight disease.

The development of novel algorithms for scRNA-seq has resulted in new approaches to infer cell-to-cell communication that aims to integrate gene expression from dissociated single-cells within a population, with available ligand–receptor information. Such methods first collating pairs of interacting ligand–receptor from curated databases (Türei et al., 2021; Türei, Korcsmáros, & Saez-Rodriguez, 2016; Vento-Tormo et al., 2018), followed by aggregating their expression values within the experiment and modeling a background enrichment level for significance testing within scRNA-seq datasets. These estimates are compared via network analysis to suggest enriched receptor–ligand pair across individual cells as a proxy for biologically meaningful interactions at the single-cell level. The cell-to-cell inference methods can be broadly classified into three types. The statistical-enrichment based methods (CellphoneDB (Efremova, Vento-Tormo, Teichmann, & Vento-Tormo, 2020), CellChat (Jin et al., 2021)) quantify the aggregated background interaction enrichment between expressed ligand–receptor pairs using a null hypothesis, followed by identification of enriched and significant interacting cells. The network-enrichment based methods (NicheNet (Browaeys, Saelens, & Saeys, 2020), Connectome (Raredon et al., 2022), NATMI (Hou, Denisenko, Ong, Ramilowski, & Forrest, 2020), CytoTalk (Hu, Peng, Gao, & Tan, 2021), Domino (Cherry et al., 2021), Scriabin (Wilk, Shalek, Holmes, & Blish, 2023)) model ligand–receptor interactions as a graph network, with edge weights to rank interactions and can also overlay regulatory information within the graphs network. The ST based methods (Garcia-Alonso et al., 2021), Giotto (Dries et al., 2021), SpaOTsc (Cang & Nie, 2020), MISTY (Tanevski, Flores, Gabor, Schapiro, & Saez-Rodriguez, 2022), stLearn (Pham et al., 2023) integrate expression based inference with available spatial information to prioritize cell-to-cell interactions between cell types in the same spatial microenvironment.

While the above inference methods have been applied to a variety of biological contexts, we would like to highlight a few examples of particular interest. Applying ST with scRNA-seq on matched primary pancreatic tumors, Moncada et al., capture spatially restricted tumor and microenvironmental niches comprising of cancer clones with distinct pancreatic and immune sub-populations (Moncada et al., 2020). Applying integrative computational mapping analysis in 10 ST pancreatic ductal adenocarcinoma (PDAC) samples, the authors uncover co-localization of cancer subpopulations with inflammatory IL-6 secretory fibroblasts, expressing signaling genes involved in stress-response. Ji et al., performed scRNA-seq and ST on cutaneous squamous cell carcinoma (cSCC) to reveal a unique tumor-specific keratinocyte (TSK) population which re-models fibrovascular niche via extensive autocrine and paracrine interaction with cancer-associated fibroblasts (CAFs) (L-R pairs: MMP9-LRP1, TNC-SDC1), endothelial cells (TFPI, FN1, THBS1), tumor-associated macrophages and MDSCs via integrins (ITGA3 and ITGB1) (Ji et al., 2020). Notably, TSK ligands (MMP9, TNC, SERPINE2, CALM1) in tumors remodel local microenvironment via induction of CAFs within a local niche (Ji et al., 2020). In a recent work, Wilk and colleagues describe Scriabin, a computational approach to integrate scRNA-seq and ST datasets for inferring cell-to-cell interaction modules within cell pairs (Wilk et al., 2023). The authors identify an interaction module between fibroblast and intestinal stem cells that is critical for maintaining stem cell niche. Applying Scriabin to longitudinal ex vivo air–liquid interface profiling of SARS-CoV2 infected human bronchial epithelial cells, the authors observed that uninfected basal, ciliated and club cells produced higher IL1B circuit-initiating ligand and pro-inflammatory cytokines which were responsible for influencing infected cells and resultant tissue remodeling. The infected basal cells subsequently amplify the inflammatory response by TGFB1-mediated upregulation of JAG1 and NOTCH signaling within single-cell pairs.

Most cell-to-cell inference methods require prior biological knowledge (ligand–receptor database) and assume that expression within single-cells is a suitable proxy for true activity between interacting cells. The scRNA-seq or ST datasets provide a static snapshot of population-level crosstalk, and often the high dynamical nature of ligand–receptor interactions can be over-interpreted. Due to the abundance of missing values (zero or NA) inherent in scRNA-seq, the inference approaches often lead to excessive target prediction as well as high false positives. Furthermore, cells simultaneously function as signal receivers and transmitters, and such multivariate dynamics are often hard to model and ignored in the inference methods.

Several complementary computational and experimental approaches such as neighbor-seq (Ghaddar & De, 2022), PIC-seq (Giladi et al., 2020) and LIPSTIC (Nakandakari-Higa et al., 2023) are developed that can isolate either physically interacting cells (doublets, multiplets) or label interacting cells within a spatial neighborhood to track their localization and migration.

4. Evolution of tumor
4.1 Clonal architecture

Clonal architecture serves as a blueprint for understanding the intricate dynamics of tumor evolution. As tumors progress, they undergo genetic and phenotypic diversification, giving rise to distinct cell populations with unique characteristics. These cellular clones, originating from a single progenitor cell, form the organizational framework of the tumor. The genetic and non-genetic alterations inherited by subsequent clones play a pivotal role in defining the aggressiveness of the tumor (Fig. 3).

The arrangement of tumor clones and the disruption of tissue architecture varies significantly between solid and liquid tumors, which in turn guides tumor evolution (Noble et al., 2021). Liquid tumors lack a defined structure and neighborhood, which leads to low ITH as the subclones do not experience competition and selective pressure. Subclones with driver events conferring growth advantage dominate with time. The spatial organization differs even among solid cancer types depending on the tissue of origin or spatial constraints. For instance, the pattern of clonal expansion differs between epithelial cancers and colon cancers. Mutated cells within the intestinal crypts lead to persistence and proliferation, while mutated cells positioned outside the niche are slowly eliminated from the population by differentiation (Flanagan et al., 2021; Van Neerven et al., 2021). The constraints set by tissue architecture lend to high heterogeneity, though only one subclone typically dominates the substructure (Noble et al., 2021). This underscores the importance of considering the specific characteristics of each tumor type when analyzing clonal architecture and its impact on tumor evolution.

The spatial organization of clones within the tissue structure holds valuable insights into tumor evolution. Computational models show that ITH is shaped based on differing spatial constraints of solid tumors (Noble et al., 2021). Further ST studies can help understand the factors shaping tumor evolution in different cancer types and the role of tissue structure in defining the varying heterogeneity of cancer cells within tissue sub-structure and invasive tumors. Invasive tumors have relatively lower heterogeneity compared to tumors within tissue structure as they compete with normal and clonal neighborhood to form dominant clones in a region with multiple different subclones. ITH further reduces in tumors not restrained by any tissue boundary or physical resistance. In breast cancer, this variability in ITH is studied using single-cell genome and transcriptome sequencing in ductal and invasive carcinomas (Lomakin et al., 2022; Yates et al., 2015).

Clonal interactions are also influenced by factors such as the TME, which play a crucial role in driving tumor progression and metastasis (Hsieh et al., 2022). Understanding the interplay between clonal architecture and TME dynamics provides a holistic view of tumor evolution and guides the

Crosstalk between tumor and microenvironment 203

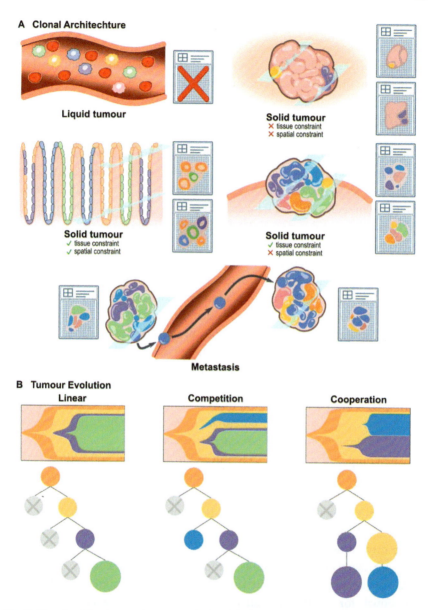

Fig. 3 Clonal architecture and tumor evolution. (A) The tumor clonal architecture can vary significantly depending on the cancer type, and spatial arrangement further shapes the intratumor heterogeneity. For solid tumors, the localization of tumor clones can influence local crosstalk and microenvironment. Spatial transcriptomics of multiple sections, samples and longitudinal/temporal profiling can provide insights into tumor evolution. (B) Tumor evolution is shaped by clonal heterogeneity and interactions between the subclones. The interactions can be neutral (linear), antagonistic (competition) or cooperative (cooperation). The driver events are accumulated over time and fitness of these subclones defines their dominance in the tumor.

development of effective therapeutic strategies tailored to target specific clones or microenvironmental cues. The disruption of normal tissue architecture by invading cancer cells and their interactions within the TME further shapes the clonal arrangement (discussed in detail in §5).

4.1.1 Role of genome

Genomic changes play a fundamental role in shaping the clonal architecture of tumors, influencing their evolution and heterogeneity. Early driver events, such as mutations or alterations in key genes, are often shared across clonal populations and initiate tumorigenesis. For instance, in myeloid neoplasms, mutations in genes like TET2 are frequently observed early in disease development, contributing to clonal expansion (Jankowska et al., 2011).

However, as cancer cells continue to divide, genomic variability is introduced into individual clones through processes like genomic instability and mutagenesis (Rowley, 1973; Stratton et al., 2009; Vogelstein et al., 2013). This leads to the accumulation of additional mutations, some of which may confer selective advantages, driving clonal evolution. Late driver mutations, acquired during tumor progression, are often observed as a result of clonal selection, further shaping the clonal architecture. This distinction between late and early mutations is not always conserved as observed in breast cancers with mutations in driver genes such as TP53, PTEN, BRCA2 or amplification of MYC and CDK6 (Yates et al., 2015). Subclonal mutations provide potential targets for therapies illustrating the need to understand ITH for personalized treatment.

The composition and characteristics of major clonal populations within a tumor can define distinct cancer subtypes. Intertumor heterogeneity influences tumor classification and treatment strategies in breast cancer based on the presence or absence of HER2 amplification (Meric-Bernstam & Hung, 2006). Similarly, in renal cell carcinoma, specific combinations of driver mutations, such as PBRM1 and SETD2 mutations or PI3K alterations, are associated with distinct tumor subtypes (Network, 2013).

Understanding the genomic landscape and clonal architecture of tumors is essential for stratifying patients and guiding personalized treatment approaches tailored to target specific clonal populations and genomic aberrations. However, achieving a comprehensive understanding often requires sophisticated methodologies such as multi-sampling and single-cell analysis. By analyzing multiple samples, researchers can uncover ITH and assess the clonal composition across the tumor landscape, providing a more accurate representation of the tumor's genomic architecture (Lomakin et al., 2022;

Yates et al., 2015). Lomakin et al. used ST on breast cancer samples to show different lineages dominating primary tumors compared to metastatic and ductal invasions. Lobular regions show mosaic clonal populations, which get more homogenous at sublobular resolution. ST-based analysis can help clonal dominance and architecture within tissue constraints which cannot be studied using bulk methods.

4.1.2 Role of heritable epigenome

The heritable nature of epigenomic changes in tumors significantly impacts clonal architecture, influencing gene expression patterns and downstream cellular pathways. Epigenomic alterations, such as DNA methylation and histone modifications, can be passed down through cell divisions, leading to stable changes in gene expression profiles within clonal populations. Mutations in regulatory regions of the genome can disrupt the normal epigenetic landscape, affecting gene expression and contributing to tumorigenesis. Interestingly, even tumors lacking mutations in well-known driver genes often exhibit alterations in regulatory regions, showing the significance of epigenomic changes in shaping clonal architecture and tumor evolution (Chatterjee, Rodger, & Eccles, 2018).

These epigenomic modifications can have profound effects on downstream signaling pathways, driving aberrant cellular behaviors and promoting tumor progression. In colorectal cancer (CRC), for example, mutations in driver genes are often accompanied by epigenomic modifications that further dysregulate key pathways involved in tumor development and growth (Bardhan & Liu, 2013). Furthermore, gene enrichment revealed the involvement of developmental genes in tumorigenesis, suggesting a link between epigenomic alterations and the reactivation of developmental programs during tumor evolution. These findings highlight the intricate interplay between epigenetic regulation, genomic alterations, and clonal architecture in driving tumorigenesis and tumor heterogeneity.

4.2 Tumor evolution types

Tumor evolution is a dynamic process driven by complex interactions among cells within the TME. These interactions give rise to ITH, where different clonal populations coexist and compete or cooperate for survival and proliferation. Understanding the various types of tumor evolution is crucial for identifying effective therapeutic strategies against cancer (Fig. 3).

In linear evolution, tumor progression follows a stepwise accumulation of driver mutations. Cells acquire mutations sequentially, with each mutation conferring a selective advantage for survival or growth. Metastatic prostate tumors originate from a dominant subclone, which then follows a linear evolution at the metastatic site (Woodcock et al., 2020). Multiple clonal populations may coexist within the tumor, each representing a distinct stage of evolution. However, older clones may diminish over time as newer, more advantageous mutations emerge. Selective pressures within the micro-environment eliminate weaker clones, leading to the dominance of stronger, more aggressive populations commonly observed in liquid cancers.

Competition plays a significant role in shaping tumor evolution, especially within the spatial constraints of the TME. Clonal populations compete for limited resources such as nutrients and oxygen. The spatial arrangement leads to competition among different subclonal cells, with some populations being eliminated or marginalized. Post-treatment, surviving subclones often dominate, leading to remission or the emergence of resistant tumors. Combinatorial drug therapies that target diverse subclonal populations are crucial for preventing tumor relapse.

Contrary to competition, some subclonal populations within tumors exhibit cooperative behaviors. These cells rely on each other for survival or proliferation, forming symbiotic relationships within the tumor. Certain phenotypic traits may only emerge in the presence of specific subclones, indicating cooperation for overall tumor growth or resistance. An example would be metastasis or drug resistance. Understanding these cooperative interactions is essential for identifying vulnerabilities in tumor ecosystems and developing targeted therapies.

4.3 Insights from spatial transcriptomics

ST is a powerful tool for unraveling the complex landscape of cancer heterogeneity within tumor tissues. Across various cancer types, including prostate cancer, PDAC (Moncada et al., 2020), CRC (Peng, Ye, Ding, Feng, & Hu, 2022), and breast cancers (Bassiouni et al., 2023; Lv et al., 2021), transcriptional differences among subclones shed light on ITH, providing insights into tumor evolution and progression.

In a study by Moncada et al., four distinct ductal subpopulations were identified in PDAC, expressing genes such as APOL1 and hypoxia-related genes (TFF1/2/3). Additionally, centroacinar ductal populations and antigen-presenting ductal cells were characterized. Antigen-presenting ductal cells may have a role to modulate the inflammatory response, while hypoxic and terminal subpopulations were associated with specific regions within the TME.

Unsupervised clustering of ST data from colorectal samples correlated with manually curated cell types (Peng et al., 2022). Interestingly, fibroblast markers expressed in one out of five clusters, including COL1A2, COL1A1, SPARC, and COL3A1 hinting at the presence of CAFs. While ITH of transcriptome is a common feature across cancer types, differential expression was also observed in adjacent normal lymphoid and stromal regions based on proximity to cancer cells in cutaneous malignant melanoma (Thrane, Eriksson, Maaskola, Hansson, & Lundeberg, 2018) and invasive micropapillary carcinoma (Lv et al., 2021), respectively.

In prostate cancer, while common transcriptional signatures across different cancer regions were identified, difference in expression profiles between central and peripheral regions was also observed (Berglund et al., 2018). The localized subclonal population consists of similar gene deletions and amplifications. Similarly, triple-negative breast cancer investigation, revealed nine distinct clusters shared across samples, highlighting the presence of ITH (Bassiouni et al., 2023). The authors also noted race-associated differences in hypoxic tumor content. Lv et al. examined invasive micropapillary carcinoma and identified clusters with high levels of lipid metabolism, along with high expression of SREBF1 in one cluster correlated with metastasis and poor prognosis. Overall, ST provides valuable insights into the complex molecular landscape of cancer, revealing diverse cellular subpopulations, gene expression patterns, and microenvironmental interactions that contribute to tumor heterogeneity and progression.

4.4 Metastasis

Metastasis, the spread of cancer from its original site to distant organs or tissues, is often associated with a poor prognosis and resistance to treatment (Christensen et al., 2022; Lomakin et al., 2022; Wang et al., 2023; Wood et al., 2023; Zhang et al., 2022) (Fig. 3). Metastatic tumor evolution develops from subclonal populations of primary tumors (Lomakin et al., 2022; Yates et al., 2015), suggesting the need for additional mutational burden before metastasizing to a secondary location. While one dominant lineage often initiates and drives systemic metastasis, the route to metastasis varies between patients (Woodcock et al., 2020).

Christensen et al. conducted a pan-cancer analysis of metastasis evolution across 25 cancer types, revealing common driver mutations associated with treatment resistance, such as ESR1 and AR in breast and prostate cancer, EGFR T790M in NSCLC, and KIT V654A in gastrointestinal cancer. In general, metastatic tumors were found to harbor more somatic mutations, increased genomic instability, an enrichment in resistance mutations, but fewer driver mutations per mutation compared to primary tumors. Mutations in TP53 (contributed to 15.7% of all driver events) and cell-cycle pathways genes were enriched in metastatic tumors.

Metastasis of CRC to the liver reveals patient prognosis is dependent on immune response (Brancolini et al., 2024; Wang et al., 2023; Wood et al., 2023). Metastases in long-term survivors exhibited adaptive immune cell populations enriched for type II IFN signaling and MHC-class II antigen presentation, while poor prognosis metastases demonstrated an increased abundance of regulatory T cells and neutrophils with enrichment of Notch and TGFb signaling pathways (Wood et al., 2023). Fibroblasts play a crucial role in prognosis with F3+ cells enriched with primary tumors contribute to poor survival, while MCAM+ cells in liver metastatic tumors promote the generation of CD8 CXCL13 and CD4 CL13 cells associated with high proliferation ability and better prognosis (Wang et al., 2023). Overexpression of tetraspanins transmembrane proteins, TSPAN9 in fibroblasts and TSPAN1/3/8/13 in tumor cells is also associated with tumor growth and metastasis (Brancolini et al., 2024).

Q. Zhang et al. highlighted the common occurrence of brain metastases in lung cancer, which correlates with poor prognosis using digital spatial profiling. The brain TME is characterized by immune dysregulation, reduced antigen presentation, and altered cellular composition, including increased neutrophils and M2-type macrophages compared to primary lung cancer. Similarly, Salachan et al. (2023) profiled spatial transcriptome of prostate cancer and neuroendocrine disease stages, revealing impaired T-cell activity in castration-resistant prostate cancer and active interactions between CAFs and immune cells in neuroendocrine prostate cancer, highlighting the complex interplay between tumor cells, stromal cells, and the immune system in metastatic disease progression. The role of TME, including CAFs and immune cells, poor prognosis, resistance to treatment and heterogeneity in molecular mechanism of metastasis are common themes across cancer types.

5. Tumor microenvironment (TME)

TME refers to the non-cancerous components of a tumor usually populated by CAFs, immune and stromal cells. TMEs regulate and release molecules that drive the progression of cancer and the reaction to treatments.

Dissection of the TMEs provides crucial insights into the biology of cancer by elucidating the dynamics of tumor infiltrating immune cells and their heterogeneity. Spatial and single-cell RNA-seq are powerful tools for characterizing gene expression across thousands of cells simultaneously, which results in comprehensive profiling of cell types and states in tumors while retaining the topology of the tumor specimens. It has proved to be a powerful tool to elaborate on the lineage dynamics and cellular architecture of tumor infiltrating immune cells, also revealed the importance of cellular crosstalk within many types of cancer. Here, we briefly introduce a few key studies applying spatial and scRNA-seq studies to explore tumor environment heterogeneity (Fig. 4).

5.1 Dynamics of cancer associated fibroblasts in cancer progression

Cancer associated fibroblasts have been identified to play principal role in shaping up tumor growth by remodeling the tissue microenvironments

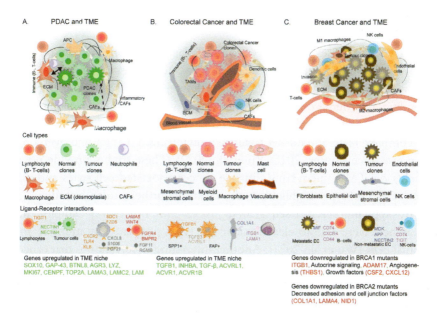

Fig. 4 **Tumor clones and crosstalk with microenvironment.** Dynamic crosstalk between tumor clones with various cell types in the microenvironment in pancreatic ductal adenocarcinoma (PDAC), colorectal Cancer (CRC) and breast cancer. Spatial and single-cell transcriptomics has led to identification and characterization of cell types of the microenvironment in (A) PDAC, (B) colorectal cancer (CRC) and (C) breast cancer. A few examples of representative cell types, inferred ligand–receptor and differentially regulated genes are also listed.

(Ishimoto et al., 2017; Sahai et al., 2020). An integrated analysis of spatial and single-cell data across six common cancer types helped identified four major functional subgroups of cancer associated fibroblasts. iCAFs and mCAFs were the most identified subtypes in all the cancer types. Their presence was also found in two rare cancer types, the MEC and EMC. Additionally, it was also observed that pCAF was involved in robust production of IFN-I, which plays the dual role of promoting as well as deterring the proliferation of cancer cells. Metabolic reprogramming mechanisms among distinct CAFs subtypes were observed, shedding light on the metabolism of cancer cells (Ma et al., 2023). In another study, scRNA-seq was performed on two clinical specimens of neuroblastoma and four sporadic cases. This study generated a compendium of the transcriptional profiles associated with familial NB and identify Fib4 as a marker that distinguishes a subpopulation of endothelial cells that differentiate into a CAF sub-type (Zhang, Liu, & Wang, 2023).

5.2 Insights about the tumor microenvironment from single cell and ST studies

The TME comprises of heterogeneous cellular populations including tumor cells and the surrounding nonmalignant cells, such as vascular, immune, and fibroblast cell types. Constant cross-talk between tumor cells and the surrounding microenvironment was associated with treatment response and survival outcomes of malignancies. ScRNA-seq is a powerful tool to explore the complex TME and further facilitate individualized therapy and overcome drug resistance (Potter, 2018; Yu et al., 2022). These studies provide novel insights on TME uncovering distance cancer associated cell states and different CAF subtypes highlighting the need for further in-depth studies.

By applying scRNA-seq analysis to 42 NSCLC biopsy samples, the researchers were able to eleven major cell types populated by immune and stromal cells (Wu et al., 2021). Another interesting observation is that cancer cells of different patients showed higher heterogeneity and patient specific gene expression signatures. Neutrophils significantly depleted/not found in LUAD patients. Prominent interaction between cancer and endothelial cells, fibroblasts and macrophages which explains the active inflammatory response within the tumor. Myofibroblasts to fibroblasts ratio in this study was unusually higher indicating that the CAF with myofibroblasts may act as an important malignant signature for advanced lung cancer. Lung squamous cell carcinoma has higher ITH than LUAD indicating higher intertumor difference in LUSC than LUAD (Wu et al., 2021).

The cellular composition of CRC from the tumor samples and adjacent tissues obtained from five non-metastatic patients were assessed by Qi et al. (2022). They found that the immune responses like IL2/STAT5 signaling and IL6/JAK/STAT3 signaling, were enriched not only in immune cell populations, but also in mesenchymal stromal cells and endothelial cells in tumor compared to normal tissues suggesting the involvement of MSCs and ECs in the response against CRC. Another interesting observation is that remodeling of stromal cells resulted in increase in a CAF subtype FAP+ depicting the role of stromal cells in CRC progression.

Even though recent scRNA-seq studies (Zhang et al., 2019) have elucidated prognostic signatures of gastric cancers, the limited number of cells sequenced in these studies have been a limiting factor in obtaining a comprehensive and novel features of gastric cancers. This led to the work from the researchers from the Duke-NUS school of medicine to come up with scRNA-seq study on a cohort of 31 patients reflecting multiple clinical stages and subtypes. They identified the expression of gastric cancer OG expression in intestinal-like epithelial cells, EpiInt1. Interestingly, EpiInt1 was also associated with high tumor-associated expression differences. These factors likely indicate the transition of epithelial cells to the state of malignancy (Kumar et al., 2022).

Breast cancer is one of the most common forms of cancer in females showed that the BRACA1/2 carrier preneoplastic epithelial cells communicate differently in their surrounding environment. Additionally they also found that expression of collagen degrading enzymes such as MMP3 and MMP8 were enriched in BRCA1 mutation carriers which is again an ideal condition for the genesis and progression of tumors (Caputo et al., 2023).

Another study which combined the ST and scRNA-seq approaches to deconvolute cell types thereby to delineate morphological features in CRC identified the co-existence of tumor cells composed different consensus molecular sub-types, a widely used gene expression-based classification of CRC (Valdeolivas et al., 2024). Most importantly using ST approach the researchers were able to correlate the growth patterns of tumors with their consensus molecular sub-types. The subpopulations in these tumors where further studied using scRNA-seq in a complementary approach, providing a higher resolution of altered global transcripts and signaling modules.

It is noteworthy that scRNA-seq and ST studies have played indispensable role in dissecting the tumor micro-environment of various cancer types retaining their cellular and molecular architecture. These studies have identified conclusive results that have been consistent with previous studies. Most commonly, the cell types identified across tumors of different cancer types have been validated through previous studies. Further, single-cell studies have also provided novel biological insights in TMEs by identifying rare cell subpopulations in a tumor as well as serving the purpose of identifying factors contributing to the progression of cancer (CAFs and inflammatory responses) and potential inhibitors of cancers (tumor infiltrating immune cells).

6. Integrating ST and scRNA-seq

Integrating scRNA-seq and ST data has emerged as a powerful approach for studying tumor evolution, offering complementary insights into the cellular composition and spatial organization of the TME. While scRNA-seq enables the study of the transcriptome (both high and medium expressed genes) at a single-cell resolution, it lacks spatial information. Conversely, ST provides spatial context (for highly expressed genes) but many of current methods lack true single-cell resolution. By integrating these datasets, researchers have designed strategies to overcome these limitations and gain a more comprehensive understanding of tumor biology.

Various computational tools have been developed for deconvoluting and inferring cell–cell interactions from scRNA-seq and ST data (Arnol, Schapiro, Bodenmiller, Saez-Rodriguez, & Stegle, 2019; Cang & Nie, 2020). However, predictions can be further refined by integrating scRNA-seq and ST data from different but similar samples obtained from the same patient. This integration can also include temporal or multi-sampled data, allowing for the investigation of dynamic changes in the TME over time.

Studies integrating scRNA-seq and ST have been conducted across different cancer types, including NSCLC (Chen et al., 2023), LUAD (Zhu et al., 2022), esophageal squamous cell carcinoma (Guo et al., 2022), CRC (Valdeolivas et al., 2024; Wang et al., 2023), rectal cancer (Qin et al., 2023), cSCC (Ou et al., 2022), and pan-cancer cohorts (Du et al., 2024). These studies aim to decipher the complexity of the TME, focusing on various cell types such as CAFs, immune cells, non-immune stromal cells, and myCAF subpopulations. Insights into ITH and metastasis can be gleaned by matching primary and metastatic tumor data, allowing for the characterization of spatial and cellular differences between these tumor stages (Fig. 4).

In addition to integrating scRNA-seq and ST data, other approaches, such as multiomic methods have been employed, where multiple omics layers are simultaneously analyzed from the same sample. This includes methods to study the transcriptome and proteome (Chen, Lake, & Zhang, 2019; Satpathy et al., 2018) or the transcriptome and epigenome (Peterson et al., 2017; Stoeckius et al., 2017) in parallel, providing a comprehensive view of tumor biology and evolution. Overall, the integration of scRNA-seq and ST data, along with other omics approaches, holds great promise for advancing our understanding of tumor evolution and guiding the development of precision cancer therapies.

7. Challenges and future directions

The field of ST is growing exponentially with development of new methods, exploration of multiple tissues and new biological insights. There are still major challenges from technical, modeling, analysis, data exploration perspective as well as accessibility, cost and broader usage. From the technical side, several methods continue to improve the resolution (sub-cellular, sub-organelle) and sensitivity (detection of lower abundance transcripts). While current ST methods have largely been applied to fresh-frozen tissue due to easier permeabilization and release of transcripts, methods suited for paraffin embedded tissues (FFPE) and preserving tissue integrity are needed. The FFPE compatible methods will enable access and profiling of decades of biological material from longitudinal clinical studies and Biobanks (Liu, Enninful, Deng, & Fan, 2020). The current ST methods profile a representative tissue section of method-specific thickness and size, which poses a challenge in highly reactive tissues with high tumor and micro environmental heterogeneity. Sequencing-based ST methods are particularly challenged due to selective capture of highly expressed transcripts, lower complexity of sequencing library (compared to scRNA-seq) and data sparsity. This is particularly relevant for different cancer types with high degree of inter-tumoral heterogeneity and ITH. The recent pan-cancer bulk and single-cell atlases across multiple cancer types are providing a glimpse into clonal states and dynamics, interaction with tumor infiltrating immune and stromal cells and cellular communication. By profiling multiple tissue slices, ST can offer a three-dimensional view of tumors, their inter-action dynamics within spatial environment and heterogeneity in spatial niches. An interesting approach has been recently applied for the 3D reconstruction of the human heart (Asp et al., 2019). The large amount of data storage and analysis poses another challenges, and the rapid adoption of new data formats, lossless compression and open-access analysis methods is exciting and paves a way for exciting developments. Although ST methods are still quite expensive, less accessibility and require specialist equipment; the community driven rapid availability of large pre-processed datasets from different tumor and matched controls fosters development of new computational analysis methods and biological insights. The next frontier is integrating multi-modal spatial studies (Vicari et al., 2023) and new computational algorithms for both static and longitudinal tissue profiling, to guide our modeling of tumor heterogeneity, clonal architecture, crosstalk with microenvironment and tumor evolution.

Acknowledgments

The research in the KNN lab is supported by the Villum Young Investigator grant (VYI#00025397) and DigitSTEM Initiative. We apologize to authors whose work could not be cited owing to space constraints.

Conflict of interest statement

The authors declare that they have no competing interests.

References

Aas, T., Børresen, A.-L., Geisler, S., Smith-Sørensen, B., Johnsen, H., Varhaug, J. E., et al. (1996). Specific P53 mutations are associated with de novo resistance to doxorubicin in breast cancer patients. *Nature Medicine, 2*, 811–814.

Alon, S., Goodwin, D. R., Sinha, A., Wassie, A. T., Chen, F., Daugharthy, E. R., et al. (2021). Expansion sequencing: Spatially precise in situ transcriptomics in intact biological systems. *Science (New York, N.Y.), 371*, eaax2656.

An, X., Tiwari, A. K., Sun, Y., Ding, P.-R., Ashby, C. R., & Chen, Z.-S. (2010). BCR-ABL tyrosine kinase inhibitors in the treatment of Philadelphia chromosome positive chronic myeloid leukemia: A review. *Leukemia Research, 34*, 1255–1268.

Armingol, E., Officer, A., Harismendy, O., & Lewis, N. E. (2021). Deciphering cell–cell interactions and communication from gene expression. *Nature Reviews. Genetics, 22*, 71–88.

Arnol, D., Schapiro, D., Bodenmiller, B., Saez-Rodriguez, J., & Stegle, O. (2019). Modeling cell-cell interactions from spatial molecular data with spatial variance component analysis. *Cell Reports, 29*(1), 202–211.

Asp, M., Giacomello, S., Larsson, L., Wu, C., Fürth, D., Qian, X., et al. (2019). A spatiotemporal organ-wide gene expression and cell atlas of the developing human heart. *Cell, 179*(7), 1647–1660.

Bai, X., Ni, J., Beretov, J., Graham, P., & Li, Y. (2018). Cancer stem cell in breast cancer therapeutic resistance. *Cancer Treatment Reviews, 69*, 152–163.

Baner, J., Nilsson, M., Mendel-Hartvig, M., & Landegren, U. (1998). Signal amplification of padlock probes by rolling circle replication. *Nucleic Acids Research, 26*, 5073–5078.

Bardhan, K., & Liu, K. (2013). Epigenetics and colorectal cancer pathogenesis. *Cancers, 5*, 676–713.

Bassiouni, R., Idowu, M. O., Gibbs, L. D., Robila, V., Grizzard, P. J., Webb, M. G., et al. (2023). Spatial transcriptomic analysis of a diverse patient cohort reveals a conserved architecture in triple-negative breast cancer. *Cancer Research, 83*, 34–48.

Berglund, E., Maaskola, J., Schultz, N., Friedrich, S., Marklund, M., Bergenståhle, J., et al. (2018). Spatial maps of prostate cancer transcriptomes reveal an unexplored landscape of heterogeneity. *Nature Communications, 9*, 2419.

Bozic, I., Antal, T., Ohtsuki, H., Carter, H., Kim, D., Chen, S., ... Nowak, M. A. (2010). Accumulation of driver and passenger mutations during tumor progression. *Proceedings of the National Academy of Sciences of the United States of America, 107*, 18545–18550.

Bozic, I., & Nowak, M. A. (2014). Timing and heterogeneity of mutations associated with drug resistance in metastatic cancers. *Proceedings of the National Academy of Sciences of the United States of America, 111*, 15964–15968.

Brancolini, C., Suwatthanarak, T., Thanormjit, K., Suwatthanarak, T., Acharayothin, O., Methasate, A., et al. (2024). Spatial transcriptomic profiling of tetraspanins in stage 4 colon cancer from primary tumor and liver metastasis. *Life (Chicago, Ill.: 1978), 14*, 126.

Browaeys, R., Saelens, W., & Saeys, Y. (2020). NicheNet: Modeling intercellular communication by linking ligands to target genes. *Nature Methods, 17*, 159–162.

Cable, D. M., Murray, E., Zou, L. S., Goeva, A., Macosko, E. Z., Chen, F., et al. (2022). Robust decomposition of cell type mixtures in spatial transcriptomics. *Nature Biotechnology, 40*(4), 517–526.

Camp, J. G., Sekine, K., Gerber, T., Loeffler-Wirth, H., Binder, H., Gac, M., et al. (2017). Multilineage communication regulates human liver bud development from pluripotency. *Nature, 546*, 533–538.

Cang, Z., & Nie, Q. (2020). Inferring spatial and signaling relationships between cells from single cell transcriptomic data. *Nature Communications, 11*, 2084.

Caputo, A., Vipparthi, K., Bazeley, P., Downs-Kelly, E., McIntire, P., Ni, Y., et al. (2023). Alterations in the preneoplastic breast microenvironment of BRCA1/2 mutation carriers revealed by spatial transcriptomics. *BioRxiv*, 2023.05.24.542078.

Chang, Y., He, F., Wang, J., Chen, S., Li, J., Liu, J., et al. (2022). Define and visualize pathological architectures of human tissues from spatially resolved transcriptomics using deep learning. *Computational and Structural Biotechnology Journal, 20*, 4600–4617.

Chatterjee, A., Rodger, E. J., & Eccles, M. R. (2018). Epigenetic drivers of tumourigenesis and cancer metastasis. *Seminars in Cancer Biology, 51*, 149–159.

Chen, A., Liao, S., Cheng, M., Ma, K., Wu, L., Lai, Y., et al. (2022). Spatiotemporal transcriptomic atlas of mouse organogenesis using DNA nanoball-patterned arrays. *Cell, 185*(10), 1777–1792.

Chen, C., Guo, Q., Liu, Y., Hou, Q., Liao, M., Guo, Y., et al. (2023). Single-cell and spatial transcriptomics reveal POSTN+ cancer-associated fibroblasts correlated with immune suppression and tumour progression in non-small cell lung cancer. *Clinical and Translational Medicine, 13*, e1515.

Chen, K. H., Boettiger, A. N., Moffitt, J. R., Wang, S., & Zhuang, X. (2015). Spatially resolved, highly multiplexed RNA profiling in single cells. *Science (New York, N.Y.), 348*, aaa6090.

Chen, S., Lake, B. B., & Zhang, K. (2019). High-throughput sequencing of the transcriptome and chromatin accessibility in the same cell. *Nature Biotechnology, 37*, 1452–1457.

Chen, S., Loper, J., Chen, X., Vaughan, A., Zador, A. M., & Paninski, L. (2021). Barcode demixing through non-negative spatial regression (BarDensr). *PLoS Computational Biology, 17*(3), e1008256.

Cherry, C., Maestas, D. R., Han, J., Andorko, J. I., Cahan, P., Fertig, E. J., et al. (2021). Computational reconstruction of the signalling networks surrounding implanted biomaterials from single-cell transcriptomics. *Nature Biomedical Engineering, 5*, 1228–1238.

Cho, C.-S., Xi, J., Si, Y., Park, S.-R., Hsu, J.-E., Kim, M., et al. (2021). Microscopic examination of spatial transcriptome using Seq-Scope. *Cell, 184*(13), 3559–3572.

Christensen, D. S., Ahrenfeldt, J., Sokač, M., Kisistók, J., Thomsen, M. K., Maretty, L., et al. (2022). Treatment represents a key driver of metastatic cancer evolution. *Cancer Research, 82*, 2918–2927.

Dees, N. D., Zhang, Q., Kandoth, C., Wendl, M. C., Schierding, W., Koboldt, D. C., et al. (2012). MuSiC: Identifying mutational significance in cancer genomes. *Genome Research, 22*, 1589–1598.

Dietz, S., Harms, A., Endris, V., Eichhorn, F., Kriegsmann, M., Longuespée, R., et al. (2017). Spatial distribution of EGFR and KRAS mutation frequencies correlates with histological growth patterns of lung adenocarcinomas. *International Journal of Cancer, 141*, 1841–1848.

Dinstag, G., & Shamir, R. (2020). PRODIGY: Personalized prioritization of driver genes. *Bioinformatics (Oxford, England), 36*, 1831–1839.

Drews, R. M., Hernando, B., Tarabichi, M., Haase, K., Lesluyes, T., Smith, P. S., et al. (2022). A pan-cancer compendium of chromosomal instability. *Nature, 606*, 976–983.

Dries, R., Zhu, Q., Dong, R., Eng, C.-H. L., Li, H., Liu, K., et al. (2021). Giotto: A toolbox for integrative analysis and visualization of spatial expression data. *Genome Biology, 22*, 78.

Du, Y., Shi, J., Wang, J., Xun, Z., Yu, Z., Sun, H., et al. (2024). Integration of pan-cancer single-cell and spatial transcriptomics reveals stromal cell features and therapeutic targets in tumor microenvironment. *Cancer Research, 84*, 192–210.

Efremova, M., Vento-Tormo, M., Teichmann, S. A., & Vento-Tormo, R. (2020). CellPhoneDB: Inferring cell-cell communication from combined expression of multi-subunit ligand-receptor complexes. *Nature Protocols, 15*, 1484–1506.

Eichenberger, B. T., Zhan, Y., Rempfler, M., Giorgetti, L., & Chao, J. A. (2021). Deep-Blink: Threshold-independent detection and localization of diffraction-limited spots. *Nucleic Acids Research, 49*(13), 7292–7297.

Fazal, F. M., Han, S., Parker, K. R., Kaewsapsak, P., Xu, J., Boettiger, A. N., et al. (2019). Atlas of subcellular RNA localization revealed by APEX-Seq. *Cell, 178*(2), 473–490.

Femino, A. M., Fay, F. S., Fogarty, K., & Singer, R. H. (1998). Visualization of single RNA transcripts in situ. *Science (New York, N.Y.), 280*, 585–590.

Flanagan, D. J., Pentinmikko, N., Luopaj¨arvi, K., Willis, N. J., Gilroy, K., Raven, A. P., et al. (2021). NOTUM from Apc-mutant cells biases clonal competition to initiate cancer. *Nature, 594*, 430–435.

Fu, X., Sun, L., Chen, J. Y., Dong, R., Lin, Y., Palmiter, R. D., et al. (2021). Continuous polony gels for tissue mapping with high resolution and RNA capture efficiency. *BioRxiv.*

Garcia-Alonso, L., Handfield, L.-F., Roberts, K., Nikolakopoulou, K., Fernando, R. C., Gardner, L., et al. (2021). Mapping the temporal and spatial dynamics of the human endometrium in vivo and in vitro. *Nature Genetics, 53*, 1698–1711.

Ghaddar, B., & De, S. (2022). Reconstructing physical cell interaction networks from single-cell data using Neighbor-seq. *Nucleic Acids Research, 50*, e82.

Ghazanfar, S., Lin, Y., Su, X., Lin, D. M., Patrick, E., Han, Z.-G., et al. (2020). Investigating higher-order interactions in single-cell data with scHOT. *Nature Methods, 17*(8), 799–806.

Giladi, A., Cohen, M., Medaglia, C., Baran, Y., Li, B., Zada, M., et al. (2020). Dissecting cellular crosstalk by sequencing physically interacting cells. *Nature Biotechnology, 38*, 629–637.

Guo, W., Zhou, B., Yang, Z., Liu, X., Huai, Q., Guo, L., et al. (2022). Integrating microarray-based spatial transcriptomics and single-cell RNA-sequencing reveals tissue architecture in esophageal squamous cell carcinoma. *eBioMedicine, 84*, 104281.

Hao, Y., Hao, S., Andersen-Nissen, E., Mauck, W. M., Zheng, S., Butler, A., et al. (2021). Integrated analysis of multimodal single-cell data. *Cell, 184*(13), 3573–3587.

Hong, D., Yang, F., Fessler, J. A., & Balzano, L. (2023). Optimally weighted PCA for high-dimensional heteroscedastic data. *SIAM Journal on Mathematics of Data Science, 5*(1), 222–250.

Hou, R., Denisenko, E., Ong, H. T., Ramilowski, J. A., & Forrest, A. R. R. (2020). Predicting cell-to-cell communication networks using NATMI. *Nature Communications, 11*, 5011.

Hsieh, W.-C., Budiarto, B. R., Wang, Y.-F., Lin, C.-Y., Gwo, M.-C., So, D. K., et al. (2022). Spatial multi-omics analyses of the tumor immune microenvironment. *Journal of Biomedical Science, 29*, 96.

Hsu, H.-C., Thiam, T. K., Lu, Y.-J., Yeh, C. Y., Tsai, W.-S., You, J. F., et al. (2016). Mutations of KRAS/NRAS/BRAF predict cetuximab resistance in metastatic colorectal cancer patients. *Oncotarget, 7*, 22257–22270.

Hu, Y., Peng, T., Gao, L., & Tan, K. (2021). CytoTalk: De novo construction of signal transduction networks using single-cell transcriptomic data. *Science Advances, 7*, eabf1356.

Ishimoto, T., Miyake, K., Nandi, T., Yashiro, M., Onishi, N., Huang, K. K., et al. (2017). Activation of transforming growth factor beta 1 signaling in gastric cancer-associated fibroblasts increases their motility, via expression of rhomboid 5 homolog 2, and ability to induce invasiveness of gastric cancer cells. *Gastroenterology, 153*(1), 191–204.

Jankowska, A. M., Makishima, H., Tiu, R. V., Szpurka, H., Huang, Y., Traina, F., et al. (2011). Mutational spectrum analysis of chronic myelomonocytic leukemia includes genes associated with epigenetic regulation: UTX, EZH2, and DNMT3A. *Blood, 118*, 3932–3941.

Ji, A. L., Rubin, A. J., Thrane, K., Jiang, S., Reynolds, D. L., Meyers, R. M., et al. (2020). Multimodal analysis of composition and spatial architecture in human squamous cell carcinoma. *Cell, 182*, 497–514.e22.

Jin, S., Guerrero-Juarez, C. F., Zhang, L., Chang, I., Ramos, R., Kuan, C.-H., et al. (2021). Inference and analysis of cell-cell communication using CellChat. *Nature Communications, 12*, 1088.

Ke, R., Mignardi, M., Pacureanu, A., Svedlund, J., Botling, J., Wählby, C., et al. (2013). In situ sequencing for RNA analysis in preserved tissue and cells. *Nature Methods, 10*, 857–860.

Kleshchevnikov, V., Shmatko, A., Dann, E., Aivazidis, A., King, H. W., Li, T., et al. (2022). Cell2location maps fine-grained cell types in spatial transcriptomics. *Nature Biotechnology, 40*(5), 661–671.

Kohno, T., Tsuta, K., Tsuchihara, K., Nakaoku, T., Yoh, K., & Goto, K. (2013). RET fusion gene: Translation to personalized lung cancer therapy. *Cancer Science, 104*, 1396–1400.

Korsunsky, I., Millard, N., Fan, J., Slowikowski, K., Zhang, F., Wei, K., et al. (2019). Fast, sensitive and accurate integration of single-cell data with harmony. *Nature Methods, 16*(12), 1289–1296.

Kueckelhaus, J., Von Ehr, J., Ravi, V. M., Will, P., Joseph, K., ... Beck, J., et al. (2020). Inferring spatially transient gene expression pattern from spatial transcriptomic studies. *BioRxiv*, 2020.10.20.346544.

Kumar, V., Ramnarayanan, K., Sundar, R., Padmanabhan, N., Srivastava, S., Koiwa, M., et al. (2022). Single-cell atlas of lineage states, tumor microenvironment, and subtype-specific expression programs in gastric cancer. *Cancer Discovery, 12*(3), 670–691.

Lawrence, M. S., Stojanov, P., Polak, P., Kryukov, G. V., Cibulskis, K., Sivachenko, A., et al. (2013). Mutational heterogeneity in cancer and the search for new cancer-associated genes. *Nature, 499*, 214–218.

Lee, J. H., Daugharthy, E. R., Scheiman, J., Kalhor, R., Yang, J. L., Ferrante, T. C., et al. (2014). Highly multiplexed subcellular RNA sequencing in situ. *Science (New York, N.Y.)*, *343*, 1360–1363.

Littman, R., Hemminger, Z., Foreman, R., Arneson, D., Zhang, G., Gómez-Pinilla, F., et al. (2021). Joint cell segmentation and cell type annotation for spatial transcriptomics. *Molecular Systems Biology*, *17*(6), e10108.

Liu, S., Punthambaker, S., Iyer, E. P. R., Ferrante, T., Goodwin, D., Fürth, D., et al. (2021). Barcoded oligonucleotides ligated on RNA amplified for multiplexed and parallel in situ analyses. *Nucleic Acids Research*, *49*, e58.

Liu, Y., Enninful, A., Deng, Y., & Fan, R. (2020). Spatial transcriptome sequencing of FFPE tissues at cellular level. *BioRxiv*, 2020.10.13.338475.

Lomakin, A., Svedlund, J., Strell, C., Gataric, M., Shmatko, A., Rukhovich, G., et al. (2022). Spatial genomics maps the structure, nature and evolution of cancer clones. *Nature*, *611*, 594–602.

Lubeck, E., Coskun, A. F., Zhiyentayev, T., Ahmad, M., & Cai, L. (2014). Single-cell in situ RNA profiling by sequential hybridization. *Nature Methods*, *11*, 360–361.

Lv, J., Shi, Q., Han, Y., Li, W., Liu, H., Zhang, J., et al. (2021). Spatial transcriptomics reveals gene expression characteristics in invasive micropapillary carcinoma of the breast. *Cell Death Disease*, *12*, 1095.

Ma, C., Yang, C., Peng, A., Sun, T., Ji, X., Mi, J., et al. (2023). Pan-cancer spatially resolved single-cell analysis reveals the crosstalk between cancer-associated fibroblasts and tumor microenvironment. *Molecular Cancer*, *22*(1), 170.

Maćkiewicz, A., & Ratajczak, W. (1993). Principal components analysis (PCA). *Computers & Geosciences*, *19*(3), 303–342.

McInnes, L., Healy, J., & Melville, J. (2018). UMAP: Uniform manifold approximation and projection for dimension reduction. *arXiv [e-print]*, *1802*, 03426.

Meric-Bernstam, F., & Hung, M.-C. (2006). Advances in targeting human epidermal growth factor receptor-2 signaling for cancer therapy. *Clinical Cancer Research*, *12*, 6326–6330.

Meyerson, M., & Pellman, D. (2011). Cancer genomes evolve by pulverizing single chromosomes. *Cell*, *144*, 9–10.

Miller, B. F., Bambah-Mukku, D., Dulac, C., Zhuang, X., & Fan, J. (2021). Characterizing spatial gene expression heterogeneity in spatially resolved single-cell transcriptomic data with nonuniform cellular densities. *Genome Research*, *31*(10), 1843–1855.

Misale, S., Yaeger, R., Hobor, S., Scala, E., Janakiraman, M., Liska, D., et al. (2012). Emergence of KRAS mutations and acquired resistance to anti-EGFR therapy in colorectal cancer. *Nature*, *486*, 532–536.

Moffitt, J. R., Lundberg, E., & Heyn, H. (2022). The emerging landscape of spatial profiling technologies. *Nature Reviews. Genetics*, *23*, 741–759.

Moncada, R., Barkley, D., Wagner, F., Chiodin, M., Devlin, J. C., Baron, M., et al. (2020). Integrating microarray-based spatial transcriptomics and single-cell RNA-seq reveals tissue architecture in pancreatic ductal adenocarcinomas. *Nature Biotechnology*, *38*, 333–342.

Moses, L., & Pachter, L. (2022). Museum of spatial transcriptomics. *Nature Methods*, *19*(5), 534–546.

Nakandakari-Higa, S., Canesso, M. C. C., Walker, S., Chudnovskiy, A., Jacobsen, J. T., Bilanovic, J., et al. (2023). Universal recording of cell-cell contacts in vivo for interaction-based transcriptomics. *BioRxiv [The Preprint Server for biology]* 2023.03.16.533003.

Noble, R., Burri, D., Sueur, C. L., Lemant, J., Viossat, Y., Kather, J. N., et al. (2021). Spatial structure governs the mode of tumour evolution. *Nature Ecology Evolution*, *6*, 207–217.

Ou, Z., Lin, S., Qiu, J., Ding, W., Ren, P., Chen, D., et al. (2022). Single-nucleus RNA sequencing and spatial transcriptomics reveal the immunological microenvironment of cervical squamous cell carcinoma. *Advanced Science*, *9*, 2203040.

Paez, J. G., Janne, P. A., Lee, J. C., Tracy, S., Greulich, H., Gabriel, S., et al. (2004). EGFR mutations in lung cancer: Correlation with clinical response to Gefitinib therapy. *Science (New York, N.Y.), 304,* 1497–1500.

Partel, G., Hilscher, M. M., Milli, G., Solorzano, L., Klemm, A. H., Nilsson, M., et al. (2020). Automated identification of the mouse brain's spatial compartments from in situ sequencing data. *BMC Biology, 18*(1), 1–14.

Partel, G., & Wählby, C. (2021). Spage2vec: Unsupervised representation of localized spatial gene expression signatures. *The FEBS Journal, 288*(6), 1859–1870.

Peng, Z., Ye, M., Ding, H., Feng, Z., & Hu, K. (2022). Spatial transcriptomics atlas reveals the crosstalk between cancer-associated fibroblasts and tumor microenvironment components in colorectal cancer. *Journal of Translational Medicine, 20,* 302.

Peterson, V. M., Zhang, K. X., Kumar, N., Wong, J., Li, L., Wilson, D. C., et al. (2017). Multiplexed quantification of proteins and transcripts in single cells. *Nature Biotechnology, 35,* 936–939.

Pham, D., Tan, X., Balderson, B., Xu, J., Grice, L. F., Yoon, S., et al. (2023). Robust mapping of spatiotemporal trajectories and cell–cell interactions in healthy and diseased tissues. *Nature Communications, 14,* 7739.

Potter, S. S. (2018). Single-cell RNA sequencing for the study of development, physiology and disease. *Nature Reviews Nephrology, 14*(8), 479–492.

Prabhakaran, S. (2022). Sparcle: Assigning transcripts to cells in multiplexed images. *Bioinformatics Advances, 2*(1), vbac048.

Qi, J., Sun, H., Zhang, Y., Wang, Z., Xun, Z., Li, Z., et al. (2022). Single-cell and spatial analysis reveal interaction of FAP+ fibroblasts and SPP1+ macrophages in colorectal cancer. *Nature Communications, 13*(1), 1742.

Qin, P., Chen, H., Wang, Y., Wang, J., Liu, S., Correspondence, H. L., et al. (2023). Cancer-associated fibroblasts undergoing neoadjuvant chemotherapy suppress rectal cancer revealed by single-cell and spatial transcriptomics. *Cell Reports. Medicine, 4*(10), 101231.

Raredon, M. S. B., Yang, J., Garritano, J., Wang, M., Kushnir, D., Schupp, J. C., et al. (2022). Computation and visualization of cell-cell signaling topologies in single-cell systems data using Connectome. *Scientific Reports, 12,* 4187.

Robinson, D. R., Wu, Y.-M., Vats, P., Su, F., Lonigro, R. J., Cao, X., et al. (2013). Activating ESR1 mutations in hormone-resistant metastatic breast cancer. *Nature Genetics, 45,* 1446–1451.

Ross, J. S., Wang, K., Chmielecki, J., Gay, L., Johnson, A., Chudnovsky, J., et al. (2016). The distribution of BRAF gene fusions in solid tumors and response to targeted therapy. *International Journal of Cancer, 138,* 881–890.

Rowley, J. D. (1973). A new consistent chromosomal abnormality in chronic myelogenous leukaemia identified by quinacrine fluorescence and Giemsa staining. *Nature, 243,* 290–293.

Sahai, E., Astsaturov, I., Cukierman, E., DeNardo, D. G., Egeblad, M., Evans, R. M., et al. (2020). A framework for advancing our understanding of cancer-associated fibroblasts. *Nature Reviews. Cancer, 20*(3), 174–186.

Salachan, P. V., Rasmussen, M., Ulhøi, B. P., Jensen, J. B., Borre, M., & Sørensen, K. D. (2023). Spatial whole transcriptome profiling of primary tumor from patients with metastatic prostate cancer. *International Journal of Cancer, 153,* 2055–2067.

Satpathy, A. T., Saligrama, N., Buenrostro, J. D., Wei, Y., Wu, B., Rubin, A. J., et al. (2018). Transcript-indexed ATAC-seq for precision immune profiling. *Nature Medicine, 24,* 580–590.

Schoenfeld, A. J., Chan, J. M., Kubota, D., Sato, H., Rizvi, H., Daneshbod, Y., et al. (2020). Tumor analyses reveal squamous transformation and off-target alterations as early resistance mechanisms to first-line osimertinib in EGFR-mutant lung cancer. *Clinical Cancer Research, 26,* 2654–2663.

Soda, M., Choi, Y. L., Enomoto, M., Takada, S., Yamashita, Y., Ishikawa, S., et al. (2007). Identification of the transforming EML4–ALK fusion gene in non-small-cell lung cancer. *Nature, 448*, 561–566.

Stewart, C. A., Gay, C. M., Xi, Y., Sivajothi, S., Sivakamasundari, V., Fujimoto, J., et al. (2020). Single-cell analyses reveal increased intratumoral heterogeneity after the onset of therapy resistance in small-cell lung cancer. *Nature Cancer, 1*, 423–436.

Stoeckius, M., Hafemeister, C., Stephenson, W., Houck-Loomis, B., Chattopad-Hyay, P. K., Swerdlow, H., et al. (2017). Simultaneous epitope and transcriptome measurement in single cells. *Nature Methods, 14*(9), 865–868.

Stratton, M. R., Campbell, P. J., & Futreal, P. A. (2009). The cancer genome. *Nature, 458*, 719–724.

Sudhakar, M., Rengaswamy, R., & Raman, K. (2022a). Multi-omic data improve prediction of personalized tumor suppressors and oncogenes. *Frontiers in Genetics, 13*, 964.

Sudhakar, M., Rengaswamy, R., & Raman, K. (2022b). Novel ratio-metric features enable the identification of new driver genes across cancer types. *Scientific Reports, 12*, 5.

Svensson, V., Teichmann, S. A., & Stegle, O. (2018). SpatialDE: Identification of spatially variable genes. *Nature Methods, 15*(5), 343–346.

Takeuchi, S., Doi, M., Ikari, N., Yamamoto, M., & Furukawa, T. (2018). Mutations in BRCA1, BRCA2, and PALB2, and a panel of 50 cancer-associated genes in pancreatic ductal adenocarcinoma. *Scientific Reports, 8*, 8105.

Tanevski, J., Flores, R. O. R., Gabor, A., Schapiro, D., & Saez-Rodriguez, J. (2022). Explainable multiview framework for dissecting spatial relationships from highly multiplexed data. *Genome Biology, 23*, 97.

The Cancer Genome Atlas Research Network. (2013). Comprehensive molecular characterization of clear cell renal cell carcinoma. *Nature, 499*, 43–49.

Thrane, K., Eriksson, H., Maaskola, J., Hansson, J., & Lundeberg, J. (2018). Spatially resolved transcriptomics enables dissection of genetic heterogeneity in stage III cutaneous malignant melanoma. *Cancer Research, 78*, 5970–5979.

Tokheim, C., & Karchin, R. (2019). CHASMplus reveals the scope of somatic missense mutations driving human cancers. *Cell Systems, 9*, 9–23 e8.

Tomasetti, C., Marchionni, L., Nowak, M. A., Parmigiani, G., & Vogelstein, B. (2015). Only three driver gene mutations are required for the development of lung and colorectal cancers. *Proceedings of the National Academy of Sciences of the United States of America, 112*, 118–123.

Türei, D., Korcsmáros, T., & Saez-Rodriguez, J. (2016). OmniPath: Guidelines and gateway for literature-curated signaling pathway resources. *Nature Methods, 13*, 966–967.

Türei, D., Valdeolivas, A., Gul, L., Palacio-Escat, N., Klein, M., Ivanova, O., et al. (2021). Integrated intra- and intercellular signaling knowledge for multicellular omics analysis. *Molecular Systems Biology, 17*, e9923.

Umemoto, K., & Sunakawa, Y. (2021). The potential targeted drugs for fusion genes including nrg1 in pancreatic cancer. *Critical Reviews in Oncology/Hematology, 166*, 103465.

Valdeolivas, A., Amberg, B., Giroud, N., Richardson, M., Gálvez, E. J., Badillo, S., et al. (2024). Profiling the heterogeneity of colorectal cancer consensus molecular subtypes using spatial transcriptomics. *Precision Oncology, 8*(1), 1–16.

Vandereyken, K., Sifrim, A., Thienpont, B., & Voet, T. (2023). Methods and applications for single-cell and spatial multi-omics. *Nature Reviews. Genetics*, 1–22.

van der Maaten, L., & Hinton, G. (2008). Visualizing data using t-SNE. *Journal of Machine Learning Research, 9*(86), 2579–2605.

Van Neerven, S. M., De Groot, N. E., Nijman, L. E., Scicluna, B. P., Van Driel, M. S., Lecca, M. C., et al. (2021). Apc-mutant cells act as supercompetitors in intestinal tumour initiation. *Nature, 594*, 436–441.

Vendramin, R., Litchfield, K., & Swanton, C. (2021). Cancer evolution: Darwin and beyond. *The EMBO Journal, 40*, e108389.

Vento-Tormo, R., Efremova, M., Botting, R. A., Turco, M. Y., Vento-Tormo, M., Meyer, K. B., et al. (2018). Single-cell reconstruction of the early maternal–fetal interface in humans. *Nature, 563*, 347–353.

Vicari, M., Mirzazadeh, R., Nilsson, A., Shariatgorji, R., Bjärterot, P., Larsson, L., et al. (2023). Spatial multimodal analysis of transcriptomes and metabolomes in tissues. *Nature Biotechnology*, 1–5. https://doi.org/10.1038/s41587-023-01937-y Epub ahead of print. PMID: 37667091.

Vickovic, S., Eraslan, G., Salmén, F., Klughammer, J., Stenbeck, L., Schapiro, D., et al. (2019). High-definition spatial transcriptomics for in situ tissue profiling. *Nature Methods, 16*(10), 987–990.

Vogelstein, B., Papadopoulos, N., Velculescu, V. E., Zhou, S., Diaz, L. A., & Kinzler, K. W. (2013). Cancer genome landscapes. *Science (New York, N.Y.), 339*, 1546–1558.

Wang, F., Long, J., Li, L., Wu, Z. X., Da, T. T., Wang, X. Q., et al. (2023). Single-cell and spatial transcriptome analysis reveals the cellular heterogeneity of liver metastatic colorectal cancer. *Science Advances, 9*(24), eadf5464.

Wang, X., Allen, W. E., Wright, M. A., Sylwestrak, E. L., Samusik, N., Vesuna, S., et al. (2018). Three-dimensional intact-tissue sequencing of single-cell transcriptional states. *Science (New York, N.Y.), 361*, eaat5691.

Wei, R., He, S., Bai, S., Sei, E., Hu, M., Thompson, A., et al. (2022). Spatial charting of single-cell transcriptomes in tissues. *Nature Biotechnology, 40*(8), 1190–1199.

Wilk, A. J., Shalek, A. K., Holmes, S., & Blish, C. A. (2023). Comparative analysis of cell–cell communication at single-cell resolution. *Nature Biotechnology, 42*, 470–483.

Wood, C. S., Pennel, K. A., Leslie, H., Legrini, A., Cameron, A. J., Melissourgou-Syka, L., et al. (2023). Spatially resolved transcriptomics deconvolutes prognostic histological subgroups in patients with colorectal cancer and synchronous liver metastases. *Cancer Research, 83*, 1329–1344.

Woodcock, D. J., Riabchenko, E., Taavitsainen, S., Kankainen, M., Gundem, G., Brewer, D. S., et al. (2020). Prostate cancer evolution from multi-lineage primary to single lineage metastases with implications for liquid biopsy. *Nature Communications, 11*, 5070.

Wooster, R., Bignell, G., Lancaster, J., Swift, S., Seal, S., Mangion, J., et al. (1995). Identification of the breast cancer susceptibility gene BRCA2. *Nature, 378*, 789–792.

Wu, F., Fan, J., He, Y., Xiong, A., Yu, J., Li, Y., et al. (2021). Single-cell profiling of tumor heterogeneity and the microenvironment in advanced non-small cell lung cancer. *Nature Communications, 12*(1), 2540.

Yan, K., Niu, L., Wu, B., He, C., Deng, L., Chen, C., et al. (2023). Copy number variants landscape of multiple cancers and clinical applications based on NGS gene panel. *Annals of Medicine, 55*, 2280708.

Yang, L., Shi, P., Zhao, G., Xu, J., Peng, W., Zhang, J., et al. (2020). Targeting cancer stem cell pathways for cancer therapy. *Signal Transduction and Targeted Therapy, 5*, 8.

Yates, L. R., Gerstung, M., Knappskog, S., Desmedt, C., Gundem, G., Loo, P. V., et al. (2015). Subclonal diversification of primary breast cancer revealed by multiregion sequencing. *Nature Medicine, 21*, 751–759.

Yu, Q., Jiang, M., & Wu, L. (2022). Spatial transcriptomics technology in cancer research. *Frontiers in Oncology, 12*, 1019111.

Yu, Y.-P., Liu, P., Nelson, J., Hamilton, R. L., Bhargava, R., Michalopoulos, G., et al. (2019). Identification of recurrent fusion genes across multiple cancer types. *Scientific Reports, 9*, 1074.

Yun, C.-H., Mengwasser, K. E., Toms, A. V., Woo, M. S., Greulich, H., Wong, K.-K., et al. (2008). The T790M mutation in EGFR kinase causes drug resistance by increasing the affinity for ATP. *Proceedings of the National Academy of Sciences of the United States of America, 105*, 2070–2075.

Zabransky, D. J., Yankaskas, C. L., Cochran, R. L., Wong, H. Y., Croessmann, S., Chu, D., et al. (2015). HER2 missense mutations have distinct effects on oncogenic signaling and migration. *Proceedings of the National Academy of Sciences of the United States of America, 112*, E6205–E6214.

Zhang, P., Yang, M., Zhang, Y., Xiao, S., Lai, X., Tan, A., et al. (2019). Dissecting the single-cell transcriptome network underlying gastric premalignant lesions and early gastric cancer. *Cell Reports, 27*(6), 1934–1947.

Zhang, Q., Abdo, R., Iosef, C., Kaneko, T., Cecchini, M., Han, V. K., et al. (2022). The spatial transcriptomic landscape of non-small cell lung cancer brain metastasis. *Nature Communications, 13*(1), 5983.

Zhang, Y., Liu, Q., & Wang, S. (2023). Single-cell transcriptome sequencing reveals tumor heterogeneity in family neuroblastoma. *Frontiers in Immunology, 14*, 1197773.

Zhao, E., Stone, M. R., Ren, X., Guenthoer, J., Smythe, K. S., Pulliam, T., et al. (2021). Spatial transcriptomics at subspot resolution with BayesSpace. *Nature Biotechnology, 39*(11), 1375–1384.

Zhao, Y., Li, Z., Zhu, Y., Fu, J., Zhao, X., Zhang, Y., et al. (2021). Single-cell transcriptome analysis uncovers intratumoral heterogeneity and underlying mechanisms for drug resistance in hepatobiliary tumor organoids. *Advanced Science, 8*, 2003897.

Zhu, J., Fan, Y., Xiong, Y., Wang, W., Chen, J., Xia, Y., et al. (2022). Delineating the dynamic evolution from preneoplasia to invasive lung adenocarcinoma by integrating single-cell RNA sequencing and spatial transcriptomics. *Experimental Molecular Medicine, 54*, 2060–2076.

CHAPTER SEVEN

Modular formation of in vitro tumor models for oncological research/therapeutic drug screening

Weiwei Wang[a,b] and Hongjun Wang[a,c,*]

[a]Department of Biomedical Engineering, Stevens Institute of Technology, Hoboken, NJ, United States
[b]School of Life Sciences, Yantai University, Yantai, Shandong, P.R. China
[c]Semcer Center for Healthcare Innovation, Stevens Institute of Technology, Hoboken, NJ, United States
*Corresponding author. e-mail address: hwang2@stevens.edu

Contents

1. Introduction	224
2. Why do we need 3D tumor models?	225
2.1 Spheroid	226
2.2 Hydrogels	227
2.3 Microfluidics	227
2.4 Polymeric scaffolds	228
3. Challenges with traditional tissue engineering strategies to form tumorous tissues	229
3.1 Comparison of top-down and bottom-up strategy	230
3.2 Modular tissue engineering	231
4. Application of modular strategy for tumor model formation	232
5. Future perspective and concluding remarks	244
Acknowledgment	245
References	246

Abstract

In recognition of the lethal nature of cancer, extensive efforts have been made to understand the mechanistic causation while identifying the effective therapy modality in hope to eradicate cancerous cells with minimal damage to healthy cells. In search of such effective therapeutics, establishing pathophysiologically relevant in vitro models would be of importance in empowering our capabilities of truly identifying those potent ones with significantly reduction of the preclinical periods for rapid translation. In this regard, wealthy progresses have been achieved over past decades in establishing various in vitro and in vivo tumor models. Ideally, the tumor models should maximally recapture the key pathophysiological attributes of their native counterparts. Many of the current models have demonstrated their utilities but also showed some noticeable limitations. This book chapter will briefly review some of the

Advances in Cancer Research, Volume 163
ISSN 0065-230X, https://doi.org/10.1016/bs.acr.2024.06.011
Copyright © 2024 Elsevier Inc. All rights are reserved, including those for text and data mining, AI training, and similar technologies.

mainstream platforms for in vitro tumor models followed by detailed elaboration on the modular strategies to form in vitro tumor models with complex structures and spatial organization of cellular components. Clearly, with the ability to modulate the building modules it becomes a new trend to form in vitro tumor models following a bottom-up approach, which offers a high flexibility to satisfy the needs for pathophysiological study, anticancer drug screening or design of personalized treatment.

1. Introduction

Cancer remains as the top lethal reasons to cause the life loss. In search of effective therapeutics, establishing pathophysiologically relevant in vitro models would be of importance in empowering our capabilities of truly identifying those potent ones with significantly reduction of the preclinical periods. As highlighted in a recent report, approximately 95% of new anti-cancer drugs demonstrating great promise in vitro testing ultimately fail in clinical trials, which poses a big financial burden to all the parties involved while significantly decelerating the new therapeutic development for life saving. It has been proposed that mismatch physiological conditions between many of the in vitro/in vivo models and human conditions might be the main contributor affecting the poor turnaround of therapeutic candidates. Furthermore, the great disparity in immune systems between humans and animals may also partially account for the failure of clinical trials on those promising drug. Thousands of researchers are devoting their careers to studying the mechanism of tumor initiation, progression, and metastasis and subsequently seeking effective treatment. In recognition of the significance of tumor models for drug screening as well as the fundamental studies, extensive efforts have been made in past decades or so to establish various tumor models like tumor spheroids and organoids, tissue-engineered tumor, microfluidic-based tumor model, xenografts and so on. In general, these models are typically generated through the cell self-assembly using a top-down approach. While recognizing the convivence of the top-down approaches, it confronts with noticeable challenges to recapitulate those structural and compositional complexities like the natural tumors. To this end, the recent endeavor has been geared toward the creation of tumor-like tissues using the bottom-up approach to replicate some of the complex structures if not fully. In this chapter, we will first briefly review several of the common in vitro tumor models and then particularly discuss the modular strategies to form complex tumor structures, followed by future perspectives and conclusion.

2. Why do we need 3D tumor models?

Traditional approaches to studying solid tumors for anticancer drug screening and testing primarily involve 2D cocultures, xenografts, or syngeneic mouse models. However, growing evidence has indicated that 2D culture is very limited for its inability to capture the complexity and heterogeneity of native tumors (Fischbach et al., 2007; Pampaloni, Reynaud, & Stelzer, 2007). Animal models, on the other hand, can better recapitulate the complexity of native conditions, but several drawbacks are also identified, for example, they are usually expensive to set up and have a limited ability to control and isolate key regulators of cell signaling due to the inherent complexity of living systems. To address these challenges, three-dimensional (3D) culture systems have received increasing attention and are expected to bridge the gap between experimental feasibility and clinical physiology. Several mainstream approaches (e.g., spheroids, hydrogel, porous scaffold, and microfluidics) have been explored to establish the 3D in vitro tumor models (see Fig. 1). A brief overview of these approaches is provided below to highlight the limitations along with the advantages.

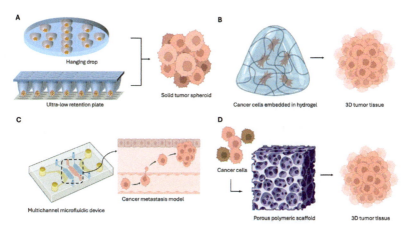

Fig. 1 Schematic illustration of the key methodology toward the formation of in vitro tumor models. (A) Cell spheroids formed through hanging drop method or in ultra-low retention plate, (B) 3D tumor model formed via embedding cancer cells in hydrogel, (C) Cancer metastasis model formed in a multichannel microfluidic device, and (D) 3D tumor tissue formed by seeding cancer cells onto porous polymeric scaffold. *Created from BioRender.com.*

2.1 Spheroid

Tumor spheroid composed of single or multiple types of cancer cells with or without the presence of stromal cells can be formed by using the hanging drop method or by growing cells in a bioreactor like a spinner flask (Fig. 1A). The former can lead to uniform spheroid formation while suffering from the involvement of extensive labors and limited scale-up capacity. The latter, on the other hand, allows for mass production, but oftentimes show a broader distribution in spheroid sizes. After initial aggregation, cells continue to grow and start to secrete their own extracellular matrix (ECM) components along with upregulated proteins mediating cell–cell interactions. In general, the resulting 3D tumor spheroids are compact and stable with noticeably decreasing gradients of proliferating cells, oxygen levels, and nutrition supplies from the periphery toward the center (Kunz-Schughart, 1999; Lee et al., 2013; Lin, Chou, Chien, & Chang, 2006; Vinci et al., 2012). Cumulative evidence has shown that cells within 3D tumor spheroids exhibit decreased sensitivity to cytotoxic agents such as 5-fluorouracil, cisplatin, or doxorubicin, better representing the in vivo responses in comparison to 2D planar culture, and as a consequence, the spheroid platform has been widely adopted for anticancer drug screening.

Despite the identified advantages, the spheroid mono-culture system lacks the typical stromal compartment of a tumor, where a basal lamina separates clumps of tumor cells from stroma. In recognition, investigators have taken the efforts to address this issue via various strategies such as mixing tumor cells with fibroblasts to form the spheroids, culturing tumor spheroids with fibroblast/monocyte suspensions, co-culturing tumor spheroids with embryonic bodies (i.e., developed from pluripotent murine ES cells that differentiate to cell types of all three germ layers) or combining tumor spheroids with a fibroblast/endothelial cell monolayer (Hauptmann, Zwadlo-Klarwasser, Jansen, Klosterhalfen, & Kirkpatrick, 1993; Seidl, Huettinger, Knuechel, & Kunz-Schughart, 2002). Some successes have been with these strategies in creating tumor models with heterogeneous structure and composition.

Although the inclusion of stromal cells within spheroids partially addresses the cellular heterogeneity limitation, the spheroid models growing in media still miss an essential component of tumor biology: the ECM. The ECM has been shown to play essential roles in providing the survival and drug resistance signals to cancer cells (Amann et al., 2014; Aoudjit & Vuori, 2001; Correia & Bissell, 2012; Huang et al., 2011;

Menendez et al., 2005; Xiong, Balcioglu, & Danen, 2013). The ECM secreted by tumor spheroids growing in media was far too simple to mimic the real complexity of tumor microenvironment (TME).

2.2 Hydrogels

Different from the spheroid method, solely relying on cell aggregation, hydrogels like Matrigel or methacrylated gelatin (GelMA) can be used to accommodate the tumor cells together with other stromal cells and provide the flexibility in designing the ECM-like environment as highlighted in Fig. 1B. These configurable hydrogel systems allow for control of biochemical and physical properties to better emulate the native tumor ECM. Theoretically, the physiochemical and biological characteristics of the hydrogel systems may be designed at the molecular level to induce the desirable cellular responses. Hydrogels based on polyethylene glycol (PEG), collagen, and Matrigel are commonly investigated to create functional and cell-instructive hydrogels (DeForest, Polizzotti, & Anseth, 2009; Kloxin, Kasko, Salinas, & Anseth, 2009; Liang et al., 2011). Despite that many of the hydrogels have been applied to stem cell differentiation, the principles are transferable to design new 3D culture for in vitro cancer models (Hutmacher, 2010), where ECM remodeling and changes in biomechanics are crucial in both primary cancers and metastatic processes. However, limited perfusibility and diffusibility through the hydrogel can cause obvious gradients of oxygen and nutrient in the experiment conditions, not fully representing the in vivo circumstances with rich and leaky vasculature around the tumors. Notably, introduction of microvascular networks into the hydrogel culture might circumvent such pathophysiologically relevant limitations (Chiew, Wei, Sultania, Lim, & Luo, 2017).

2.3 Microfluidics

With the channel size (i.e., $10-100\,\mu m$) close to microvascular diameter, microfluidic culture platforms offer sufficient circulation for mass exchange during the tissue formation, which has been explored to create in vitro tumor model with the hope to solve the perfusion issues. Besides enhanced perfusion through the cultured cells, the microfluidic platforms also enable to deliver various biological reagents including the therapeutic molecules in a time-resolved manner while inducing the shear force to the cells by modulating the flow rates. As noted, microfluidic devices or micro-bioreactors are efficient but small-scale with precise control of cell microenvironment. Recent advances in microfluidic technology have boosted the development of novel in vitro drug testing/screening platforms compatible with high-throughput applications (Tsui, Lee, Pun, Kim, & Kim, 2013) and the development of "organs-on-chips," composed of biologically functional tissue mimetics, in which inter-tissue interactions can be reconstituted by connecting individual "tissue/organ-on-a chip" modules (see Fig. 1C) (Yum, Hong, Healy, & Lee, 2014). The microenvironment within the microfluidic devices can be readily controlled by precisely adjusting the fluid flows, allowing for perfusion operation modes and resulting in shear stress, and by modulating the influx medium supplements to deliver various chemical stimuli (Li et al., 2022; Paterson, Zanivan, Glasspool, Coffelt, & Zagnoni, 2021; Xie, Appelt, & Jenkins, 2021; Zhang et al., 2022). In recognition of the above-mentioned advantages, there are some limitations and the major ones are associated with the low scalability, setup complexity and the difficult-to-operate nature.

2.4 Polymeric scaffolds

Distinct from hydrogel, oftentimes exhibiting limited nutrition diffusion throughout the culture, and spheroid technologies, typically forming small size aggregates, 3D polymeric scaffolds with a high porosity (>50%) and interconnected pores support the formation of tumor tissues with the shape and size defined by the scaffolds. Such scaffolds can be made from synthetic polymeric biomaterials such as polycaprolactone, poly(lactic-*co*-glycolic acid) (PLGA), inorganic ceramics such as hydroxyapatite or bioglass, or natural materials such as collagen, silk fibroin (Ozkan, Kalyon, & Yu, 2010; Ozkan, Kalyon, Yu, McKelvey, & Lowinger, 2009; Pathi, Kowalczewski, Tadipatri, & Fischbach, 2010; Tasso et al., 2009). Many of them are still under extensive investigation for possible formation of tumor tissues. In 2007, a 3D model was first established in polymeric scaffolds by culturing oral squamous cell carcinoma (Fischbach et al., 2007). As demonstrated, this model displayed many attributes much more closely to the tumors formed in vivo. Cells in this model were also less sensitive to chemotherapy and yielded tumors with enhanced malignancy. Since then, porous collagen scaffolds, porous silk scaffolds, and nanoparticle-incorporated scaffolds have been explored to better emulate the tumor activity (Blanco, Mantalaris, Bismarck, & Panoskaltsis, 2010; Chen et al., 2012; Choi et al., 2011; Da Silva, Lautenschlager, Sivaniah, & Guck, 2010; Kang & Bae, 2009; Senturk-Ozer, Aktas, He, Fisher, & Kalyon, 2017; Tan, Aung, Toh, Goh, & Nathan, 2011; Zhang et al., 2010). Compared to hydrogel-based culture, synthetic polymeric scaffolds allow to better control over the properties of the scaffolds and provide the ability to tailor the properties as required. However, the biocompatibility requirement has limited the usable material types as well as their utilities for forming the tumor models.

3. Challenges with traditional tissue engineering strategies to form tumorous tissues

As elaborated in the use of polymeric scaffolds for tumor tissue formation, better called as tumor tissue engineering, 3D scaffolds provide the cells with a temporary platform to attach, migrate, proliferate and form the tumor mass. Apart from the scaffolding functions to support cell growth, these scaffolds also define an influential microenvironment to induce the corresponding cellular responses via cell–materials interactions and subsequently regulate the synthesis of new tissue matrix for various functions. Ideally, the 3D scaffolds for in vitro tumor formation should maximally replicate the natural growing environment not only for cancer cells, but also for the cancer-related stromal cells such as cancer-associated fibroblasts and endothelial cells to induce desirable cell phenotypes and tissue matrix synthesis (Williams, 2014). However, due to the high complexity and heterogeneity of native tumor tissue, many of the in vitro 3D tumor models established through traditional tissue-engineering methods are far from the pathophysiological relevance. The tumor complexity can be briefly categorized in three aspects. First, tumors are integrated systems. Tumor progression and metastasis depend on the reciprocal changes between tumor cells and their local microenvironment, including genetic changes, cell–cell interactions, soluble signaling, and biochemical and biophysical properties of ECM such as the fibrotic formation and the stiffness alteration. It becomes essential to mimic the TME and tumor–stromal interactions in order to better achieve the ideal 3D in vitro tumor model. Second, tumors are always heterogeneous as characterized by different growth rates and differential responses to the therapeutic treatment, and this heterogeneity refers to not only their original diversity but also the genetic mutation of the same origin (Faupel-Badger et al., 2013). Third, tumors, whether primary or secondary, require the recruitment of new blood vessels to support the continuous growth into a large mass while increasing the probability of harmful mutations. More importantly, the presence of an adjacent blood vessel also facilitates the process of metastasis (Folkman & Kalluri, 2004; Gimbrone, Leapman, Cotran, & Folkman, 1972).

In other words, the expected 3D in vitro tumor model not only comprises the tumor cells, but also needs to: (1) include the cancer-associated stromal cells such as cancer-associated fibroblasts, endothelial cells, pericytes, and immune cells; (2) develop abundant tumorous ECM; (3) enable the spatial organization of various cells; (4) grow into a similar tissue size to that of the in vivo tumor; and (5) form tumorous vasculature networks. To accomplish such goals, more than one method should be adopted to fabricate 3D tumor models. In this regard, the fabrication strategy would benefit from the shift of traditional top-down strategy to more flexible ones such as bottom-up methods.

3.1 Comparison of top-down and bottom-up strategy

For in vitro tissue formation, two possible strategies can be considered, including a top-down and/or bottom-up method. As a matter of fact, the top-down one is most popular, which has been widely adopted in the traditional tissue engineering endeavor. Typically, it involves the seeding of cells onto a prefabricated large scaffold with a defined shape and delicate structure such as 3D printed scaffolds. During the top-down tissue regeneration, cells seeded onto the scaffolds are expected to distribute throughout the scaffolds without distinct control of their spatial organization, instead of self-organizing and growing into the scaffolds to develop their own ECM and micro-architecture with the help of perfusion (Niklason et al., 1999), growth factors (Gooch et al., 2002), and/or mechanical stimulation (Boublik et al., 2005; Gopalan et al., 2003). However, despite the advances in surface design to induce the desirable cell attachment or the use of more biomimetic scaffolds to match the tissue native environment, many of the top-down approaches oftentimes fail to recapitulate the complex microstructural characteristics of native tissue due to the difficulties of controlling spatially cell attachment with desired seeding density (Nichol & Khademhosseini, 2009) (Table 1).

To overcome the above-mentioned drawbacks of top-down approach, a bottom-up strategy is accordingly explored in order to better control cellular distribution. In fact, the layer-by-layer bioprinting of cells suspended in the bioink (e.g., hydrogel) into 3D structure represents a good example of the bottom-up modality (Murphy & Atala, 2014). In general, the bottom-up strategy focuses on building tissues or organs from the building units, which can be as small as cells or cell aggregates, upon assembly of them into large constructs with the support of matrices like hydrogel or molds. As noted, this strategy can better mimic the natural development of tissues and organs taking place in the body, where cells self-organize and interact to form

Table 1 Comparison between top-down and bottom-up approach.

	Top-down	Bottom-up
Key procedural steps	• Fabrication of large 3D scaffolds with defined shape • Seeding and culture of desired Cells onto the scaffolds	• Creation of cell-laden building modules • Assembly of modules into larger constructs
Advantages	• Relatively easy to operate • Defined size and shape	• Control of spatial cell distribution • Accommodation of compositional heterogeneity and structural complexity
Limitations	• Limited control of cell distribution, uniformity and density • Limited control of spatial distribution of multiple cell types	• Involvement of additional step • Prolonged experimental time

complex structures. In recognition, the bottom–up tissue regeneration offers several immediate advantages, including the ability to create complex, multicellular tissues with spatial organization and functionality much close to native tissues. However, foreseeable challenges are also expected including scalability, vascularization, and immune compatibility for widespread clinical translation. Notably, continued advancements in biomaterials, cell biology, and tissue engineering techniques would enable to address these challenges.

3.2 Modular tissue engineering

As part of the bottom–up approaches, modular tissue engineering is accordingly developed, involving the fabrication and assembly of small tissue blocks/ modules to create larger, more complex tissues or organs. This approach is inspired by the idea of building complex structures from simpler components. The tissue modules are designed in such a way to typically mimic specific functions or characteristics of natural tissues. For example, efforts have been made to design such structural units on the microscale. Several ways have been adopted to fabricate the tissue modules such as self-assembled aggregation (Dean, Napolitano, Youssef, & Morgan, 2007), cell sheets (L'Heureux, Paquet, Labbe, Germain, & Auger, 1998), cell-laden hydrogels (Yeh et al., 2006), or direct printing of tissues (Mironov, Boland, Trusk, Forgacs, & Markwald, 2003). Among them, 3D bioprinting, microfluidics, and self-assembly are the most popular ones for precise control over their composition, structure, and function. These building blocks can be assembled through different manners like random packing (McGuigan & Sefton, 2007; McGuigan, Leung, & Sefton, 2006), stacking of layers (L'Heureux, McAllister, & De La Fuente, 2007) or directed assembly (Du, Lo, Ali, & Khademhosseini, 2008), and the assembly process can be achieved manually or with the help of robotic systems. After assembly, the constructs can be further cultured in vitro under controlled conditions, allowing the cells to proliferate and differentiate and leading to integration of the modules with each other and with the surrounding environment (Du et al., 2008; L'Heureux et al., 2007; McGuigan & Sefton, 2007; McGuigan et al., 2006). Although more steps are involved and thus prolong the experimental time, the resulting structure is more controllable and predictable.

In contrast to extensive progress in creation of various tissues like cartilage via the modular tissue engineering (Ford et al., 2018), growing attentions have been seen to form tumor tissues with the modular strategy. In the endeavor to form in vitro 3D tumor models through a bottom-up strategy, cell aggregation and cell-laden hydrogel are very commonly used to build modules. Cell aggregation refers to simple tumor spheroids; cell-laden hydrogel refers to those microtumors formed in different hydrogels that can help to provide abundant ECM such as collagen gel and Matrigel. There are other methods commonly used as tumor model building blocks such as polymeric microspheres and 3d printed microfibers/microtubes.

As mentioned previously, the complexity of a tumor and its microenvironment make it difficult to rebuild through the top-down approaches. In particular, the heterogeneity of tumors cannot be easily achieved even through the simple bottom-up approaches. In this regard, the application of modular strategy in building up 3D in vitro tumor models often requires more than one type of module and involves more complicated assembling methods, which are further elaborated with detailed examples below.

4. Application of modular strategy for tumor model formation

Researchers have made extensive progress toward the development of tumor spheroids or other small-scale models, such as cell-encapsulated hydrogel microspheres and organoids, to study the pathophysiology of various types of cancer (Pradhan, Clary, Seliktar, & Lipke, 2017; Zhang et al., 2021). After decades of hard work, these small-scale models can now be regarded as successful microtumor tissues. They can include multiple cell types, modest TME, and a certain degree of tumor architecture, such as different cell distributions from core to shell. However, the scale of these microtumors is unrealistic compared to the native tumor tissue. For example, the size of breast cancer usually varies from 2 cm (stage I) to 5 cm (stage IV); lung cancer tumors have a size from less than 3 cm (stage I) to larger than 7 cm (stage IV), etc. Many groups choose to use modular strategies to solve this problem. In recognition of such size-related disparity, emerging efforts are made to connect such microtumors into large ones. For example, Woodfield group reported a modular assembly approach to the fabrication of well-defined and reproducible ovarian cancer tumor models (Mekhileri et al., 2018, 2023). The building modules used for assembly were heterotypic ovarian carcinoma spheroids and cell-encapsulated GelMA microspheres. Both modules contained multicellular components (a mixture of ovarian carcinoma cells and foreskin fibroblasts); the microspheres not only could be reproduced easily but also contained a hydrogel matrix that could partially represent the TME. The modules were then bioassembled into the void space of a large 3D-printed scaffold ($3\times3\times2.64$ mm with 0.7-mm interfiber spacing for spheroid segment and $3\times3\times1.76$ mm with 1 mm interfiber spacing for microsphere segment) of PEG-PBT copolymers (poly(ethylene glycol)-terephthalate-poly(butylene terephthalate) with 55:45 wt% PEGT:PBT and PEG at 300 g/mol) by an automated device composed of singularization and injection to form a relevant size that can better mimic in vivo complex extracellular organization and architecture of the TME (see Fig. 2). In this case, the 3D printed scaffolds served mainly as the scaffolding frameworks to support the bioassembly while defining the shape of the final constructs. To demonstrate the flexibility of this bioassembly approach and the potential utilities, Woodfield and his colleagues compared the cell behaviors and their drug response profiles between the larger tumor models constructed from multicellular spheroid or microsphere modules and the tumor models established with traditional methods like standard 2D monolayer culture, individual multicellular 3D spheroid, and individual hydrogel microsphere models. The results clearly showed that there were significant advantages in using large models over small-scale models.

As a matter of fact, the ovarian cancer models established by Woodfield group share many advantages similar to other tumor models developed

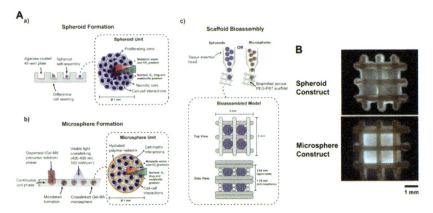

Fig. 2 Formation of ovarian cancer model via the bioassembly of spheroids and cell-laden microspheres. (A) Cancer 3D in vitro model overview. (a) Spheroid module formation using the liquid-overlay technique to produce tailorable multicellular spheroids of tunable size through differential cell seeding. The resulting spheroids have distinct biomolecular gradients and model cell–cell interactions. (b) Microsphere module formation using a droplet microfluidics system, coupled with cell-friendly visible-light photo-polymerization. Cancer cells are encapsulated within a spherical hydrated polymer network (e.g., GelMA hydrogel) and demonstrate both cell–cell and cell–matrix interactions. Microsphere and spheroid modules can replicate heterotypic interactions in the tumor microenvironment through modifying the cellular make-up during assembly (e.g., cocultures with fibroblastic cells). (c) Spheroid and microsphere tumor modules can be bioassembled into tumor-scale composite constructs for screening, drug discovery, or personalized medicine using an automated biofabrication platform which singularizes and inserts tumor modules into bioplotted thermoplastic polymer framework. (B) Darkfield images of spheroid and microsphere bioassembled constructs. The schematic image of Woodfield's model and its cell culture results. *Reproduced with permission from Mekhileri et al. (2023). Copyright 2023, Wiley.*

through the modular strategies, that is Efficiently reproducible modules that are less time-consuming and more cost-efficient; larger tissue size that can lead to lower drug sensitivities; tunable cell population and matrix molecules that closely mimic in vivo microenvironment, and a few more. Besides all the mentioned above, tumor models created through the modular strategies can also better recapture the physical architecture of native tumor tissue. With that said, the spatial organization of each module is not randomly packed into a bulk but is distributed more intentionally. Back to 2009, Morgan group was the first to report a fusion study to investigate the factors governing the fusion of microtissues and the cell sorting that occurred after fusion (Rago, Dean, & Morgan, 2009). They first looked into the fusion of normal human fibroblasts (NHF) spheroids by combining the spheroids cultured for 1, 4, or 7 days in trough shaped recesses. As demonstrated, the spheroids could fuse into a rod shape microtissue over the 24-h culture period. Notably, varying the amount of spheroid culture time prior to fusion (1–7 days) can lead to the variation in the rate of fusion, the coherence of the building units (as measured by fusion angle) and the steady state length of the structure. Longer pre-culture times for the spheroids resulted in slower fusion, less coherence and increased length of rod microtissues. Apparently, cell–cell interactions and cell–matrix interactions along with new ECM deposition all contribute to the spheroid fusion. In their study, further efforts were also made to fabricate the spheroids from the coculture of NHF and rat hepatocytes (H35s cell line) and use as building blocks to create tumor models with varying cell portions and physical architectures. By controlling the cell sorting and cell position in the building blocks, the fusion of spheroids, as well as the cell position in the tumor microtissue, could be well modulated (see Fig. 3).

In view of the essence to control cell portion and spatial distribution of various types of cells within the 3D tumor tissue model, further efforts have been extended to include other stromal cells such as vascular endothelial cells, fibroblasts and adipose stromal cells (Agarwal et al., 2017; Wang et al., 2023). Adequate vascularization is a key factor in determining the success

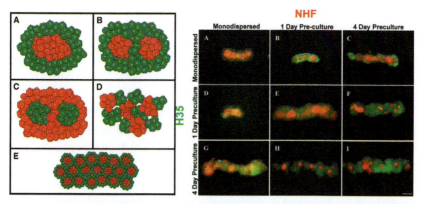

Fig. 3 Schematic representation of microtissues that can be generated by controlling the modules before assembly and fluorescent image of cell sorting in the cocultured microtissues. Schematic representation of microtissues that can be generated by controlling the pre-culture time of the building unit before assembly: (A) core-coating; (B) separate, but completely engulfed; (C) inside-out; (D) random unengulfed; (E) individually coated spheroids. *Reproduced with permission from Rago et al. (2009). Copyright 2009, Wiley.*

of a 3D in vitro tumor model. Unfortunately, the lack of vasculature in many current in vitro models implies that they still cannot fully recapture the key features of native solid tumors, typically with the presence of very rich microvascular networks as a result of the hypoxia-induced high level of pro-angiogenic factors. As a matter of fact, tumors, whether primary or secondary, require the recruitment of new blood vessels to grow to a large mass. In the absence of blood supply, a tumor can grow into a mass of about 10^6 cells, roughly a sphere of 2 mm in diameter (Brancato et al., 2017). At this point, division of cells on the outside of the tumor mass is balanced by the death of those in the center due to inadequate supply of nutrients. Such tumors, unless they secrete hormones, cause minimal problems in vivo. However, most tumors induce new blood vessel formation via autocrine secretion or paracrine secretion of proangiogenic growth factors to stimulate the angiogenesis, the process of new blood-vessel formation from the growth and sprouting of existing blood vessels. For example, basic fibroblast growth factor (bFGF), transforming growth factor α, and vascular endothelial growth factor (VEGF), secreted by many tumor cells, all have angiogenic properties. These new vessels nourish the growing tumor for large size while facilitating metastasis and increasing the probability for additional unwanted mutations. (Folkman & Kalluri, 2004; Gimbrone et al., 1972). In recognition of the crucial role of angiogenesis in cancer progression and metastasis, therapeutics have also been designed to target the abnormal angiogenesis with the intention to inhibit tumor growth and minimize the metastatic event. Thus, in the course of screening and testing of such anti-angiogenic therapeutic molecules, it becomes necessary to have those microvascular networks with the in vitro tumor models as well.

In an effort to vascularize the in vitro tumor model, He group took a "bottom-up" approach to fabricate a 3D vascularized human tumor with controlled formation of a complex 3D vascular network and then studied the effect of vascularization on cancer drug resistance. (Agarwal et al., 2017). They first encapsulated and cultured cancer cells in core–shell microcapsules ($< \sim 400\,\mu m$ in diameter) consisting of a type-I collagen-rich core enclosed with a semipermeable alginate hydrogel shell to form cell aggregates or microtumors (μtumors) with the assistance of a high-throughput nonplanar polydimethylsiloxane (PDMS) microfluidic encapsulation device (Fig. 4A(a)). Within a PDMS-glass microfluidic perfusion device ($1\times5\times0.5$ mm), the μtumors as the building blocks were assembled with human umbilical vein endothelial cells (HUVEC) and adipose-derived stem cells suspended collagen

In vitro tumor models for oncological research 237

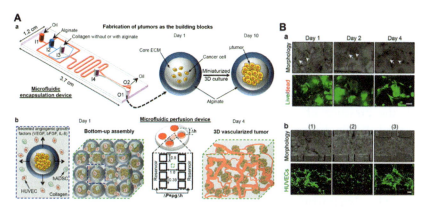

Fig. 4 The formation of vascularized tumor model. (A) Schematic illustration of major process involved. (a) A nonplanar microfluidic encapsulation device is used for encapsulating cancer cells in core–shell microcapsules, and the cells are cultured in the microcapsules for 10 days to form microtumors (μtumors, < ~200 μm in radius). Mineral oil infused with calcium chloride, aqueous sodium alginate solution (to form the microcapsule shell), aqueous collagen solution (with or without cells) to form the microcapsule core, and aqueous extraction solution are pumped into the device via inlets I1, I2, I3, and I4, respectively. The aqueous phase (containing core–shell microcapsules) and oil exit the device from outlets O1 and O2, respectively. (b) A microfluidic perfusion device is used to assemble the μtumors and stromal cells including endothelial cells for perfusion culture to form 3D vascularized tumor. The μtumors in core–shell microcapsules are assembled with human umbilical vein endothelial cells (HUVECs) and human adipose-derived stem cells (hADSCs) in collagen hydrogel in the microfluidic perfusion device. The alginate shell of the microcapsules is dissolved to allow cell–cell interactions and the formation of 3D vascularized tumor in the microfluidic perfusion device under perfusion driven by hydrostatic pressure. Units for the dimensions of micropillars and sample chamber: mm; P: pressure; ρ: density; g: gravitational acceleration; and h: height of medium column linked to the reservoirs. (B) Assembly of μtumors in microfluidic perfusion device to form vascularized 3D tumors. (a) Live/dead staining of the vascularized tumor on different days showing high cell viability. HUVECs without green fluorescence protein (GFP) were used. Arrow and arrowhead represent microcapsule shell and μtumor, respectively. (b) Phase contrast and fluorescence images (4×) showing vessel formation on day 4 with (1) μtumors encapsulated in microcapsules, (2) empty microcapsules, and (3) μtumors without microcapsules (the alginate hydrogel shell was dissolved by perfusing the samples with 75 mM sodium citrate solution for 5 min after 1 day of culturing the sample). HUVECs with GFP were used. Scale bar: (a) 100 μm and (b) 200 μm. *Reproduced with permission from Agarwal et al. (2017). Copyright 2017, American Chemical Society.*

hydrogel and then cultured under perfusion conditions to form millimeter-sized 3D vascularized tumor cancer cells (Fig. 4A(b)). It was expected that μtumors not only provided geometric support, but also offered the physicochemical guidance (secretion of angiogenic factors like bFGF, VEGF, IL8) to the HUVEC cells to form a complex 3D vascular network around the μtumors to mimic the vascular and cellular configuration of in vivo tumors (Fig. 4B). To demonstrate its utility, this study conducted drug tests to compare the drug sensitivity between the vascularized model with the avascular microtumors and monolayer of cancer cells. The results proved that with the presence of a vasculature network, drug resistance significantly increased. While demonstrating the possibility of forming vascularized miniaturized tumor tissues, it remains a challenge to make a larger size (e.g., ~cm in size) with a high flexibility to accommodate tumor heterogeneity typically composed of multiple mutated cancer cells and stromal cells at designated ratios. Thereby, it would be beneficial to develop a more robust and flexible platform allowing to better replicate such complicated circumstances in concurrent with the opportunity to develop microvascular networks. In this regard, Wang group recently conducted a study to use cell-laden polymeric microspheres as the modules to build up 3D vascularized tumor models with tunable cell portions and controllable spatial distributions of cancer cells and stromal cells (Fig. 5B) (Wang et al., 2023). As highlighted in Fig. 5A, the microsphere-enabled modular formation of breast cancer models was realized by following three key steps, including: (1) fabrication of PLGA microspheres (150–200 μm in diameter) with varying pore sizes (micropores vs. macropores) for different cell types (breast cancer cells, fibroblasts, and endothelial cells), (2) creation of cell-laden microsphere modules populated with respective cells under dynamic cultivation within spinner flasks, and (3) assembly of cell-laden microspheres into 3D constructs within a customized microfluidic chamber and formation of tumor tissue models with the presence of vasculature-like structures. This study not only focused on the cell-laden microspheres enabled-modular strategy to build up tumor models but also paid much attention to introducing endothelial cell-laden microspheres into cancer cell-laden microspheres at different ratios, aiming to induce angiogenesis within the culture to yield vascularized tumor. The established tumor models were successfully used to evaluate two known therapeutic molecules (i.e., doxorubicin and cediranib) for their anticancer efficacy. Cediranib, a small molecule inhibitor of all VEGF receptors, is now regarded as a potential anticancer drug because of its anti-angiogenesis function. As demonstrated, the vascularized tumor models showed the responses correlated with in vivo outcomes, which was similarly reported by Perch and Torchilin in the use of cancer spheroid models (Perche & Torchilin, 2012), confirming the potentials of the tumor models as the platform for anticancer drug testing. Meanwhile, considering the high flexibility to customize the

In vitro tumor models for oncological research   239

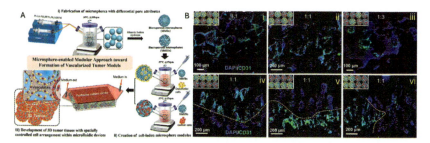

Fig. 5 Microsphere-enabled modular approach to forming breast cancer tissue models with structural and compositional complexity. (A) Schematic illustration of the key steps involved in microsphere-based formation of tumor model. (B) Fluorescence images showing tissue formation with the control of spatial distribution of various cells. *Reproduced with permission from Wang et al. (2023). Copyright 2023, Wiley.*

cell types within the tissue model, the model could be potentially used for pathophysiological study or design of personalized treatment. Notably, the porous microsphere-assisted creation of in vitro breast cancer models offers several benefits, including: (1) separate design of microspheres with physicochemical properties tailored for individual cell type, (2) uniform cell distribution within individual microspheres via dynamic cell seeding, (3) flexibility to combine different cell types in the models at desirable ratios by simply mixing the cell-laden microspheres, (4) creation of inter-sphere channels for mass/gas exchange upon the assembly of cell-laden microspheres into 3D constructs, (5) capability of forming arbitrarily shaped constructs via assembly of the cell-laden microspheres in corresponding molds, and (6) spatial organization of cell-laden microspheres in the assembled construct. In addition, such a strategy can be adapted to create other tissue models with a high scale-up potential.

Based on the findings from Wang and He groups, it is evident that microfluidic perfusion culture is favorable to support vascularization of the assembled building blocks along with tumor-like tissue formation. Taking advantage of the perfusion device setup, Lammers group also applied a perfusion culture to establish the vascularized tumor-like tissue (De Lorenzi et al., 2024). Different from the above-mentioned approaches, Lammers group included the avascular microtumors within a perfusion device containing pre-made vessels (see Fig. 6I). Vascular strands (gelatin mixed with endothelial cells, HUVEC) were bioprinted inside the perfusion device to form an oval space surrounding a tumor spheroid in ECM substitute containing other stromal cells. The ECM substitute used was fibrin-collagen

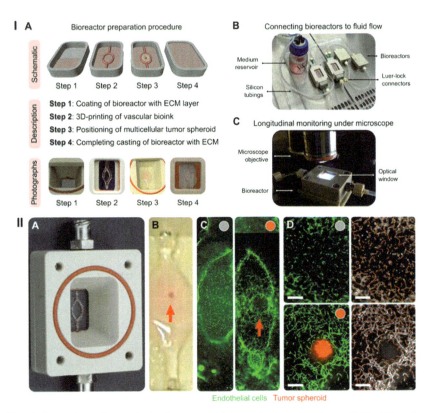

Fig. 6 Formation of vascularized tumor within perfusion chambers. (I) Workflow for creating perfused mesoscopic artificial tumors. (A) The process started with the placement of an ECM layer at the bottom of the bioreactor to increase the adherence of the hydrogel to the glass slide (step 1). Vascular strands (gelatin mixed with EC) were bioprinted to form an oval space in which a tumor spheroid in ECM containing either fibroblasts or fibroblasts and EC was positioned (steps 2 and 3). Finally, the bioreactor was filled with ECM (step 4). (B) Gelatin was removed and sterile perfusion was applied using a peristaltic pump in a recirculating flow loop. (C) EC organization and tumor spheroid growth were monitored by fluorescence microscopy. (II) The presence of tumor spheroids enhances vascularization of ECM hydrogel substitutes. (A, B) Bright field images display a single-inlet glass bottom bioreactor where branched gelatin strands are 3D-bioprinted to create an oval-shaped space. In the displayed example, a fibrinogen-collagen hydrogel blend containing endothelial cells (GFP-HUVEC), stromal cells (fibroblasts and mesenchymal stem cells), and a tumor spheroid (DiD-labeled A431, human skin squamous cell carcinoma) was cast. The position of the tumor spheroid within the hydrogel blend is indicated by the red arrow. (C, D) After seven days of dynamic cultivation, fluorescence microscopy images show a denser and more ramified vascular network between the two branched feeding vessels in the presence of tumor spheroids. Scale bar = 200 μm. *Reproduced with permission from De Lorenzi et al. (2024). Copyright 2024, Wiley.*

hydrogel blend, which could be modulated to yield the optimal mechanical properties by crosslinking the fibrinogen with thrombin and stabilizing with transglutaminase, antifibrinolytics tranexamic acid, and aprotinin. Furthermore, fibrin matrix itself could support the angiogenesis. The multicellular tumor spheroid was made of cancer (A431 human squamous carcinoma) and stromal cells (dermal fibroblasts). The dynamic medium flow (at a flow speed of 0.1 mL/min, imparting wall shear stress values of $\approx 0.01-0.03$ dyn/cm^2). went through the pre-made vascular strands, and the capillary network formation and microtumor growth can be easily monitored under the fluorescence microscope. As noted, the presence of cancer cells, secreting high level of VEGF, promoted the formation of a vascular network in the space comprised by the bioprinted vascular strands (Fig. 6II) after one week of dynamic cultivation. Different from all above-mentioned modular models, in this study there was no really assembly procedure but only one module (i.e., tumor spheroid) was included in the model. But this module could be changed to any other type of spheroid or microtumor unit, and the ECM surrounding vessel strands and microtissue unit could also be controlled. Such flexibility and visualization made this model a great candidate for studying tumor pathology.

Noteworthily, Lammers group used bioprinting to pre-make vessel strands within their tumor model. This technique has become popular in the creation of 3D in vitro tissue models. Bioprinting, also called direct cell printing or organ printing, creates 2-D arrays of cells and ECM/polymer (Zhang et al., 1999), which can be built layer-by-layer into 3D large structures (Fedorovich et al., 2007; Fedorovich, De Wijn, Verbout, Alblas, & Dhert, 2008; Roth et al., 2004). Typically, it deposits cells in small clusters using the technology similar to that of traditional printing systems but replaces the ink with cells suspended in aqueous solution of ECM or self-assembled ECM mimics, which are usually called bio-ink. The advantage of this technique is the high control level of cell and ECM placement and the potential to create engineered tissues with a wide array of properties and geometries. The resultant cell aggregates/structures are then cultured for an extended period, allowing the cells and ECM to fuse into one integrated tissue piece. This technique can be used to create individually patterned cell clusters or more intricate 2D arrays of cells (Zhang et al., 1999). In addition, it is possible to print subsequent layers on top of the printed layers once they are stabilized, creating 3D arrays of cells and tissues. Bioprinting has the capability to solve some of the challenges that traditional top-down methods are facing, e.g., spatial organization of cells.

However, the need of bio-inks to support the cell viability throughout the entire printing and beyond suggest that more efforts are necessary to develop more robust bio-inks suitable for different cell types. Moreover, to achieve the layer-to-layer binding and the integrity of ultimately printed structure, the bio-ink needs to meet certain viscosity and fluidity at 37 °C or below (Theus et al., 2020). On the other hand, with the aim of maintaining the geometry and structure of the desired tissues, the bio-ink must be able to solidify quickly to retain the printed structure and maintain the low fluidity right after the printing. Furthermore, the cell viability needs to be kept at a high level (>80% viability) during printing, which means the bio-ink should be very biocompatible.

Applying bioprinting in the modular strategy to create 3D in vitro models has been conducted by quite some groups. In particular, Kilian group developed a versatile approach to multi-domain tissue mimetics by extruding peptide-conjugated alginate filaments under the controlled flow rates, which could produce a variety of structures in a single simple device (Grolman, Zhang, Smith, Moore, & Kilian, 2015). With the single fluidic extrusion step, it is possible to integrate multiple cell types in distinct and controllable spatial domains of the extruded filaments as illustrated in Fig. 7, allowing for modeling tissue-mimetic interactions in vitro. Human breast cancer cells (MDA-MB-231) and mouse macrophage cells (RAW 264.7) were respectively mixed into pre-prepared 3.2% weight alginate and $0.046\,g/mL$ $CaCl_2$ in Dulbecco's Modified Eagles' media to formulate the respective bio-ink. After being placed in syringe pumps, the solutions were extruded through the microfluidic device with desired geometry and domain configuration as illustrated in Fig. 7A and B and collected in $45\,mg/mL$ $CaCl_2$ aqueous solution. These filaments were then cut into pieces suitable for mounting as flowable tissue culture chips or used in 96-well plates for long term culture. Depending on the outer fluid/inner fluid ratio, the architecture and pattern amplitude spacing could be changed. Therefore, Kilian fulfilled the aim to control the spatial positioning of 3D tumor-mimetic microenvironments containing multiple cell types with a simple concentric flow device in a single step. They used this method to demonstrate several geometric architectures, and the migration of segregated tumor cells and macrophages was also explored by using drugs that inhibited heterotypic interactions.

Combining bioprinting with other techniques could help build 3D in vitro tumor models with better replication of in vivo conditions. As highlighted in one of the early reviews on cancer organoids, cancer organoids

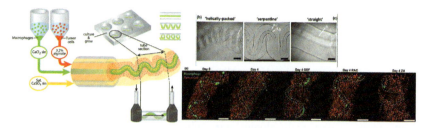

Fig. 7 Schematic of dual-cultured hydrogel fiber production. (A) 3D reconstructions of macrophages in green and tumor cells (231) in red stained with CellTracker and the location of the hollow inner channel indicated in dotted gray lines. The scale bar is 400 μm. (B) Optical images of select patterned fiber structures with 200 μm scale bars. Reproduced with permission from Grolman et al. (2015). Copyright 2015, Wiley.

have been considered as a major technological breakthrough and received significant attention recently because of their capability to retain the heterogeneity of original tumors (Fatehullah, Tan, & Barker, 2016). As a matter of fact, cancer organoids have been established as an essential tool in many basic biology and clinical applications, such as an accurate study of many in vivo biological processes including tissue renewal, stem cell/niche functions and tissue responses to drugs, mutation, or damage. More recent endeavors have been geared to establish the patient-derived in vitro cancer organoids, closely recapitulating the pathophysiological features of natural tumorigenesis and metastasis (Neal et al., 2018; Vlachogiannis et al., 2018). Meanwhile, cancer organoids have recently been utilized in the discovery of personalized anti-cancer therapy and prognostic biomarkers (Aberle et al., 2018). However, cancer organoid is still regarded as a newly emerging technique, and the protocols for establishment and batch-to-batch quality control are not fully standardized. And its heterogeneity can have negative effects in drug testing due to poor reproducibility (Karthaus et al., 2014; Tuysuz et al., 2017). Bioprinting can help to solve these problems and conveniently produce arrays of organoids with consistent shapes, sizes, and compositions. Therefore, the use of 3D bioprinting for organoid generation would enable the organoids with more complicated structures. Sachs group explored the incorporation of 3D bioprinting for generating tumor organoids with a self-gelling hydrogel that solely contained ECM from decellularized rat or human breast tissue (Mollica et al., 2019). Since such ECM hydrogel was able to faithfully recapitulate the compositional features of the TME and preserved various signaling molecules for the organoid growth, Sachs group successfully created large organoids with the assistance of a 3D bioprinter.

Utilization of native ECM to support the formation of tumor organoids determined their suitability not only for cancer mechanistic study but also for identifying the personalized treatment recipe. Personalized treatment/precision medicine for cancer imply the medical intervention or therapeutic cocktails could be tailored for each patient based on the individual characteristics of the cancer (Savoia et al., 2017). The personalized treatment apparently relies on the advancement of new technologies, e.g., whole body genomes, to understand the unique molecular and genetic profiles susceptible to certain disease developments and, therefore, to suggest the most effective treatment for each patient (Akhondzadeh, 2014). With increasing interest in "personalized medicine" for cancer therapies, especially considering the drastic variation in response to therapeutics between individual patients, the modular strategy offers great potentials to fabricate patient-specific 3D in vitro tumor models to screen various therapeutic cocktails for personalized therapy. This is especially true because the modular tumor formation has the flexibility in assembly and scalability, allowing for the construction of homogeneous or heterogeneous tumor tissues in terms of varying shape, size, and cell types.

5. Future perspective and concluding remarks

The modular strategy provides an effective avenue to form 3D in vitro tumor models toward biologically relevant sizes with precise control over the architectural arrangement of modules and spatial compartmentalization of cell populations, biomaterials, and native ECM microenvironments. Clearly, these modular tumor models provide opportunities to better understand drug/nutrient diffusion profiles, much more representative of the native tumor conditions to advance cancer research and therapeutic development, e.g., preclinical drug screening capacities. Thus, the tumor models via modular approach holds significant promise for advancing personalized medicine, enhancing drug development and screening, facilitating the understanding of tumor heterogeneity, investigating immune response, discovering new biomarkers and accelerating clinical translation.

Modular in vitro tumor models can be used for high-throughput drug screening to identify novel anti-cancer compounds and predict drug responses. By incorporating patient-derived cells or tumor fragments into the modular platforms, drug efficacy and toxicity can be assessed in a more physiologically relevant context, potentially accelerating the drug development process. Furthermore, validating the efficacy of novel drugs and treatment strategies in modular systems that closely mimic human tumors can enhance the predictability of preclinical studies and therefore increase the likelihood of successful clinical translation. However, modular tumor models also bring challenges with compatible analysis tools and technical difficulties. To prove valuable for high-throughput screening, the formation of these tumors must be reproducible, scalable, and automated (Nichol & Khademhosseini, 2009). Tumors are typically composed of heterogeneous cell populations with varying genetic and phenotypic profiles. The modular in vitro tumor models allow to create customized models that closely mimic the genetic and phenotypic characteristics of individual patients' tumors, thereby enabling the testing of various therapeutic interventions to identify the most effective treatment strategies for each patient. Along with potentials for drug screening and testing, the modular in vitro tumor models can provide a platform for studying tumor heterogeneity and the dynamics of tumor–stroma interactions. By incorporating different cell types and microenvironmental components into the modular systems, researchers can elucidate the underlying mechanisms driving tumor progression, metastasis, and therapeutic resistance. Furthermore, the TME plays a crucial role in modulating immune responses and shaping tumor progression. Modular in vitro tumor models can be used to interrogate the complex interactions between cancer cells, immune cells, and the ECM. Integrating immune components into the modular tumor platforms enables to investigate immune evasion mechanisms employed by tumors and develop immunotherapeutic strategies to enhance anti-tumor immunity. Besides the utilities for therapy, modular in vitro tumor models can be used to identify novel biomarkers associated with tumor progression, metastasis, and therapeutic response. By analyzing the molecular and cellular changes within the modular systems, potential diagnostic and prognostic markers can be identified for different cancer types, thus, facilitating early detection and personalized treatment approaches.

Overall, the modular approach to forming tumor models represents a versatile and powerful tool for advancing our understanding of cancer biology, identifying new therapeutic targets, and developing more effective cancer treatments. Continued research and technological advancements in this field are expected to further enhance the utility and applicability of modular tumor models in cancer research and clinical practice.

Acknowledgment

The authors are thankful for the partial financial support by the National Science Foundation (NSF-GCR award number 2219014) and Army Medical Research and Material Command with award number W81XWH2211044.

References

Aberle, M. R., Burkhart, R. A., Tiriac, H., Olde Damink, S. W. M., Dejong, C. H. C., Tuveson, D. A., et al. (2018). Patient-derived organoid models help define personalized management of gastrointestinal cancer. *The British Journal of Surgery, 105*(2), e48–e60.

Agarwal, P., Wang, H., Sun, M., Xu, J., Zhao, S., Liu, Z., et al. (2017). Microfluidics enabled bottom-up engineering of 3D vascularized tumor for drug discovery. *ACS Nano, 11*(7), 6691–6702.

Akhondzadeh, S. (2014). Personalized medicine: A tailor made medicine. *Avicenna Journal of Medical Biotechnology, 6*(4), 191.

Amann, A., Zwierzina, M., Gamerith, G., Bitsche, M., Huber, J. M., Vogel, G. F., et al. (2014). Development of an innovative 3D cell culture system to study tumour–stroma interactions in non-small cell lung cancer cells. *PLoS One, 9*(3), e92511.

Aoudjit, F., & Vuori, K. (2001). Integrin signaling inhibits paclitaxel-induced apoptosis in breast cancer cells. *Oncogene, 20*(36), 4995–5004.

Blanco, T. M., Mantalaris, A., Bismarck, A., & Panoskaltsis, N. (2010). The development of a three-dimensional scaffold for ex vivo biomimicry of human acute myeloid leukaemia. *Biomaterials, 31*(8), 2243–2251.

Boublik, J., Park, H., Radisic, M., Tognana, E., Chen, F., Pei, M., et al. (2005). Mechanical properties and remodeling of hybrid cardiac constructs made from heart cells, fibrin, and biodegradable, elastomeric knitted fabric. *Tissue Engineering, 11*(7–8), 1122–1132.

Brancato, V., Garziano, A., Gioiella, F., Urciuolo, F., Imparato, G., Panzetta, V., et al. (2017). 3D is not enough: Building up a cell instructive microenvironment for tumoral stroma microtissues. *Acta Biomaterialia, 47*, 1–13.

Chen, L., Xiao, Z., Meng, Y., Zhao, Y., Han, J., Su, G., et al. (2012). The enhancement of cancer stem cell properties of MCF-7 cells in 3D collagen scaffolds for modeling of cancer and anti-cancer drugs. *Biomaterials, 33*(5), 1437–1444.

Chiew, G. G., Wei, N., Sultania, S., Lim, S., & Luo, K. Q. (2017). Bioengineered three-dimensional co-culture of cancer cells and endothelial cells: A model system for dual analysis of tumor growth and angiogenesis. *Biotechnology and Bioengineering, 114*, 1865–1877.

Choi, J., Kim, K., Kim, T., Liu, G., Bar-Shir, A., Hyeon, T., et al. (2011). Multimodal imaging of sustained drug release from 3-D poly(propylene fumarate) (PPF) scaffolds. *Journal of Controlled Release: Official Journal of the Controlled Release Society, 156*(2), 239–245.

Correia, A. L., & Bissell, M. J. (2012). The tumor microenvironment is a dominant force in multidrug resistance. *Drug Resistance Updates: Reviews and Commentaries in Antimicrobial and Anticancer Chemotherapy, 15*(1-2), 39–49.

Da Silva, J., Lautenschlager, F., Sivaniah, E., & Guck, J. R. (2010). The cavity-to-cavity migration of leukaemic cells through 3D honey-combed hydrogels with adjustable internal dimension and stiffness. *Biomaterials, 31*(8), 2201–2208.

De Lorenzi, F., Hansen, N., Theek, B., Daware, R., Motta, A., Breuel, S., et al. (2024). Engineering mesoscopic 3D tumor models with a self-organizing vascularized matrix. *Advanced Materials, 36*(5), e2303196.

Dean, D. M., Napolitano, A. P., Youssef, J., & Morgan, J. R. (2007). Rods, tori, and honeycombs: The directed self-assembly of microtissues with prescribed microscale geometries. *The FASEB Journal, 21*(14), 4005–4012.

DeForest, C. A., Polizzotti, B. D., & Anseth, K. S. (2009). Sequential click reactions for synthesizing and patterning three-dimensional cell microenvironments. *Nature Materials, 8*(8), 659–664.

Du, Y., Lo, E., Ali, S., & Khademhosseini, A. (2008). Directed assembly of cell-laden microgels for fabrication of 3D tissue constructs. *Proceedings of the National Academy of Sciences of the United States of America, 105*(28), 9522–9527.

Fatehullah, A., Tan, S. H., & Barker, N. (2016). Organoids as an in vitro model of human development and disease. *Nature Cell Biology, 18*(3), 246–254.

Faupel-Badger, J. M., Arcaro, K. F., Balkam, J. J., Eliassen, A. H., Hassiotou, F., Lebrilla, C. B., et al. (2013). Postpartum remodeling, lactation, and breast cancer risk: Summary of a National Cancer Institute-sponsored workshop. *Journal of the National Cancer Institute, 105*(3), 166–174.

Fedorovich, N. E., Alblas, J., de Wijn, J. R., Hennink, W. E., Verbout, A. J., & Dhert, W. J. (2007). Hydrogels as extracellular matrices for skeletal tissue engineering: State-of-the-art and novel application in organ printing. *Tissue Engineering, 13*(8), 1905–1925.

Fedorovich, N. E., De Wijn, J. R., Verbout, A. J., Alblas, J., & Dhert, W. J. (2008). Three-dimensional fiber deposition of cell-laden, viable, patterned constructs for bone tissue printing. *Tissue Engineering. Part A, 14*(1), 127–133.

Fischbach, C., Chen, R., Matsumoto, T., Schmelzle, T., Brugge, J. S., Polverini, P. J., et al. (2007). Engineering tumors with 3D scaffolds. *Nature Methods, 4*(10), 855–860.

Folkman, J., & Kalluri, R. (2004). Cancer without disease. *Nature, 427*(6977), 787.

Ford, A. C., Chui, W. F., Zeng, A. Y., Nandy, A., Liebenberg, E., Carraro, C., et al. (2018). A modular approach to creating large engineered cartilage surfaces. *Journal of Biomechanics, 67,* 177–183.

Gimbrone, M. A., Jr., Leapman, S. B., Cotran, R. S., & Folkman, J. (1972). Tumor dormancy in vivo by prevention of neovascularization. *The Journal of Experimental Medicine, 136*(2), 261–276.

Gooch, K. J., Blunk, T., Courter, D. L., Sieminski, A. L., Vunjak-Novakovic, G., & Freed, L. E. (2002). Bone morphogenetic proteins-2, -12, and -13 modulate in vitro development of engineered cartilage. *Tissue Engineering, 8*(4), 591–601.

Gopalan, S. M., Flaim, C., Bhatia, S. N., Hoshijima, M., Knoell, R., Chien, K. R., et al. (2003). Anisotropic stretch-induced hypertrophy in neonatal ventricular myocytes micropatterned on deformable elastomers. *Biotechnology and Bioengineering, 81*(5), 578–587.

Grolman, J. M., Zhang, D., Smith, A. M., Moore, J. S., & Kilian, K. A. (2015). Rapid 3D extrusion of synthetic tumor microenvironments. *Advanced Materials, 27*(37), 5512–5517.

Hauptmann, S., Zwadlo-Klarwasser, G., Jansen, M., Klosterhalfen, B., & Kirkpatrick, C. J. (1993). Macrophages and multicellular tumor spheroids in co-culture: A three-dimensional model to study tumor-host interactions. Evidence for macrophage-mediated tumor cell proliferation and migration. *The American Journal of Pathology, 143*(5), 1406–1415.

Huang, C., Park, C. C., Hilsenbeck, S. G., Ward, R., Rimawi, M. F., Wang, Y. C., et al. (2011). beta1 integrin mediates an alternative survival pathway in breast cancer cells resistant to lapatinib. *Breast Cancer Research: BCR, 13*(4), R84.

Hutmacher, D. W. (2010). Biomaterials offer cancer research the third dimension. *Nature Materials, 9*(2), 90–93.

Kang, S. W., & Bae, Y. H. (2009). Cryopreservable and tumorigenic three-dimensional tumor culture in porous poly(lactic-co-glycolic acid) microsphere. *Biomaterials, 30*(25), 4227–4232.

Karthaus, W. R., Iaquinta, P. J., Drost, J., Gracanin, A., van Boxtel, R., Wongvipat, J., et al. (2014). Identification of multipotent luminal progenitor cells in human prostate organoid cultures. *Cell, 159*(1), 163–175.

Kloxin, A. M., Kasko, A. M., Salinas, C. N., & Anseth, K. S. (2009). Photodegradable hydrogels for dynamic tuning of physical and chemical properties. *Science (New York, N.Y.), 324*(5923), 59–63.

Kunz-Schughart, L. A. (1999). Multicellular tumor spheroids: Intermediates between monolayer culture and in vivo tumor. *Cell Biology International, 23*(3), 157–161.

L'Heureux, N., McAllister, T. N., & De La Fuente, L. M. (2007). Tissue-engineered blood vessel for adult arterial revascularization. *The New England Journal of Medicine, 357*(14), 1451–1453.

L'Heureux, N., Paquet, S., Labbe, R., Germain, L., & Auger, F. A. (1998). A completely biological tissue-engineered human blood vessel. *The FASEB Journal, 12*(1), 47–56.

Lee, J. M., Mhawech-Fauceglia, P., Lee, N., Parsanian, L. C., Lin, Y. G., Gayther, S. A., et al. (2013). A three-dimensional microenvironment alters protein expression and chemosensitivity of epithelial ovarian cancer cells in vitro. *Laboratory Investigation; a Journal of Technical Methods and Pathology, 93*(5), 528–542.

Li, Y., Fan, H., Ding, J., Xu, J., Liu, C., & Wang, H. (2022). Microfluidic devices: The application in TME modeling and the potential in immunotherapy optimization. *Frontiers in Genetics, 13*, 969723.

Liang, Y., Jeong, J., DeVolder, R. J., Cha, C., Wang, F., Tong, Y. W., et al. (2011). A cell-instructive hydrogel to regulate malignancy of 3D tumor spheroids with matrix rigidity. *Biomaterials, 32*(35), 9308–9315.

Lin, R. Z., Chou, L. F., Chien, C. C., & Chang, H. Y. (2006). Dynamic analysis of hepatoma spheroid formation: Roles of E-cadherin and beta1-integrin. *Cell and Tissue Research, 324*(3), 411–422.

McGuigan, A. P., Leung, B., & Sefton, M. V. (2006). Fabrication of cell-containing gel modules to assemble modular tissue-engineered constructs [corrected]. *Nature Protocols, 1*(6), 2963–2969.

McGuigan, A. P., & Sefton, M. V. (2007). Design and fabrication of sub-mm-sized modules containing encapsulated cells for modular tissue engineering. *Tissue Engineering, 13*(5), 1069–1078.

Mekhileri, N. V., Lim, K. S., Brown, G. C. J., Mutreja, I., Schon, B. S., Hooper, G. J., et al. (2018). Automated 3D bioassembly of micro-tissues for biofabrication of hybrid tissue engineered constructs. *Biofabrication, 10*(2), 024103.

Mekhileri, N. V., Major, G., Lim, K., Mutreja, I., Chitcholtan, K., Phillips, E., et al. (2023). Biofabrication of modular spheroids as tumor-scale microenvironments for drug screening. *Advanced Healthcare Materials, 12*(14), e2201581.

Menendez, J. A., Vellon, L., Mehmi, I., Teng, P. K., Griggs, D. W., & Lupu, R. (2005). A novel CYR61-triggered 'CYR61-alphavbeta3 integrin loop' regulates breast cancer cell survival and chemosensitivity through activation of ERK1/ERK2 MAPK signaling pathway. *Oncogene, 24*(5), 761–779.

Mironov, V., Boland, T., Trusk, T., Forgacs, G., & Markwald, R. R. (2003). Organ printing: Computer-aided jet-based 3D tissue engineering. *Trends in Biotechnology, 21*(4), 157–161.

Mollica, P. A., Booth-Creech, E. N., Reid, J. A., Zamponi, M., Sullivan, S. M., Palmer, X. L., et al. (2019). 3D bioprinted mammary organoids and tumoroids in human mammary derived ECM hydrogels. *Acta Biomaterialia, 95*, 201–213.

Murphy, S. V., & Atala, A. (2014). 3D bioprinting of tissues and organs. *Nature Biotechnology, 32*(8), 773–785.

Neal, J. T., Li, X., Zhu, J., Giangarra, V., Grzeskowiak, C. L., Ju, J., et al. (2018). Organoid modeling of the tumor immune microenvironment. *Cell, 175*(7), 1972–1988.e16.

Nichol, J. W., & Khademhosseini, A. (2009). Modular tissue engineering: Engineering biological tissues from the bottom up. *Soft Matter, 5*(7), 1312–1319.

Niklason, L. E., Gao, J., Abbott, W. M., Hirschi, K. K., Houser, S., Marini, R., et al. (1999). Functional arteries grown in vitro. *Science (New York, N.Y.), 284*(5413), 489–493.

Ozkan, S., Kalyon, D. M., & Yu, X. (2010). Functionally graded beta-TCP/PCL nano-composite scaffolds: In vitro evaluation with human fetal osteoblast cells for bone tissue engineering. *Journal of Biomedical Materials Research. Part A, 92*(3), 1007–1018.

Ozkan, S., Kalyon, D. M., Yu, X., McKelvey, C. A., & Lowinger, M. (2009). Multifunctional protein-encapsulated polycaprolactone scaffolds: Fabrication and in vitro assessment for tissue engineering. *Biomaterials, 30*(26), 4336–4347.

Pampaloni, F., Reynaud, E. G., & Stelzer, E. H. (2007). The third dimension bridges the gap between cell culture and live tissue. *Nature Reviews. Molecular Cell Biology, 8*(10), 839–845.

Paterson, K., Zanivan, S., Glasspool, R., Coffelt, S. B., & Zagnoni, M. (2021). Microfluidic technologies for immunotherapy studies on solid tumours. *Lab on a Chip, 21*(12), 2306–2329.

Pathi, S. P., Kowalczewski, C., Tadipatri, R., & Fischbach, C. (2010). A novel 3-D mineralized tumor model to study breast cancer bone metastasis. *PLoS One, 5*(1), e8849.

Perche, F., & Torchilin, V. P. (2012). Cancer cell spheroids as a model to evaluate chemotherapy protocols. *Cancer Biology & Therapy, 13*(12), 1205–1213.

Pradhan, S., Clary, J. M., Seliktar, D., & Lipke, E. A. (2017). A three-dimensional spheroidal cancer model based on PEG-fibrinogen hydrogel microspheres. *Biomaterials, 115*, 141–154.

Rago, A. P., Dean, D. M., & Morgan, J. R. (2009). Controlling cell position in complex heterotypic 3D microtissues by tissue fusion. *Biotechnology and Bioengineering, 102*(4), 1231–1241.

Roth, E. A., Xu, T., Das, M., Gregory, C., Hickman, J. J., & Boland, T. (2004). Inkjet printing for high-throughput cell patterning. *Biomaterials, 25*(17), 3707–3715.

Savoia, C., Volpe, M., Grassi, G., Borghi, C., Agabiti Rosei, E., & Touyz, R. M. (2017). Personalized medicine—A modern approach for the diagnosis and management of hypertension. *Clinical Science (London, England: 1979), 131*(22), 2671–2685.

Seidl, P., Huettinger, R., Knuechel, R., & Kunz-Schughart, L. A. (2002). Three-dimensional fibroblast-tumor cell interaction causes downregulation of RACK1 mRNA expression in breast cancer cells in vitro. *International Journal of Cancer, 102*(2), 129–136.

Senturk-Ozer, S., Aktas, S., He, J., Fisher, F. T., & Kalyon, D. M. (2017). Nanoporous nanocomposite membranes via hybrid twin-screw extrusion-multijet electrospinning. *Nanotechnology, 28*(2), 025301.

Tan, P. H., Aung, K. Z., Toh, S. L., Goh, J. C., & Nathan, S. S. (2011). Three-dimensional porous silk tumor constructs in the approximation of in vivo osteosarcoma physiology. *Biomaterials, 32*(26), 6131–6137.

Tasso, R., Augello, A., Carida, M., Postiglione, F., Tibiletti, M. G., Bernasconi, B., et al. (2009). Development of sarcomas in mice implanted with mesenchymal stem cells seeded onto bioscaffolds. *Carcinogenesis, 30*(1), 150–157.

Theus, A. S., Ning, L., Hwang, B., Gil, C., Chen, S., Wombwell, A., et al. (2020). Bioprintability: Physiomechanical and biological requirements of materials for 3D bioprinting processes. *Polymers (Basel), 12*(10).

Tsui, J. H., Lee, W., Pun, S. H., Kim, J., & Kim, D. H. (2013). Microfluidics-assisted in vitro drug screening and carrier production. *Advanced Drug Delivery Reviews, 65*(11–12), 1575–1588.

Tuysuz, N., van Bloois, L., van den Brink, S., Begthel, H., Verstegen, M. M., & Cruz, L. J. (2017). Lipid-mediated Wnt protein stabilization enables serum-free culture of human organ stem cells. *Nature Communications, 8*, 14578.

Vinci, M., Gowan, S., Boxall, F., Patterson, L., Zimmermann, M., Court, W., et al. (2012). Advances in establishment and analysis of three-dimensional tumor spheroid-based functional assays for target validation and drug evaluation. *BMC Biology, 10*, 29.

Vlachogiannis, G., Hedayat, S., Vatsiou, A., Jamin, Y., Fernandez-Mateos, J., Khan, K., et al. (2018). Patient-derived organoids model treatment response of metastatic gastrointestinal cancers. *Science (New York, N.Y.), 359*(6378), 920–926.

Wang, W., Zhang, L., O'Dell, R., Yin, Z., Yu, D., Chen, H., et al. (2023). Microsphere-enabled modular formation of miniaturized in vitro breast cancer models. *Small (Weinheim an der Bergstrasse, Germany), 20*, e2307365.

Williams, D. F. (2014). The biomaterials conundrum in tissue engineering. *Tissue Engineering. Part A, 20*(7–8), 1129–1131.

Xie, H., Appelt, J. W., & Jenkins, R. W. (2021). Going with the flow: Modeling the tumor microenvironment using microfluidic technology. *Cancers (Basel), 13*(23), 6052.

Xiong, J., Balcioglu, H. E., & Danen, E. H. (2013). Integrin signaling in control of tumor growth and progression. *The International Journal of Biochemistry & Cell Biology, 45*(5), 1012–1015.

Yeh, J., Ling, Y., Karp, J. M., Gantz, J., Chandawarkar, A., Eng, G., et al. (2006). Micromolding of shape-controlled, harvestable cell-laden hydrogels. *Biomaterials, 27*(31), 5391–5398.

Yum, K., Hong, S. G., Healy, K. E., & Lee, L. P. (2014). Physiologically relevant organs on chips. *Biotechnology Journal, 9*(1), 16–27.

Zhang, J., Tavakoli, H., Ma, L., Li, X., Han, L., & Li, X. (2022). Immunotherapy discovery on tumor organoid-on-a-chip platforms that recapitulate the tumor microenvironment. *Advanced Drug Delivery Reviews, 187*, 114365.

Zhang, S., Yan, L., Altman, M., Lassle, M., Nugent, H., Frankel, F., et al. (1999). Biological surface engineering: A simple system for cell pattern formation. *Biomaterials, 20*(13), 1213–1220.

Zhang, T., Zhang, H., Zhou, W., Jiang, K., Liu, C., Wang, R., et al. (2021). One-step generation and purification of cell-encapsulated hydrogel microsphere with an easily assembled microfluidic device. *Frontiers in Bioengineering and Biotechnology, 9*, 816089.

Zhang, Y., Cai, X., Choi, S. W., Kim, C., Wang, L. V., & Xia, Y. (2010). Chronic label-free volumetric photoacoustic microscopy of melanoma cells in three-dimensional porous scaffolds. *Biomaterials, 31*(33), 8651–8658.

CHAPTER EIGHT

Unraveling the complexity: Advanced methods in analyzing DNA, RNA, and protein interactions

Maria Leonor Peixoto[a,b] and Esha Madan[c,d,e,*]

[a]Champalimaud Center for the Unknown, Lisbon, Portugal
[b]Instituto Superior Técnico, Universidade de Lisboa, Lisbon, Portugal
[c]Department of Surgery, Virginia Commonwealth University, School of Medicine, Richmond, VA, United States
[d]Massey Comprehensive Cancer Center, Virginia Commonwealth University, Richmond, VA, United States
[e]VCU Institute of Molecular Medicine, Department of Human and Molecular Genetics, Virginia Commonwealth University, School of Medicine, Richmond, VA, United States
*Corresponding author. e-mail address: Esha.Gogna@vcuhealth.org

Contents

1. Introduction	252
2. Intramolecular DNA-DNA and RNA-RNA interactions	254
2.1 DNA	256
2.2 RNA	257
2.3 X-Ray crystallography	258
2.4 Nuclear magnetic resonance spectroscopy	259
2.5 Cryo-electron microscopy	260
2.6 Circular dichroism	261
2.7 Single-molecule fluorescence resonance energy transfer (smFRET)	261
2.8 Further techniques	262
2.9 In silico predictions and combined approaches	263
3. Intermolecular DNA-RNA interactions	263
3.1 DNA-RNA interactions	263
3.2 Protein-DNA interactions	271
3.3 Protein-RNA interactions	278
4. Conclusion	284
Acknowledgments	286
References	286

Abstract

Exploring the intricate interplay within and between nucleic acids, as well as their interactions with proteins, holds pivotal significance in unraveling the molecular complexities steering cancer initiation and progression. To investigate these interactions, a diverse array of highly specific and sensitive molecular techniques has been developed.

Advances in Cancer Research, Volume 163
ISSN 0065-230X, https://doi.org/10.1016/bs.acr.2024.06.010
Copyright © 2024 Elsevier Inc. All rights are reserved, including those for text and data mining, AI training, and similar technologies.

The selection of a particular technique depends on the specific nature of the interactions. Typically, researchers employ an amalgamation of these different techniques to obtain a comprehensive and holistic understanding of inter- and intramolecular interactions involving DNA-DNA, RNA-RNA, DNA-RNA, or protein-DNA/RNA. Examining nucleic acid conformation reveals alternative secondary structures beyond conventional ones that have implications for cancer pathways. Mutational hotspots in cancer often lie within sequences prone to adopting these alternative structures, highlighting the importance of investigating intra-genomic and intra-transcriptomic interactions, especially in the context of mutations, to deepen our understanding of oncology. Beyond these intramolecular interactions, the interplay between DNA and RNA leads to formations like DNA:RNA hybrids (known as R-loops) or even DNA:DNA:RNA triplex structures, both influencing biological processes that ultimately impact cancer. Protein-nucleic acid interactions are intrinsic cellular phenomena crucial in both normal and pathological conditions. In particular, genetic mutations or single amino acid variations can alter a protein's structure, function, and binding affinity, thus influencing cancer progression. It is thus, imperative to understand the differences between wild-type (WT) and mutated (MT) genes, transcripts, and proteins. The review aims to summarize the frequently employed methods and techniques for investigating interactions involving nucleic acids and proteins, highlighting recent advancements and diverse adaptations of each technique.

1. Introduction

Within cells, the establishment of intramolecular interactions driven by forces inherent to nucleic acids and proteins induces the adoption of atypical secondary structures (Baek et al., 2024; Bochman, Paeschke, & Zakian, 2012; Spitale & Incarnato, 2023). Concurrently, molecules incessantly partake in intramolecular and intermolecular interactions with their counterparts, thereby fostering a complex molecular landscape. Understanding these molecular interplays within cancer cells serves as a critical avenue for elucidating the mechanisms orchestrating cancer initiation and progression, and identification of potential therapeutic targets (Ghersi & Singh, 2013; Li et al., 2022; Tateishi-Karimata & Sugimoto, 2021).

In the realm of intramolecular interactions, the non-canonical secondary structures inherent to nucleic acids have been documented as influential players in pathways impacting the hallmarks of cancer. Furthermore, cancer-associated mutational hotspots consistently exhibit an enrichment in sequences predisposed to form secondary structures (Bansal, Kaushik, & Kukreti, 2022). Notably, secondary structures increase the likelihood of formation of novel mutations across the genome (Georgakopoulos-Soares, Morganella, Jain, Hemberg, & Nik-Zainal, 2018). Therefore, studying intramolecular interactions spanning the entire genome and transcriptome, particularly in the context of mutational events, contributes significantly to advancing our understanding of oncological processes. This knowledge bears paramount importance for drugvivo design, offering a basis in molecular structures and interactions for the development of targeted therapeutic interventions (Amato, Iaccarino, Randazzo, Novellino, & Pagano, 2014; Banerjee, Panda, & Chatterjee, 2022).

DNA exhibits the capacity to interact with RNA molecules, forming structures such as DNA:RNA hybrids (R-loops) (Zhang et al., 2024) and DNA:DNA:RNA triplexes (Yue Li, Syed, & Sugiyama, 2016). These configurations have been documented to impact various biological processes, ultimately influencing cancer. Notably, DNA:RNA hybrids have been proposed as potentially efficient therapeutic targets in human cancer (Boros-Oláh et al., 2019). Furthermore, some authors have suggested that DNA:DNA:RNA triplex formation can be exploited to design anti-cancer strategies, such as suppressing the expression of oncogenes (Li et al., 2022). Concurrently, interactions between proteins and nucleic acids constitute intrinsic phenomena within cellular environments, occurring under both physiological and pathological conditions. For instance, abnormal expression of transcription factors is linked with various diseases such as cancer (Shiroma, Takahashi, Yamamoto, & Tahara, 2020). Particularly, genetic mutations and single amino acid variations (SAVs) can yield proteins with distinct structures and functions, altering binding affinities and influencing cancer progression (Liu et al., 2021). Consequently, an imperative aspect of research involves scrutinizing variations in interactions between wild-type (WT) and mutated (MT) genes, transcripts, and proteins. For example, the tumor suppressor p53, acknowledged as the guardian of the genome, plays a pivotal role in cellular responses to diverse stresses (Fusée et al., 2023; Lane, 1992). Operating through sequence-specific interactions with DNA response elements, p53 initiates cascades of downstream events. Conversely, missense mutations in p53, prevalent in cancer cells, result in altered p53 proteins that impact their interaction with DNA molecules (Stiewe & Haran, 2018). Therefore, investigating protein–nucleic acid interactions and characterizing the distinctions between WT and MT forms in cancer cells are critical, given their association with drug resistance and adverse prognoses (Stiewe & Haran, 2018). This multifaceted interplay underscores the complexity of molecular relationships in cancer biology and highlights the significance of investigating such interactions for therapeutic interventions.

A diverse array of techniques has undergone continuous development and refinement to unravel intramolecular and intermolecular interactions. Namely, spectroscopic modalities, including Circular Dichroism (CD), Surface Plasmon Resonance (SPR), and Nuclear Magnetic Resonance (NMR), have been instrumental (Bishop & Chaires, 2003; Del Villar-Guerra, Trent, & Chaires, 2018; Mei et al., 2021; Plavec, 2022; Subramani & Kim, 2023). Electrophoretic techniques typified the Electrophoretic Mobility Shift Assay (EMSA), and contemporary capillary-based methods like QIAxcel, have found application (Leisegang et al., 2022; Mondal et al., 2015). Biochemical methodologies, such as immunoprecipitation (IP) and protein pull-down assays (Underwood et al., 2013), have also been employed. Microscopic techniques including Fluorescence Resonance Energy Transfer (FRET), as well as conventional fluorescence and confocal microscopy techniques (Mondal et al., 2015) have further allowed to study such interactions. In addition, in silico analyses have allowed researchers to predict the formation of intramolecular and intermolecular structures (Dans et al., 2019; Ho, Ellison, Quigley, & Rich, 1986; Jie Xiao, 2008). The selection of a specific method depends upon the distinctive nature of interactions and the requisite information. Researchers often integrate diverse approaches to attain a holistic comprehension of interaction characteristics and properties, as some techniques are complementary to each other (Brünger, 1997; Leisegang et al., 2022; Mondal et al., 2015; Postepska-Igielska et al., 2015).

This review endeavors to outline frequently utilized techniques for scrutinizing interactions within nucleic acids and proteins, providing insights into recent advancements and method-specific variations. The review includes concise depictions of these techniques, coupled with illustrative examples of their applications in oncology research.

2. Intramolecular DNA-DNA and RNA-RNA interactions

In the genomic and transcriptomic landscapes, persistent and dynamic intramolecular interactions occur, leading to the formation of non-canonical secondary structures in nucleic acids (Fig. 1). These structural arrangements exert discernible influences on both transcriptional and post-transcriptional processes (Bochman et al., 2012; Tang, 2020). Specifically, regions characterized by open chromatin, gene promoters, 5' untranslated regions (5'UTRs), and 3' untranslated regions (3' UTRs) have been identified as sites harboring motifs conducive to the formation of secondary DNA or RNA structures (Esnault et al., 2023; Georgakopoulos-Soares, Chan, Ahituv, & Hemberg, 2022; Lago et al., 2021). Various techniques have been developed and improved to determine and characterize such structures (Fig. 2).

Unraveling the complexity 255

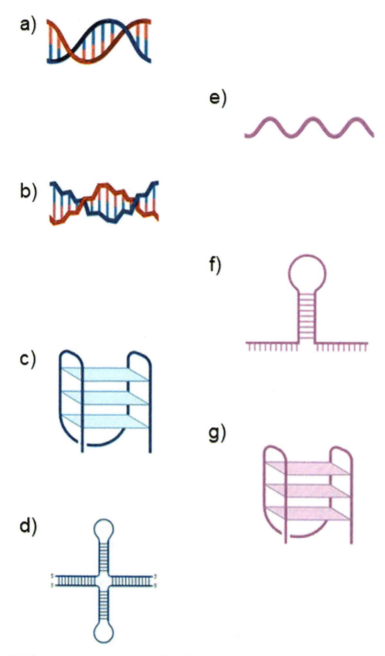

Fig. 1 Schematic representations of nucleic acid structures. (A) Canonical B-DNA: a right-handed double helix formed by two antiparallel strands held together by hydrogen bonding between complementary bases (B) Z-DNA: a left-handed double

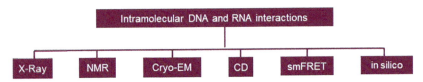

Fig. 2 Diagram illustrating techniques for investigating intramolecular DNA and RNA interactions. NMR: Nuclear Magnetic Resonance; Cryo-EM: Cryogenic Electron Microscopy; CD: Circular Dichroism; smFRET: Single-molecule Fluorescence Resonance Energy Transfer.

2.1 DNA

DNA typically adopts the canonical conformation known as B-DNA, elucidated by Watson and Crick in 1953 (Fig. 1A). This double-stranded helical structure is held by hydrogen bonds and remains stable under physiological conditions (Watson & Crick, 1953). In living cells, B-DNA is the prevalent form of DNA, manifested as a right-handed canonical helix with approximately 10.5 residues per turn (Belmont, Constant, & Demeunynck, 2001).

Under specific conditions, certain sequence motifs can assume alternative DNA conformations, referred to as non-canonical secondary structures (Bansal et al., 2022). These alternative DNA structures usually situated at promoter regions exhibit specificity in recognition and binding by certain transcription factors, thereby influencing gene expression (Bochman et al., 2012; Raiber, Kranaster, Lam, Nikan, & Balasubramanian, 2012). Moreover, these structures may serve as physical impediments that hinder nucleosome formation (Westin, Blomquist, Milligan, & Wrange, 1995). Additionally, sequences predisposed to adopting these non-canonical structures are associated with RNA polymerase II pausing, thereby affecting exon recognition, splicing events, and transcriptional termination (Szlachta et al., 2018). These processes have been implicated in various diseases, including cancer (Georgakopoulos-Soares et al., 2018).

The A-DNA structure is characterized by a right-handed double helix with 11 residues per turn, typically forming at GC-rich regions (Dickerson & Ng, 2001). In contrast, Z-DNA exhibits a left-handed double helical structure, consisting of purine and pyrimidine sequences (McLean, Lee, & Wells, 1988) (Fig. 1B). Z-DNA, less energetically favorable and stable than B-DNA, was identified through X-ray diffraction patterns and Circular Dichroism (CD) spectroscopy (Mitsui et al., 1970; Wang et al., 1979).

helix configuration of DNA. (C) DNA G-quadruplexes (G4s): formed by the stacking of G-quartets held together by Hoogsteen base pairing. (D) Hairpin loops and cruciform structures: formed by DNA sequences containing palindromic repeats. (E) Canonical single-stranded RNA. (F) RNA hairpin structure. (G) RNA G-quadruplex structure. Created with BioRender.com.

While sparsely present in living cells, Z-DNA has been implicated in gene regulation and associated with cancer progression and other diseases (Ravichandran, Subramani, & Kim, 2019). Bioinformatic tools, such as Z-hunt (Ho et al., 1986) and Z-catcher (Jie Xiao, 2008), facilitate the prediction of Z-DNA-forming regions across the genome. Analytical data indicates their enrichment in upstream and downstream regions of transcriptional start sites (TSS), suggesting a role in transcriptional regulation (Jie Xiao, 2008).

Guanine (G)-quadruplexes (G4s) represent another non-canonical DNA secondary structure commonly found in promoter regions, exerting both positive and negative regulation on gene expression (Varshney, Spiegel, Zyner, Tannahill, & Balasubramanian, 2020). G-quadruplexes, formed in guanine-rich regions, involve intramolecular and intermolecular structures bound by Hoogsteen hydrogen bonds between four guanines, constituting a G-quartet (Zyner et al., 2022). The arrangement of multiple G-quartets gives rise to a G-quadruplex (Chu, Zhang, & Paukstelis, 2019) (Fig. 1C). These structures are induced by the presence of guanines that bind to each other in a parallel, antiparallel, or hybrid manner (Biver, 2022). Promoters of genes implicated in cancer, such as KRAS, BCL2, VEGFA (Jana et al., 2017), and MYC (Chen, Dickerhoff, Sakai, & Yang, 2022), have been reported to be enriched in G-quadruplexes.

Beyond G4s, inverted repeated sequences, or palindromic regions, can form hydrogen bonds leading to hairpin structures (Ganapathiraju, Subramanian, Chaparala, & Karunakaran, 2020). Additionally, cruciform structures, composed of two hairpins and a four-way junction (Inagaki et al., 2013), have been observed to accumulate at high levels in human cancer cell lines with microsatellite instability (MSI) (Matos-Rodrigues et al., 2022) (Fig. 1D). Moreover, H-DNA is a secondary structure resulting from the binding of a DNA strand to double-stranded DNA via Hoogsteen or reverse-Hoogsteen interactions, forming a triple-stranded DNA.

Single-molecule Fluorescence Resonance Energy Transfer (FRET) microscopy emerged as a valuable tool for studying various secondary structures including DNA hairpins, G-quadruplexes, and i-motifs (Bandyopadhyay & Mishra, 2021).

2.2 RNA

While RNA molecules are inherently stable, they exhibit remarkable dynamism within cells, with single-stranded sequences capable of adopting both canonical and non-canonical structural conformations, thus exerting diverse regulatory functions and giving rise to distinct cellular phenotypes (Spitale & Incarnato, 2023). Beyond the canonical single-stranded RNA molecule (Fig. 1E), and analogous to DNA, RNA displays the formation of secondary structures including hairpins (Fig. 1F), G-quadruplexes (Fig. 1G) (Mei et al., 2021; Wan, Kertesz, Spitale, Segal, & Chang, 2011), and Z-RNA (Nichols, Krall, Henen, Vögeli, & Vicens, 2023).

Commonly employed methods such as X-ray crystallography and cryo-electron microscopy have contributed to the structural elucidation of molecules (Filius, Kim, Severins, & Joo, 2021). The investigation of these non-canonical secondary structures extends to methodologies like Nuclear Magnetic Resonance (Davis, Adamiak, & Tinoco, 1990; Jeong et al., 2014) and Circular Dichroism (Mei et al., 2021), particularly for studying RNA-quadruplexes. Techniques for identifying Z-RNA in cells involve pull-downs utilizing Z-RNA-specific antibodies or Zα-containing proteins, albeit with undisclosed specific binding sites (Nichols, Krall, Henen, Vogeli, & Vicens, 2023). Furthermore, investigations into intermolecular interactions encompass different DNA (Eisenstein & Shakked, 1995; Rau & Parsegian, 1992; Yoo & Aksimentiev, 2016) and RNA-RNA interactions (Romero-López, Roda-Herreros, Berzal-Herranz, Ramos-Lorente, & Berzal-Herranz, 2023; Van Treeck & Parker, 2018).

Coupling methods and advanced techniques enhance the precision in unraveling the nature and function of intramolecular interactions, facilitating the development of innovative anti-cancer strategies (Kosiol, Juranek, Brossart, Heine, & Paeschke, 2021).

2.3 X-Ray crystallography

X-ray crystallography is a pivotal scientific method employing X-ray diffraction patterns to elucidate the intricate three-dimensional configurations of molecules, including nucleic acids (Bragg, 1962). For successful application, the substance under scrutiny must crystallize, necessitating exploration of various crystallization conditions (Thomas, 2012). In the 1970s, X-ray techniques were instrumental in unveiling the structure of transfer RNA (tRNA) (Blake, Fresco, & Langridge, 1970). Advanced methodologies utilizing X-ray have emerged to address non-canonical nucleic acid structures like DNA and RNA G-quadruplexes (Zhang et al., 2020). However, limitations exist as X-ray crystallography cannot capture information under conditions closely resembling the native state. Notably, small-angle X-ray scattering (SAXS), despite operating at a lower resolution, has proven to be an effective complementary technique in advancing the understanding of

DNA and RNA structures (Burke & Butcher, 2012). Its utility extends to larger RNA molecules and heterogeneous samples (Chen & Pollack, 2016; Fang, Stagno, Bhandari, Zuo, & Wang, 2015), thus broadening its applicability in structural biology.

2.4 Nuclear magnetic resonance spectroscopy

Nuclear Magnetic Resonance (NMR) spectroscopy, initially delineated in 1938 by Rabi and colleagues (Webb, 2008), constitutes a powerful analytical technique rooted in the scrutiny of the chemical milieu enveloping atomic nuclei. This method involves the absorption of radio-frequency electromagnetic radiation upon the imposition of a robust magnetic field. In the realm of chemistry, NMR has proven instrumental for the precise determination of molecular structures and interactions, extending its application to elucidate the secondary structures of DNA and RNA molecules (Plavec, 2022).

NMR offers distinctive advantages over alternative techniques, providing atom-specific insights into RNA structures, a feat shared with methods like X-ray crystallography and cryo-electron microscopy (Marušič, Toplishek, & Plavec, 2023). Overcoming certain limitations inherent in other approaches, NMR proves particularly valuable for characterizing macromolecular systems and delineating alterations occurring during the interactions of DNA and RNA molecules with proteins, targets, or ligands. The amalgamation of various NMR types enhances the capacity to discern the intricacies of these interactions, further advancing our comprehension of molecular dynamics in biological systems (Marušič et al., 2023).

NMR spectroscopy serves as a versatile technique, furnishing intricate structural insights into molecular interactions and facilitating the characterization of double-stranded DNA molecules (Imeddourene et al., 2016; Parvathy et al., 2002; Sathyamoorthy, Sannapureddi, Negi, & Singh, 2022). This capability extends to the identification of sequence-specific variations within the DNA structure, enhancing our understanding of genetic intricacies (Sathyamoorthy et al., 2022). Additionally, NMR has proven invaluable in the analysis of G-quadruplexes and their interactions with other molecules under physiological conditions (Lin, Dickerhoff, & Yang, 2019). Further, collaborative efforts with X-ray crystallography have enabled the elucidation of the structures of both right- and left-handed hybrid G-quadruplexes through NMR (Winnerdy et al., 2019). The utility of NMR is further exemplified in its exploration of DNA triple helix formation, underscoring its applicability in unraveling diverse nucleic acid structures and interactions (Coman & Russu, 2004). Beyond DNA NMR serves as a crucial tool for determining and characterizing RNA intramolecular interactions, elucidating structural nuances, and deciphering associated kinetics (Barnwal, Yang, & Varani, 2017; Chen, Bellaousov, & Turner, 2016; Kennedy, 2016). In addition, the versatility of NMR extends to the analysis of the impact of nucleotide alterations *in vitro* and *in vivo* (Marušič et al., 2023). Noteworthy adaptations of *in vitro* techniques enable the study of systems under physiological intracellular conditions, providing biologically relevant insights into molecular dynamics (Marušič et al., 2023).

2.5 Cryo-electron microscopy

Cryo-electron microscopy (cryo-EM) is an advanced technique in structural biology utilized to unveil the three-dimensional architectures of biomacromolecules, including nucleic acids. Over recent years, cryo-EM has experienced notable advancements, leading to a paradigm shift within the structural biology domain (Guaita, Watters, & Loerch, 2022). Unlike conventional methods like X-ray crystallography and NMR, cryo-EM offers advantages by maintaining samples in a hydrated state and close to their native biological conditions through the freezing process and measurements (Doerr, 2017). Single-particle cryo-electron microscopy (cryo-EM) has facilitated the determination of diverse DNA nanostructures (Kato, Goodman, Erben, Turberfield, & Namba, 2009) and chromatin architectures, shedding light on their cellular functions (Takizawa & Kurumizaka, 2022). Despite significant progress, cryo-EM has encountered limited application in the study of RNA molecules (Ma, Jia, Zhang, & Su, 2022).

Cryo-EM has been successfully employed to resolve DNA G-quadruplexes (Monsen, Chua, Hopkins, Chaires, & Trent, 2023). Nevertheless, challenges persist, including the absence of computational methods tailored specifically for nucleic acid structure modeling (Wang, Terashi, & Kihara, 2023). CryoREAD, a deep learning-based approach, has been developed to assist in the de novo modeling of DNA and RNA atomic structures (Wang et al., 2023). This method identifies the positions of phosphate groups, sugar moieties, and bases within cryo-EM maps, facilitating the construction of three-dimensional structures.

The Electron Microscopy Data Bank (EMDB) database serves as a repository for cryo-EM 3D maps and associated information contributed by researchers, providing a valuable resource for the scientific community.

2.6 Circular dichroism

Circular Dichroism (CD) spectroscopy is another important technique for measuring absorption of left and right circularly polarized light by optically active molecules, particularly chiral molecules (Cotton, 1895). It is widely applied in the study of nucleic acid secondary structures, is instrumental in unraveling various nucleic acid conformations, including non-B DNA structures (Kypr, Kejnovska, Renciuk, & Vorlickova, 2009), namely H-DNA (Raghavan et al., 2005), Z-DNA (Subramani & Kim, 2023), DNA G-quadruplexes (G4s) (Del Villar-Guerra et al., 2018). In addition, CD is employed in the study of RNA G-quadruplexes (Asha, Green, Esposito, Santoro, & Improta, 2023). In the case of double-stranded B-DNA, CD reveals distinct signatures with a positive peak around 275 nm and a negative peak around 245 nm, offering a means to differentiate structural features (Josué Carvalho, 2017). For G4s, CD facilitates the discrimination of antiparallel conformations, yielding a positive peak around 290 nm and a weak negative band around 260 nm, from parallel structures that generate positive and negative signals around 260 nm and 240 nm, respectively (Paramasivan, Rujan, & Bolton, 2007).

While CD spectroscopy proves powerful in monitoring changes in the secondary and tertiary structure of nucleic acids due to variations in pH, temperature, and other conditions (Bishop & Chaires, 2003), it is constrained to providing overall structural characteristics, lacking sequence-specific details (Sosnick, 2001). To overcome this limitation, CD has been synergistically combined with other techniques such as nuclear magnetic resonance (NMR) and X-ray crystallography, culminating in a more comprehensive understanding of macromolecular interactions (Ravichandran, Razzaq, Parveen, Ghosh, & Kim, 2021; Winnerdy et al., 2019).

The publicly accessible Nucleic Acid Circular Dichroism Database (NACDDB) serves as a repository for CD data related to nucleic acids, providing relevant experimental metadata (Cappannini et al., 2023). This database consolidates CD data emanating from both intramolecular and intermolecular interactions involving nucleic acids.

2.7 Single-molecule fluorescence resonance energy transfer (smFRET)

Single-molecule Fluorescence Resonance Energy Transfer (smFRET), an application of Förster Resonance Energy Transfer (FRET) serves as a molecular nanoscopic "ruler," proficient in measuring distances between two labeled molecules within the 3–9 nm range (Lerner et al., 2018). This innovative technique involves the transfer of energy between two fluorophores—a donor and an acceptor—exemplified by pairs like Cy3 and Cy5 (Patel, Yilmaz, Bhatia, Biswas-Fiss, & Biswas, 2018). Upon excitation by a light source, the resultant readout is the transferred energy from the donor to an acceptor in close proximity. Consequently, smFRET has emerged as a potent tool for real-time investigations into molecular dynamics, enabling the precise determination of intramolecular distances. The variations in smFRET signals serve as indicators for conformational alterations, as well as the proximity or separation of the labeled sequences (Qiao, Luo, Long, Xing, & Tu, 2021; Sasmal, Pulido, Kasal, & Huang, 2016). Critical specifications of this technique include the use of low molar concentrations to ensure the presence of a single molecule within the analyzed volume at a given time (Sasmal et al., 2016). This technique serves as a dynamic tool, offering insights into transitions between various secondary DNA structures, spanning DNA hairpins, holiday junctions, G-quadruplexes, and i-motifs (Bandyopadhyay & Mishra, 2021; Hu et al., 2018). Moreover, smFRET provides a lens to observe the formation of H-DNA triplexes, shedding light on dynamics under different concentrations that influence triplex formation (Lee, Lee, Lee, & Hong, 2012). The versatility of smFRET extends to monitoring real-time conformational changes and folding during RNA transcription (Uhm, Kang, Ha, Kang, & Hohng, 2018). The precision of smFRET make it an invaluable tool for unraveling the dynamic intricacies of molecular structures and interactions at the single-molecule level.

2.8 Further techniques

The introduction of S1-END-seq, a recently developed methodology that uses S1 endonuclease provides researchers with a robust tool for obtaining a high-resolution insight into DNA secondary structures *in vivo* within human cells (Matos-Rodrigues et al., 2022). Complementing this, specific authors have pioneered a permanganate/S1 nuclease footprinting technique (Lahnsteiner et al., 2023). Utilizing potassium permanganate, this method identifies single-stranded non-B DNA structures. The subsequent application of S1 nuclease digestion and adapter ligation enables the comprehensive detection of genome-wide non-canonical structures in live cells (Lahnsteiner et al., 2023). Additionally, a genome-wide profiling initiative has been undertaken to explore RNA secondary structures within human cells from individuals representing different generations, including a mother, father, and child (Wan et al., 2014).

2.9 In silico predictions and combined approaches

In recent years, there has been a proliferation of computational tools aimed at predicting non-canonical DNA and RNA structures. These tools encompass various methodologies designed to modulate and simulate DNA and RNA structures, complemented by bioinformatic analyses (Binet, Padiolleau-Lefèvre, Octave, Avalle, & Maffucci, 2023; Dans et al., 2019). For instance, RRDNAnalyser has been engineered to forecast DNA secondary structures (Afzal, Shahid, Shehzadi, Nadeem, & Husnain, 2012), while RNAsnp has been developed to detect alterations in RNA secondary structures induced by single nucleotide polymorphisms (SNPs) (Sabarinathan et al., 2013). RS3D integrates RNA secondary structures with small-angle X-ray scattering (SAXS) data to elucidate tertiary RNA structures (Bhandari et al., 2017). Additionally, Mxfold2 employs deep learning techniques, incorporating thermodynamic data to predict RNA secondary structures (Sato, Akiyama, & Sakakibara, 2021).

Collectively, these cutting-edge methodologies significantly advance our comprehension of non-canonical secondary structures at a genomic scale, thereby contributing to the elucidation of underlying mechanisms associated with health and disease, including cancer. The insights derived from deciphering these secondary structures hold considerable promise for guiding the identification of drug targets and the development of innovative therapeutic strategies (Childs-Disney et al., 2022; Tan & Lan, 2020).

3. Intermolecular DNA-RNA interactions
3.1 DNA-RNA interactions

The intricate interplay between DNA and RNA molecules encompasses a spectrum of diverse structures (Yan et al., 2019). One prevalent interaction involves the hybridization of an RNA strand with a double-stranded DNA template, resulting in the formation of a three-stranded structure known as an R-loop or RNA-DNA hybrid loop (Fig. 3A) (Thomas, White, & Davis, 1976). Frequently found in sequence-favored regions, R-loops can extend up to a kilobase in length (Yu, Chedin, Hsieh, Wilson, & Lieber, 2003) and demonstrate heightened thermodynamic favorability compared to canonical DNA:DNA base-pairing (Roberts & Crothers, 1992). These RNA-DNA hybrids are distributed widely across the human genome and play crucial roles in various cellular processes, including transcription, where RNA is synthesized from a DNA template (Belotserkovskii, Tornaletti, D'Souza, & Hanawalt, 2018). While DNA:RNA hybrids are pivotal for gene expression

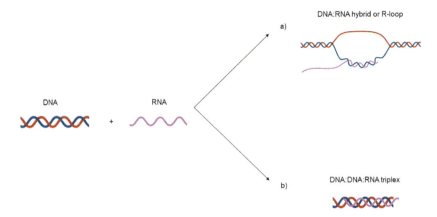

Fig. 3 Intermolecular interactions involving DNA and RNA. (A) DNA:RNA hybrid structures, also known as R-loops. (B) DNA:DNA:RNA triplexes. *Created with BioRender.com*

(Aguilera & García-Muse, 2012), their excessive or prolonged formation is associated with DNA damage, genomic instability, and various diseases, including cancer (Richard & Manley, 2017).

Beyond R-loops, DNA and RNA can form DNA:DNA:RNA triplexes (Fig. 3B) (Warwick et al., 2022), stabilized by Hoogsteen or reverse Hoogsteen hydrogen bonds (Maldonado, Filarsky, Grummt, & Längst, 2018). These triplexes selectively recruit proteins, including methylases (Schmitz, Mayer, Postepska, & Grummt, 2010), histone acetyltransferases (Postepska-Igielska et al., 2015), and protein complexes (Leisegang et al., 2022) to specific genomic sequences, thereby influencing their activation or repression. DNA:DNA:RNA triplexes play crucial roles in cell viability, metabolism, immune response, and biological processes during development, aging, and tumorigenesis (Warwick, Brandes, & Leisegang, 2023). Additionally, DNA and RNA can engage in the formation of DNA:RNA hybrid G-quadruplexes (Zhang, Zheng, Xiao, Hao, & Tan, 2014), non-canonical structures linked to genomic instability and implicated in various pathological diseases, including cancer. Importantly, these structures, along with their associated binding proteins, emerge as potential targets for anti-cancer therapeutic interventions (Boros-Oláh et al., 2019).

The diverse interactions between DNA and RNA molecules, particularly significant in oncology, are amenable to analysis through various techniques such as electrophoretic mobility shift assay (EMSA), microscopy, spectroscopy and immunostaining-based techniques, nuclear magnetic resonance

(NMR) spectroscopy, surface plasmon resonance (SPR), and pull-down assays (Mondal et al., 2015; O'Leary et al., 2015). The synergistic application of these methods provides unique insights into the structural and dynamic characteristics of DNA-RNA interactions, fostering a comprehensive understanding of their roles and functions, particularly in the context of oncological research (Li et al., 2016). An overview of some techniques commonly employed to study and characterize DNA-RNA interactions has been provided in Fig. 4.

3.1.1 Electrophoretic mobility shift assay (EMSA)

The electrophoretic mobility shift assay EMSA assay (Fried & Crothers, 1981; Garner & Revzin, 1981), has evolved into a widely employed technique providing valuable insights into macromolecular interactions, including those involving DNA and RNA (Hellman & Fried, 2007). In this assay, nucleic acids, either alone or in combination with others, are loaded onto a polyacrylamide or agarose gel and subjected to an electric field. This induces the migration of single molecules and molecular complexes, leading to the analysis of distinct shifts in the gel. Subsequent to electrophoresis, the gel containing migrated nucleic acids is typically analyzed under UV light (Cunha, de Souza, Lanza, & Lima, 2020; "Electrophoretic mobility shift assays," 2005). Detection is facilitated using covalently or non-covalently

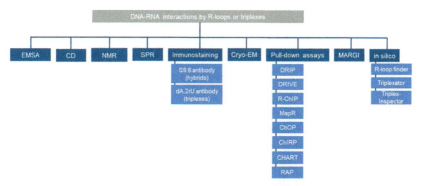

Fig. 4 Schematic representation of the techniques used for studying DNA and RNA hybrids and triplexes. EMSA: Electrophoresis Mobility Shift Assay; CD: Circular Dichroism spectroscopy; NMR: Nuclear Magnetic Resonance; SPR: Surface Plasmon Resonance; Cryo-EM: Cryogenic Electron Microscopy; DRIP: DNA-RNA Immunoprecipitation; DRIVE: DNA:RNA in vitro enrichment; ChIP: Chromatin Immunoprecipitation; ChOP: Chromatin Oligo-affinity Precipitation; ChIRP: Chromatin Isolation by RNA Purification; CHART: Capture Hybridization Analysis of RNA Targets; RAP: RNA Antisense Purification; MARGI: Mapping RNA-Genome Interactions.

bound fluorophores or biotin. For enhanced sensitivity, radioisotope-labeled nucleic acids can be employed, enabling the detection of molecules at nanomolar concentrations (Hellman & Fried, 2007).

EMSA technique proves versatile in detecting molecules with varied sizes and structures, spanning single-stranded, double-stranded, quadruplex, triplex, hairpin, and circular forms. Protocols for EMSA derivations have been comprehensively reviewed (Hellman & Fried, 2007). While often employed for qualitative purposes, the intensity of bands obtained in EMSA assays can be densitometrically quantified. This quantification enables comparisons between untreated and treated samples, such as those treated with RNAse A and H, utilizing software like ImageJ (NIH) (Haeusler et al., 2014), QuantityOne (BioRad), Intelligent or Advanced Quantifier (Bio Image), and Densitometric Image Analysis Software (Van Oeffelen, Peeters, Nguyen Le Minh, & Charlier, 2014).

In an EMSA assay, the formation of R-loops or DNA:DNA:RNA triplex structures results in decreased migration in the gel compared to unbound nucleic acids (Yuan et al., 2019) (Fig. 5A). Owing to their heterogeneity, these hybrids migrate as a smear. To confirm the nature of the DNA:RNA structure, treatment with RNases can be employed (Yuan et al., 2019). Specifically, RNase H cleaves DNA:RNA hybrids, while RNase A selectively removes ssRNA without affecting R-loop formation (Haeusler et al., 2014).

3.1.2 Circular dichroism

Circular Dichroism has emerged as a pivotal tool for studying DNA-RNA interactions (Mondal et al., 2015). Specifically, the distinctive pattern exhibited in circular dichroism experiments allows for the elucidation of DNA:DNA:RNA triplex formation. This pattern manifests as a positive small peak around 220 nm, two negative bands peaking at approximately 210 nm and 240 nm, and another positive peak at around 270 nm (Leisegang et al., 2022). Notably, these spectral features distinguish themselves from those generated by DNA duplexes or DNA:RNA heteroduplexes (Leisegang et al., 2022) (Fig. 5B).

3.1.3 Surface plasmon resonance

Surface plasmon resonance (SPR) stands out as a highly sensitive tool capable of measuring real-time changes in the refractive index at the surface of a sensor chip during molecular interactions (Majka & Speck, 2007). In essence, one of the interacting molecules under scrutiny is immobilized on the surface of the SPR chip, while the other molecule is introduced into

Unraveling the complexity

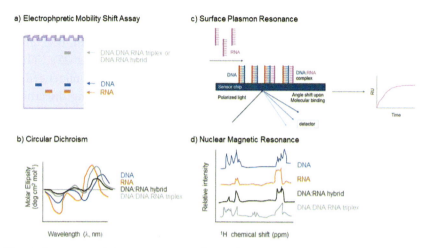

Fig. 5 Illustration of diverse strategies for analyzing interactions between DNA and RNA: (A) Electrophoresis Mobility Shift Assay (EMSA): Different structures like DNA:RNA hybrids (R-loops) and DNA:DNA:RNA triplexes manifest a distinct shift in the gel (gray), contrasting with free DNA (blue) and RNA (orange) molecules; (B) Circular Dichroism (CD) Spectra: Enables differentiation among double-stranded DNA, single-stranded RNA, DNA:RNA hybrids, and DNA:DNA:RNA triplex structures. Each structure, when free or bound, exhibits unique spectral peaks; (C) Surface Plasmon Resonance (SPR): Facilitates examination of oligonucleotide interactions. Probes fixed on a sensor chip bind molecules passing through, causing a change in the light source angle upon binding; (D) Nuclear Magnetic Resonance (NMR) Spectra: Representative spectra showcase specific peaks for double-stranded DNA, single-stranded RNA, DNA:RNA hybrids, or DNA:DNA:RNA triplex, aiding in the identification of novel structures.

the flow cell chamber. The experimental outcome is presented as a curve illustrating the increase in mass over time, expressed in resonance units (RU), providing insights into the interaction dynamics between the two molecules (Stahelin, 2013). This method facilitates the assessment of interaction kinetics, binding affinity, melting temperature and the determination of association and dissociation rate constants. A notable advantage of SPR lies in the absence of the need for labeling molecules (Majka & Speck, 2007), and the capability to reuse sensor chips in multiple cycles, thereby enhancing the method's throughput. This versatile technique has found application in the investigation of DNA:RNA hybrids (Zhou et al., 2015) and DNA:DNA:RNA triplexes (O'Leary et al., 2015) (Fig. 5C).

3.1.4 Nuclear magnetic resonance (NMR)
NMR emerges as a pivotal tool for investigating the interactions involving single-stranded RNAs (ssRNAs) and double-stranded DNA

(Trembinski et al., 2020). It plays a critical role in delineating the characteristics of both DNA:RNA hybrids (Znosko, Barnes, Krugh, & Turner, 2003) and DNA:DNA:RNA triplexes (Leisegang et al., 2022). These interactions and resulting complex formations are intricately influenced by experimental parameters such as pH and temperature, yielding discernible peaks. In hybrids, the appearance of imino proton NMR peaks within the range of 12 to 14/15 ppm signifies Watson-Crick base pairing, as observed in studies by Gudanis et al. (Gudanis et al., 2021; Pilch, Levenson, & Shafer, 1990; Trembinski et al., 2020). Conversely, peaks falling within the range of 9–12 ppm in triplexes suggest the presence of Hoogsteen base pairing (Leisegang et al., 2022) (Fig. 5D).

3.1.5 Immunostaining analyses

R-Loop structures can be detected by immunostaining using the S9.6 monoclonal antibody (Boguslawski et al., 1986; Bou-Nader, Bothra, Garboczi, Leppla, & Zhang, 2022; Skourti-Stathaki, 2022) Immunostaining employing the anti-triplex dA.2rU antibody, designed to discern triplex structures, proves efficacious in situating triplexes *in vivo* and unraveling their subcellular localization (Mondal et al., 2015). To validate the nature of *in vivo* triplex structures, differentiating between hydrogen bond-formed hybrids and those sustained by Hoogsteen-base pairing, treatment with RNase H—specifically designed for cleaving DNA-RNA hybrids—can be judiciously employed (Mondal et al., 2015).

3.1.6 Electron microscopy

In the quest for visualizing three-dimensional (3D) structures of R-loops in their native environment, advanced microscopy techniques, including electron microscopy (EM) (Reddy et al., 2010) and cryo-electron microscopy (cryo-EM) single-particle analysis (SPA) (Li et al., 2022), have proven invaluable.

3.1.7 Pull-down assays

DNA:RNA Immunoprecipitation (DRIP) is another important tool used to study DNA-RNA interactions. The *in vivo* application of the previously developed DNA:RNA hybrid-specific S9.6 monoclonal antibody (Boguslawski et al., 1986; Bou-Nader et al., 2022), in conjunction with high-throughput DNA sequencing (DRIP-seq), facilitates the comprehensive genome-scale mapping of R-loops in human cells (Sanz & Chédin, 2019; Stork et al., 2016). With the emergence of next-generation sequencing, numerous methods are developed to elucidate the distribution, size, and dynamic alterations of R-loops (Khan & Danckwardt, 2022). However, this technique lacks information on the strand-specificity or directionality of R-loop formation. To address these limitations, several protocol variants have been developed.

One such variant, DNA:RNA immunoprecipitation followed by cDNA conversion coupled to high-throughput sequencing (DRIPc-seq) (Hartono et al., 2018; Sanz et al., 2016), represents a high-resolution, strand-specific method for characterizing R-loops. In this strategy, immunoprecipitated fragments using the S9.6 monoclonal antibody undergo digestion with DNase I, enabling the recovery of purified RNA. Subsequently, the RNA is converted into cDNA and analyzed by RNA-seq, providing a strand-specific map of R-loops. Another method addressing limitations of DRIP technique, involves sonication for genome fragmentation at the initial step (sDRIP-seq) rather than using a restriction enzyme (Sanz, Castillo-Guzman, & Chédin, 2021). Additionally, bisulfite-based DRIP-seq (bisDRIP-seq), a combination of immunoprecipitation (IP) using S9.6 mAb with bisulfite sequencing of DNA sequence (Dumelie & Jaffrey, 2017), discriminates the R-loop sequence and the flanking non-R-loop sequence, establishing precise boundaries for this structure.

Other DRIP variants include DNA: RNA immunoprecipitation followed by hybridization on tiling microarray (DRIP-chip), utilizing the S9.6 mAb but employing hybridization of immunoprecipitated sequences rather than sequencing (Chan et al., 2014). While this method is rapid, it is limited to chip microarray probes and lacks DNA sequence information. Nadel et al., have optimized the conventional protocol to create RDIP, which involves RNAse I pre-treatment, sonication, S9.6 IP, and sequencing of RNA strands (Nadel et al., 2015).

Single-strand DRIP-seq (ssDRIP-seq) employs a cocktail of restriction enzymes for DNA fragmentation, followed by S9.6 IP of DNA:RNA hybrids and ligation of a single-stranded DNA adaptor for sequencing (Xu et al., 2017, 2022). RNASEH1, with increased affinity for RNA–DNA hybrids, has been utilized in the DNA: RNA *in vitro* enrichment (DRIVE) technique, employing the human RNASEH1 mutant protein in affinity pull-down assays, primarily chosen for *in vivo* studies (Bhatia et al., 2014; Tresini et al., 2015).

Moreover, MapR, a technique independent of antibodies and relying on the intrinsic specificity of RNase H, has been devised to identify native R-loops throughout the genome (Yan & Sarma, 2020). Upon R-loop recognition, cleavage is triggered, leading to the liberation of R-loops, which can subsequently undergo sequencing. This method, characterized by its high sensitivity and rapidity, is applicable across various cell types, not limited to stable ones. Notably, BisMapR integrates nucleases to isolate R-loops and employs non-denaturing bisulfite chemistry to comprehensively characterize these structures at a genome-wide scale, thereby contributing to elucidating their role in gene expression (Wulfridge & Sarma, 2021).

R-ChIP-seq maps R-loops genome-wide using catalytically inactive RNAse H1 (dRNASEH1) (Chen et al., 2017). This technique requires the exogenous expression of dRNASEH1 in cells, binding to hybrids, followed by IP (ChIP) and the generation of a strand-specific library for sequencing. facilitating the binding to hybrids, followed by IP (ChIP) and the generation of a strand-specific library for subsequent sequencing. This approach involves establishing the stable expression of the modified enzyme for approximately three weeks, in addition to the R-ChIP protocol taking five days (Chen, Zhang, Fu, & Chen, 2019). Notably, qDRIP emerged as a technique to quantify RNA:DNA hybrid formation across the genome (Crossley, Bocek, Hamperl, Swigut, & Cimprich, 2020).

3.1.8 Chromatin oligo-affinity precipitation (ChOP)

Chromatin oligo-affinity precipitation (ChOP), a modified adaptation of chromatin immunoprecipitation, has emerged as a valuable tool for studying interactions between long non-coding RNAs (lncRNA) and specific gene loci (Mariner et al., 2008). This technique involves an initial cross-linking step to stabilize nucleic acid and protein interactions. Subsequent steps include cellular lysis, chromatin sonication, hybridization of biotinylated anti-sense oligonucleotide probes with RNA-DNA-protein complexes, and pull-down facilitated by anti-biotin antibodies. Following stringent washing, RNA-associated chromatin is eluted with RNase, and subsequent treatment with proteinase K enables protein removal. The purified DNA is then subjected to analysis through quantitative polymerase chain reaction (qPCR) or sequencing (Mariner et al., 2008).

Similar methodologies, including chromatin isolation by RNA purification (ChIRP), capture hybridization analysis of RNA targets (CHART), and RNA antisense purification (RAP), empower researchers to analyze long non-coding RNAs bound to DNA sequences (Kuo et al., 2019). For a comprehensive genome-wide analysis of chromatin-associated long non-coding RNAs, chromatin RIP followed by high-throughput sequencing (ChIRP-seq) can be performed using antibodies targeting H3K27me3 and EZH2 in human cells (Mondal et al., 2015).

3.1.9 In silico predictions

Bioinformatic tools including R-loop tracker (Brázda, Havlík, Kolomazník, Trenz, & Šťastný, 2021), R-loopBase (Lin et al., 2021), Triplex Domain Finder (Kuo et al., 2019), Triplex-Inspector (Buske, Bauer, Mattick, & Bailey, 2013), and Triplexator (Buske, Bauer, Mattick, & Bailey, 2012), play a pivotal role in providing a comprehensive genome-wide prediction of DNA:RNA interactions.

3.1.10 Other chromatin interaction mapping techniques

The innovative MARGI (Mapping RNA-Genome Interactions) technique introduces a holistic strategy for elucidating native RNA-chromatin interactions throughout the entire genome (Sridhar et al., 2017). Unlike traditional methods limited to analyzing one RNA per experiment, MARGI discerns chromatin-enriched RNAs and their associated genomic loci within a single experimental setup. This involves the ligation of chromatin-associated RNAs with target DNA sequences through proximity ligation, subsequently followed by sequencing (Sridhar et al., 2017). Two MARGI variants, namely pxMARGI and diMARGI, have been devised. While pxMARGI lacks the ability to differentiate direct from indirect interactions, diMARGI precisely identifies direct interactions of proteins or RNA with chromatin (Sridhar et al., 2017). Another potent tool for the *in situ* examination of chromatin-associated coding and non-coding RNAs in human, mouse, and Drosophila cells is GRID-seq (Li et al., 2017).

3.2 Protein-DNA interactions

Protein interactions with DNA represent a cornerstone in the orchestration of various biological phenomena, spanning replication, transcription, and DNA repair (Guo et al., 2023). Simultaneously, the intricate interplay of proteins and DNA influences pivotal cellular functions, including transport, translation, splicing, and transcriptional silencing.

The engagement of proteins with nucleic acids encompasses an array of sophisticated mechanisms, incorporating hydrogen bonding, electrostatic interactions (especially salt bridges), hydrophobic interactions, and base stacking. Conventionally, the negatively charged phosphate groups of DNA and RNA intricately interact with the positively charged amino acid residues in proteins (Corley, Burns, & Yeo, 2020; Lin & Guo, 2019; Lunde, Moore, & Varani, 2007; Ohlendorf & Matthew, 1985). Noteworthy instances include the formation of electrostatic forces and hydrogen bonds between DNA and positively charged histones (Davey, Sargent, Luger, Maeder, &

Richmond, 2002; Luger, Mäder, Richmond, Sargent, & Richmond, 1997). These protein-DNA relationships may manifest in a nucleotide-specific or non-sequence-specific manner, with DNA-binding domains assuming diverse structures such as Zinc fingers, helix-turn-helix, helix-loop-helix, winged helix, and leucine zipper (Yesudhas, Batool, Anwar, Panneerselvam, & Choi, 2017). The relentless pursuit of understanding these intricate interactions has seen contributions from both *in vitro* and *in vivo* methodologies, unraveling the identities of DNA-binding proteins across diverse physiological and pathological scenarios. An overview of these techniques is outlined in Fig. 6.

The choice of methodology adopted is contingent upon the specific objectives at hand, whether they pertain to retrieving and characterizing DNA-binding proteins, identifying binding sites, investigating binding kinetics and affinity, or conducting a genome-wide analysis. Factors such as desired precision and accuracy, as well as the availability of technical resources, are crucial contributes when selecting the most suitable approach.

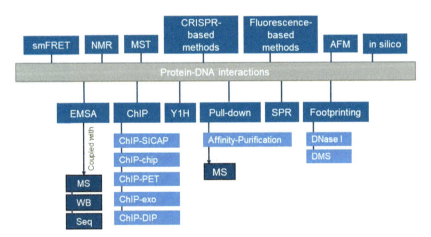

Fig. 6 Overview of techniques utilized for investigating interactions between proteins and nucleic acids: EMSA: Electrophoresis Mobility Shift Assay; ChIP: Chromatin Immunoprecipitation; Y1H: Yeast One Hybrid; MS: Mass Spectrometry; WB: Western Blotting; SDS/PAGE: Sodium Dodecyl Sulfate Polyacrylamide Gel Electrophoresis; ChIP: Chromatin Immunoprecipitation; SICAP: Selective Isolation of Chromatin-Associated Proteins; PET: Protein Expression Technology; DIP: Done In Parallel; SPR: Surface Plasmon Resonance; DMS: Dimethyl Sulfate; smFRET: Single-molecule Fluorescence Resonance Energy Transfer; NMR: Nuclear Magnetic Resonance; MST: MicroScale Thermophoresis; AFM: Atomic Force Microscopy.

3.2.1 Electrophoresis mobility shift assay (EMSA)

Electrophoresis Mobility Shift Assay (EMSA) serves a straightforward, cost-effective, and expeditious *in vitro* technique extensively utilized to study protein-nucleic acid interactions, even in the presence of low abundance of proteins (Hellman & Fried, 2007). Noteworthy for its ability to analyze both purified proteins and whole-cell extracts, EMSA reveals a slower-migrating band in protein-nucleic acid complexes compared to free molecules. While primarily qualitative, EMSA offers insights into the affinity and kinetics of interactions. Importantly, it can be combined with various techniques such as western blotting (Chen & Chang, 2001; Demczuk, Harbers, & Vennström, 1993; Granger-Schnarr, Lloubes, De Murcia, & Schnarr, 1988; Singh, LeBowitz, Baldwin, & Sharp, 1988) and SDS-PAGE and mass spectrometry (Stead, Keen, & McDowall, 2006; Woo, Dods, Susanto, Ulgiati, & Abraham, 2002; Onel, Wu, Sun, Lin, & Yang, 2019), for the identification of unknown binding proteins. EMSA is also well-suited for examining protein interactions with non-canonical structures, such as G-quadruplex-protein complexes, and determining their binding affinity (Haeusler et al., 2014; Onel et al., 2019). While this method is valuable and commonly utilized, its applicability is confined to *in vitro* analysis, and it lacks the capability for precise quantification.

3.2.2 Chromatin immunoprecipitation (ChIP) and its variants

Chromatin Immunoprecipitation (ChIP), first developed by Lis and Gilmour in 1984, and its various adaptations, constitute an *ex vivo* protocol designed to enable the isolation and identification of interactants within a protein-nucleic acid complex. Originating post the establishment of the EMSA assay, ChIP has emerged as a pivotal advancement, significantly enriching our comprehension of protein-nucleic acid interactions (Gilmour, Pflugfelder, Wang, & Lis, 1986). In its fundamental process, DNA undergoes cross-linking with proteins utilizing chemical cross-linkers or UV radiation. Subsequently, chromatin is extracted and fragmented through either restriction enzymes or sonication. Protein-DNA complexes of interest are selectively precipitated using a specific antibody. The immunoprecipitated DNA is then purified, the cross-links are eradicated, and subsequent analytical methodologies, including PCR, qPCR, hybridization to arrays or cloning, and sequencing, can be judiciously applied.

ChIP, when synergistically combined with high-throughput sequencing (ChIP-seq), emerges as the preferred choice for genome-wide analysis, proffering data with expansive coverage and exceptional resolution

(Grosselin et al., 2019; Park, 2009). The evolution of ChIP protocols has seen integration with additional techniques, further augmenting our understanding of protein-DNA interactions occurring within living cells. Noteworthy variants encompass CHIP-SICAP (Selective Isolation of Chromatin-Associated Proteins), Chromatin Immunoprecipitation followed by Microarray Hybridization (ChIP-chip) (Lee, Johnstone, & Young, 2006), and ChIP-exo, a fusion of ChIP-seq with exonuclease digestion, facilitating studies at single-base pair resolution of genomic regions bound by DNA-binding proteins (Yeh & Rhee, 2023). Moreover, ChIP-DIP (ChIP Done In Parallel) has been innovated to facilitate the concurrent analysis of numerous proteins, spanning from histone modifications to transcription factors and RNA polymerases, on a genome-wide scale within a single protocol. (Andrew, Isabel, Mario, Jimmy, & Mitchell, 2023).

3.2.3 Pull-down assay; affinity purification methods

The genesis of methods investigating protein-nucleic acid interactions lies in the inherent affinity of proteins for specific oligonucleotide sequences, and pull-down assays have been employed to analyze specific protein-DNA interactions in samples, usually using biotin-labeled DNA (Müller & Engeland, 2021; Tsuji, 2023). The eluted proteins recovered can be detected *via* Western blot or analyzed through mass spectrometry (Kriel et al., 2020; Müller & Engeland, 2021). Alternatively, the protein itself, rather than the DNA, can be tagged, or the protein-DNA complex can be retrieved using a specific antibody targeting the protein of interest. Subsequently, the bound DNA fragments can be analyzed through Southern blotting or PCR.

This approach entails incubating protein extracts with an insoluble support housing chemically or affinity-immobilized oligonucleotide probes (Cozzolino, Iacobucci, Monaco, & Monti, 2021). The proteins that interact with the oligonucleotides are retained after washing, elution, and identification by MS (Müller & Engeland, 2021). The efficacy of this technique hinges on the thoughtful design and synthesis of the probes, considering factors such as probe length tailored to the nucleotide sequence's length of interest and the specific proteins under investigation. The pH, concentration, and composition of the binding buffer exert profound effects on the success of retrieving protein-nucleic acid interactions. The application of DNA-affinity purification coupled with MS has proven instrumental in delineating DNA-protein interactions, notably in the field of oncology (Müller & Engeland, 2021).

In the quest to discriminate specific from non-specific interactions between proteins and nucleic acids, innovative strategies have emerged. Stable isotope labeling with amino acids in cell culture (SILAC), a mass spectrometry technique, distinguishes specific protein-DNA interactions from non-specific ones (Butter, Scheibe, Mörl, & Mann, 2009; Mittler, Butter, & Mann, 2009). Another recent method, leveraging affinity purification, tandem mass tag (TMT) labeling, and mass spectrometry, offers a proteome-wide investigation of proteins-DNA interactions (Makowski et al., 2018). The determination of apparent dissociation constants (KdApp) through this technique stands as a valuable alternative to established methods such as isothermal titration calorimetry, Surface Plasmon Resonance (SPR) (O'Leary et al., 2015), and FRET (Makowski et al., 2018). The CRAPome repository, compiling contaminants common in affinity purification experiments and specific cell lines, enhances the precision of interaction analyses (Mellacheruvu et al., 2013).

3.2.4 Yeast one-hybrid (Y1H) assay

The yeast one-hybrid (Y1H) assay is a genetic methodology designed to identify proteins interacting with specific DNA sequences *in vivo* (Li & Herskowitz, 1993; Wang & Reed, 1993). Distinguished by its ability to illustrate protein-DNA interactions without triggering transcription, Y1H offers a distinct advantage over the luciferase assay. Moreover, in contrast to ChIP, which may lack isoform-specific information, Y1H excels in discerning DNA-isoform-specific interactions. However, it is crucial to note that Y1H does not provide insights into the functional aspects of the studied interaction. Various Y1H protocols have been developed over the last years (Reece-Hoyes & Marian Walhout, 2012), namely to analyze the interaction between transcription factor with specific DNA sequences (Tsai et al., 2021). For instance, a recent advancement, the paired Y1H (pY1H) (Berenson et al., 2023), extends the method's utility by shedding light on how transcription factors may collaborate or antagonize each other within specific DNA regions. This evolution opens avenues for comprehending the impact of protein occupancy within DNA regions (Berenson et al., 2023).

3.2.5 Footprinting techniques

Footprinting approaches are frequently employed by researchers to unravel the precise sequences bound to proteins. While several variants exist, they commonly entail labeling analyzed nucleic acids with 32 P. For example, DNase I protection mapping, initially outlined in 1978 (Galas & Schmitz, 1978), has been utilized for such analyses. Notably, this method facilitated the conduct of an exhaustive analysis of transcription factor-DNA interactions across 243 human cell and tissue types (Vierstra et al., 2020).

Dimethyl sulfate (DMS) footprinting, while characterizing DNA G-quadruplexes, also provides crucial insights into their interactions with proteins (Onel et al., 2019). This method unveils the localization of DNA-binding sites and delineates the contribution of each nucleotide to the binding affinity. Additionally, *in vivo* footprinting techniques have been successfully implemented (Wells, Hughes, Igel, & Ares, 2000). This technique has been also employed to analyze RNA with various sizes and RNA-protein complexes (Tijerina, Mohr, & Russell, 2007). The development of DMS-seq allows for the comprehensive characterization and mapping of *in vivo* protein-DNA interactions on a genome-wide scale (Umeyama & Ito, 2017).

3.2.6 Surface plasmon resonance (SPR)

Surface plasmon resonance is widely employed in the study of protein-DNA interactions (Stockley & Persson, 2009; Teh, Peh, Su, & Thomsen, 2007), It is a sensitive technique that allows for the assessment of the kinetics and association and dissociation rate constants of such interactions (Bondeson, Frostellkarlsson, Fagerstam, & Magnusson, 1993; Guille & Kneale, 1997). Furthermore, it is more suitable than other methods such as EMSA to compare wild-type *versus* mutant proteins (Cai & Huang, 2012). For instance, SPR was utilized to characterize the real-time kinetic parameters of the interaction between wild-type and mutant p53 with a gene promoter in human cancer cells (Song, Lee, Kang, Park, & Sim, 2020). Nonetheless, this method does present certain limitations, including the method's susceptibility to high binding affinity and strong electrostatic forces introduces technical complexities (Wang, Poon, & Wilson, 2015).

3.2.7 Nuclear magnetic resonance (NMR)

The application of Nuclear Magnetic Resonance (NMR) has proven invaluable in unraveling the intricacies of protein-DNA interactions (Campagne, Gervais, & Milon, 2011; Iwahara, Zweckstetter, & Clore, 2006; Kang, 2019).

3.2.8 MicroScale thermophoresis (MST)

MicroScale thermophoresis (MST) emerges as a fitting choice for assessing quantitative biomolecular interactions in solution on a microliter scale (Mueller et al., 2017). Relying on the thermophoresis phenomenon induced by an infrared laser, this method demonstrates rapidity, high sensitivity, and precision. MST facilitates the study of molecular parameters such as size and conformation, with the mobility of molecules measured using covalently attached or intrinsic fluorophores. This technique has proven valuable in characterizing protein–DNA interactions (Jerabek-Willemsen et al., 2014).

3.2.9 Single-molecule FRET (SmFRET)
Single-molecule FRET (SmFRET) has emerged as an invaluable method for understanding DNA interactions with proteins, including DNA and RNA polymerases (Farooq, Fijen, & Hohlbein, 2014).

3.2.10 Integrated techniques
The recently developed CRISPR-based Chromatin Affinity Purification coupled with Mass Spectrometry (CRISPR-ChAP-MS) provides a means to isolate and study proteins bound to a specific genomic region *in vivo* (Waldrip et al., 2014). Employing the CRISPR/Cas9 system, comprising the endonuclease Cas9 and guide RNA (gRNA), this technique orchestrates the recruitment of the enzyme to the target gene locus. After cross-linking and sonication, Cas9 is immunoprecipitated in conjunction with the associated chromatin, and mass spectrometry orchestrates a comprehensive analysis of proteins. Another pioneering approach, leveraging a hybrid of BirA* and Cas9 named CasID, enables the study of chromatin composition across the genome in a nucleotide-specific manner (Schmidtmann, Anton, Rombaut, Herzog, & Leonhardt, 2016). CasID biotinylates proteins at a specific gene, and subsequent streptavidin-based precipitation coupled with MS, facilitates protein identification (Schmidtmann et al., 2016). A unique RNA-centric method, CRISPR-based RNA Proximity Proteomics (CBRPP), employs Cas13 fused with a proximity-based labeling (PBL) enzyme. PBL identifies proteins in the proximity of the RNA of interest, subsequently analyzed by MS (Li et al., 2021). Leveraging Cas13's distinctive targeting and editable mechanisms, it has found application in real-time *in vivo* RNA imaging, enhancing resolution and elucidating RNA-protein relationships (Leisegang et al., 2022). Importantly, these methods obviate the necessity for labeling the RNA of interest or utilizing antisense probes.

3.2.11 Atomic force microscopy
Atomic Force Microscopy, pioneered by Binning et al. in the 1980s (Binnig, Quate, & Gerber, 1986; Binnig et al., 1986), has emerged as a valuable technique for investigating protein-nucleic acid interactions, particularly with gene

promoters (Suzuki, Endo, & Sugiyama, 2015). This method involves analyzing the topography of each sample with high resolution. AFM offers the advantage of requiring significantly smaller amounts of specimen compared to other techniques. Moreover, it obviates the need for additional preparation protocols such as labeling, and importantly, the specimens remain intact throughout the process (Heath et al., 2021). AFM enables the measurement of bond strength in protein-nucleic acid interactions and the mapping of sample surfaces. However, some limitations include the restriction to analyzing only the surface (Dufrêne et al., 2017; Suzuki et al., 2015).

3.2.12 Fluorescence-based methods

Recent fluorescence methodologies, gauging fluorescence intensity emission during molecular interactions, prove adept at dynamic interaction analysis, surpassing limitations associated with stable and fixed scenarios. Protein-Induced Fluorescence Enhancement (PIFE) (Hwang & Myong, 2014; Rashid et al., 2019), relying on the heightened local viscosity due to the presence of protein residues, emerges as a fluorescence-based tool for investigating pro-tein-DNA interactions. PIFE, utilizing only one dye attached to DNA or RNA, circumvents challenges associated with FRET, such as the intricate fluorescent labeling of proteins and potential alterations in protein function due to tagging. Fluorescence Polarization or Fluorescence Anisotropy, measuring alterations in the orientation of specific molecules, has found application in studying DNA-protein interactions (Yang & Moerner, 2018). While unsuitable for larger macromolecules (>30 kDa), the innovative Selective Time-Resolved Anisotropy with Reversibly Switchable States (STARSS) (Volpato et al., 2023) surmounts this constraint, accounting for the rotational mobility of molecular complexes within cells and demonstrating applicability across the entire human proteome.

3.2.13 In silico predictions

In silico predictions can offer valuable insights into the interactions between macromolecules. For instance, the recently developed RoseTTAFoldNA is a machine learning-guided method that generates three-dimensional structure models for protein–DNA and protein–RNA complexes, potentially aiding in the design of RNA and DNA binding proteins (Baek et al., 2024).

3.3 Protein-RNA interactions

The context of protein-RNA interactions introduces the concept of RNA Binding Domains within RNA molecules, binding specifically to distinct proteins. The dynamic exploration of RNA and RNA-binding protein interactions, and their pivotal regulatory roles in gene expression, has emerged as a focal point of scientific inquiry in recent years. Advancements in sequencing technologies have propelled investigations into the transcriptomic realm, casting light upon the complex realm of ribonucleoprotein complexes (RNPs). These intricate molecular alliances, forged between RNA molecules and proteins, exert pivotal influences on mRNA processing and non-coding RNA (ncRNA)-mediated mechanisms. Unraveling the repertoire of RNAs associated with RNA-binding proteins (RBPs) within cellular contexts enhances our comprehension of RNA's multifaceted functions across diverse biological processes, with notable implications in cancer. The significance of this research is underscored by the implication of mutations in genes encoding RNA-binding proteins (RBPs) in neurological diseases and cancer. Consequently, RNA-binding proteins (RBPs), commonly implicated in cancer-associated cellular behaviors, represent promising candidates for therapeutic intervention in cancer treatment (Kang, Lee, & Lee, 2020).

The immunoprecipitation-based methodologies employed to dissect RNA-protein interactions can be categorized into two strategic approaches: one centered around proteins and the other focusing on RNA (Mattay, 2023). The former involves the purification of specific proteins or classes of proteins, subsequent analysis of associated RNAs, and comprehensive mapping of RNA-binding proteins (RBPs) across the transcriptome. Conversely, the latter strategy revolves around capturing specific RNAs or groups of RNAs, followed by the recovery and analysis of associated proteins, often utilizing sophisticated techniques such as mass spectrometry (MS). The techniques employed to analyze protein-RNA interactions have been summarized in Fig. 7.

3.3.1 RNA immunoprecipitation and its variants

Pioneering experiments, notably pulldown assays, have provided avenues for studying proteins interacting with specific RNAs (Trembinski et al., 2020). Commonly, mRNA targets were discerned through cross-linking methodologies, employing ultraviolet light, nitrocellulose filter binding, RNA electromobility shift assays (REMSAs), pull-down assays, and RNAse H assays. However, these conventional methods possess limitations in recognizing unknown RNA targets of a specific RNA-binding protein.

Novel techniques for identifying RNA-protein interactions *in vivo* are rooted in RNA immunoprecipitation (RIP), utilizing specific antibodies for protein immunoprecipitation. This method categorizes into native

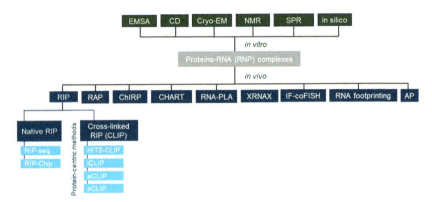

Fig. 7 Overview of methodologies utilized for studying protein-RNA interactions. RIP: RNA Immunoprecipitation; RAP: RNA Antisense Purification; CLIP: Cross-linked RNA Immunoprecipitation; ChIRP: Chromatin Isolation by RNA Purification; CHART: Capture Hybridization Analysis of RNA Targets; PLA: Proximity Ligation Assay; IF-coFISH: Immunofluorescence coupled with RNA FISH; AP: Affinity Purification; EMSA: Electrophoresis Mobility Shift Assay; CD: Circular Dichroism; Cryo-EM: Cryogenic Electron Microscopy; NMR: Nuclear Magnetic Resonance; SPR: Surface Plasmon Resonance.

RIP, encompassing RNP immunoprecipitation-microarray (RIP-Chip) (Keene, Komisarow, & Friedersdorf, 2006), RIP-seq (Zhao et al., 2010), and cross-linked RIP (CLIP) (Niranjanakumari, Lasda, Brazas, & Garcia-Blanco, 2002; Ule et al., 2003), utilizing physical or chemical agents to crosslink RNA and protein interactions (Velema & Lu, 2023). While native methods detect RNA-protein complexes under physiological conditions they carry the caveat of potential RNP re-associations post-cell lysis, possibly deviating from *in vivo* interactions. In contrast, cross-linking agents, exemplified by formaldehyde, firmly fix RNA-protein interactions, precluding re-associations. UV, another cross-linking agent, though beneficial, lacks reversibility.

Both native and cross-linked RIP methods require specific and efficient antibodies generated against the protein of interest. Following covalent cross-linking of RNA to RNA-binding proteins, immunoprecipitation or affinity purification of protein-RNA complexes is performed. Noncovalently bound RNA undergoes elimination *via* sodium dodecyl sulfate-polyacrylamide gel electrophoresis (SDS-PAGE), and the purified RNA-protein complexes are subject to proteinase K treatment, ensuring protein degradation while preserving RNA integrity. The ensuing reverse transcription of RNA yields cDNA, scrutinized through polymerase chain reaction (RT-PCR). Data analysis, often complemented with bioinformatic tools, is a precursor to genomic exploration, akin to RNA immunoprecipitation using microarrays (RIP-Chip) or next-generation sequencing (NGS) methods (RIP-Seq), reminiscent of the ChIP-Seq paradigm, amalgamating immunoprecipitation with high-throughput sequencing to scrutinize RNA molecules tethered to specific proteins.

Variants of CLIP encompass individual-nucleotide resolution CLIP (iCLIP), infrared CLIP (irCLIP), and enhanced CLIP (eCLIP) (Van Nostrand et al., 2016), each employing unique methods for purification and cDNA library preparation (Hafner et al., 2021). eCLIP further evolves into single-end enhanced CLIP (seCLIP), coupled with sequencing (seCLIP-seq) (Blue et al., 2022). Antibody-barcode eCLIP, featuring DNA-barcode antibodies and proximity ligation of DNA oligonucleotides to RNA fragments, streamlines the eCLIP protocol's efficiency, facilitating simultaneous processing of multiple RNA-binding proteins (Lorenz et al., 2023). A diverse array of CLIP-derived techniques, such as Throughput Sequencing of RNA isolated by CLIP (HITS-CLIP), cross-linking and sequencing of hybrids (CLSH), Photoactivatable ribonucleoside-enhanced CLIP (PAR-CLIP) (Hafner et al., 2010), Proximity-CLIP, and hiCLIP, complement other methodologies in mapping transcriptomic features and interactions. These techniques have not only enabled the exploration of RNA hybrids and targeted regions of secondary structures but have also laid the groundwork for binding maps of ribonucleoprotein complexes across the transcriptome, thereby unraveling novel mechanisms influencing gene regulation (Nichols, Krall, Henen, Vogeli, et al., 2023). A recently developed method called EasyCLIP allows for the quantification of RNA cross-links per protein (Porter et al., 2021). Moreover, chromatin isolation by RNA purification coupled with mass spectrometry (ChIRP-MS) facilitates the broad identification of RNA-binding proteins (Chu et al., 2015). Additionally, RNA antisense purification-mass spectrometry (RAP-MS) enables the purification of specific lncRNA complexes and the identification of associated proteins through quantitative mass spectrometry (McHugh & Guttman, 2018).

Techniques like the identification of direct RNA interacting proteins (iDRiP) are based on isolating the protein interactome of a specific RNA, enabling a comprehensive analysis of RNA-protein complexes (Chu et al., 2017; Minajigi et al., 2015). These methods collectively contribute to generating binding maps of ribonucleoprotein (RNP) complexes across the transcriptome, revealing novel mechanisms influencing gene regulation. Notably, studies have reported interactions between long-non-coding RNAs and proteins, such as the Polycomb-repressive complex 2 (PRC2) protein (Khalil et al., 2009).

Beyond ChIRP, Capture Hybridization Analysis of RNA Targets (CHART) provides a valuable tool for studying DNA and proteins associated with lncRNAs. CHART-seq, an extension involving RNA-seq post-CHART, has unveiled a plethora of interactions for lncRNAs NEAT1 and MALAT1 (West et al., 2014). CHART-MS, developed to identify proteins linked with RNAs *in vivo* (Simon et al., 2013; West et al., 2014), expands the toolkit for a comprehensive analysis.

In summary, the continuous evolution of methodologies for investigating RNA-protein complexes has systematically unraveled the intricacies inherent in these interactions, thereby significantly enriching our understanding of gene regulation and paving the way for innovative therapeutic interventions.

3.3.2 In situ proximity ligation assay (PLA)

The In Situ Proximity Ligation Assay (PLA), a well-established immune staining technique for unveiling protein-protein interactions within cells, has undergone modification with the introduction of RNA-PLA (George, Mittal, Kadamberi, Pradeep, & Chaluvally-Raghavan, 2022). This innovative adaptation enables the specific detection of RNA-protein interactions in fixed cells. The technique employs a DNA oligonucleotide that hybridizes with the complementary RNA, coupled with an antibody targeting the RNA-associated protein. Operating in fixed cells, RNA-PLA is adept at identifying both stable and transient Ribonucleoprotein (RNP) complexes, generating fluorescent signals within cellular compartments (Zhang, Xie, Shu, Steitz, & DiMaio, 2016). Additionally, the introduction of XRNAX presents a novel method for the extraction and purification of cross-linked RNA, encompassing both coding and non-coding (ncRNAs), as well as other non-polyadenylated RNA molecules (Trendel et al., 2019). Downstream applications include sequencing and mass spectrometry (MS). Notably, XRNAX has seamlessly integrated with CLIP-seq, providing validation for the interactions of ncRNAs with proteins (Trendel et al., 2019).

3.3.3 Fluorescence techniques

The challenge of localizing specific RNA-protein interactions within cells is met by fluorescence techniques. Although Fluorescence in Situ Hybridization (FISH) and immunofluorescence have conventionally served this purpose, their limitations, including time consumption, technical intricacies, and high background noise with low-resolution images, have spurred the development of enhanced methodologies. The amalgamation of immunofluorescence and RNA FISH (IF-coFISH) emerges as a strategic solution, facilitating precise colocalization of RNA with histone modifications (Chu et al., 2015).

3.3.4 Circular dichroism

Circular dichroism (CD) finds application in the study of proteins-RNA interactions, particularly through the CD melting assay. This assay empowers researchers to discern the capacity of certain proteins to stabilize RNA upon interaction (Mei et al., 2021). Furthermore, Surface Plasmon Resonance (SPR) has been enlisted to delve into these interactions, providing valuable insights into corresponding kinetic parameters (Katsamba, Park, & Laird-Offringa, 2002).

3.3.5 Nuclear magnetic resonance

In recent years, there has been an increasing inclination towards resolving high molecular weight protein-RNA complexes (>50 kDa) through the integration of NMR spectroscopy with complementary methods like small-angle neutron scattering (Carlomagno, 2014; Lapinaite et al., 2013). An exemplification of this trend is evidenced in the elucidation of the structure of a 390 kDa protein-RNA complex, featuring the archaeal Box C/D ribonucleoprotein bound to RNA.

3.3.6 Footprinting techniques

Another technique that not only facilitates the examination of RNA's secondary and tertiary structures but also provides insights into protein-RNA interactions is dimethyl sulfate (DMS) footprinting (Tijerina, Mohr, & Russell, 2007). Additionally, endoribonuclease footprinting offers detailed observations of RNA secondary structures and protein-RNA interactions at a nucleotide-specific level. This is accomplished by leveraging RNA's varied susceptibility to enzymatic cleavage based on its structural features or interactions, as well as delineating the location of protein binding sites and the consequential effects of protein binding on RNA structure (Peng, Soper, & Woodson, 2012).

3.3.7 Pull-down assays

RNA-affinity chromatography (such as Biotin pull-down) combined with mass spectrometry (MS) has revealed RNA-binding proteins potentially influencing the translation of caspase-2, thus impacting drug resistance in human colorectal carcinoma cancer cells (Nasrullah et al., 2019). Notably, the introduction of AptA–MS (aptamer affinity–mass spectrometry) (Ray et al., 2020) enhances sensitivity in the study of RNA-protein interactions. In this method, interactors of a GFP-tagged protein are identified using a high-affinity RNA aptamer against Green Fluorescent Protein (GFP), followed by MS analysis.

3.3.8 Cryo-electron microscopy

Cryo-electron microscopy (Cryo-EM) has become increasingly instrumental in unraveling various protein-RNA interactions that were previously elusive. Liu et al. provided insights into the structural dynamics of the interaction between the human protein Dicer and its cofactor, the transactivation response element RNA-binding protein (TRBP) (Liu et al., 2018). The study revealed the spatial arrangement of multiple domains within human Dicer and delineated two distinct structures of a Dicer-TRBP complex interacting with a microRNA precursor, thus elucidating the functional role of Dicer ribonucleases (Liu et al., 2018). Notably, Cryo-EM facilitated the determination of the replicating SARS-CoV-2 polymerase in association with RNA molecules (Chen et al., 2020; Hillen et al., 2020; Wang et al., 2020).

4. Conclusion

The intricate orchestration of molecular interactions within cellular environments involves a complex symphony of regulatory processes, notably featuring interactions within DNA and RNA molecules—intramolecular interactions—and between nucleic acids or proteins and nucleic acids—intermolecular interplays. An in-depth understanding of these interactions *in vitro* and *in vivo* is pivotal for comprehending the intricacies of cellular regulation, and various cutting-edge techniques have been indispensable in this scientific endeavor.

Intramolecular DNA and RNA interactions offer profound insights into structural alterations and binding sites within the nucleic acids, and some methods have allowed to study in a real-time context. However, challenges persist in discerning dynamic interactions and unraveling higher-order chromatin structures. Despite these challenges, understanding intramolecular interactions remains crucial for maintaining genome stability and regulation, although aberrations in these interactions contribute significantly to the progression of cancer.

Investigating intermolecular DNA:RNA interactions, by utilizing techniques such as EMSA, CD NMR and RNA pull-down assays, unveils the intricate interplay between coding and non-coding RNAs with DNA. These interactions wield considerable influence over gene expression and contribute to observed dysregulations in various cancers. The advent of RNA-sequencing technologies has empowered researchers to conduct comprehensive analyses of the transcriptomic landscape, shedding light on RNA's pivotal roles in cancer initiation and progression.

The interactions between proteins and nucleic acids are pivotal in various biological processes. Despite significant progress in predicting protein structures, accurately predicting those of protein-nucleic acid complexes remains a challenging task. Protein:DNA and protein:RNA interactions, explored through a diverse array of methodologies such as Chromatin Immunoprecipitation (ChIP) and its derivatives, provide insights into the dynamic associations between proteins and genomic DNA or RNA. High-throughput approaches like ChIP-seq facilitate genome-wide analyses, providing valuable data for understanding transcriptional regulation. Dysregulation of these interactions is a hallmark of cancer, with mutations in DNA-binding proteins contributing to oncogenesis.

Despite the unprecedented insights offered by these techniques, each method presents its distinct advantages and limitations. Advanced fluorescence techniques, proximity ligation assays, and circular dichroism contribute to a comprehensive understanding of these interactions but grapple with challenges such as technical complexity and the demand for sophisticated instrumentation. Notably, in silico predictions have yielded valuable insights into the dynamic structures of intramolecular and intermolecular interactions. In the realm of cancer research, the elucidation of these interactions provides a roadmap for identifying biomarkers, unraveling oncogenic mechanisms, and developing targeted therapeutic interventions. The promise lies in therapeutic strategies targeting intramolecular DNA or RNA interactions, along with DNA:RNA, protein:DNA, and protein:RNA interactions, offering avenues for precision medicine and transformative advancements in cancer treatment.

In summary, the arsenal of techniques utilized to investigate molecular interactions has significantly transformed our understanding of cellular mechanisms, providing valuable insights into the complex molecular dynamics governing both physiological and pathological states. Navigating this scientific landscape mandates the amalgamation of diverse methodologies, presenting a holistic view crucial for advancing cancer research and propelling therapeutic interventions into the forefront of precision medicine.

Acknowledgments

This study is supported by VCU Massey Comprehensive Cancer Center and School of Medicine start-up funds, Tina's Wish Foundation Award, La Caixa Funding LCF/BQ/PR20/ 11770006 to E. Madan.

All authors declare that they have no conflict of interests.

References

Afzal, M., Shahid, A. A., Shehzadi, A., Nadeem, S., & Husnain, T. (2012). RDNAnalyzer: A tool for DNA secondary structure prediction and sequence analysis. *Bioinformation, 8*(14), 687–690. https://doi.org/10.6026/97320630008687.

Aguilera, A., & García-Muse, T. (2012). R loops: From transcription byproducts to threats to genome stability. *Molecular Cell, 46*(2), 115–124.

Amato, J., Iaccarino, N., Randazzo, A., Novellino, E., & Pagano, B. (2014). Noncanonical DNA secondary structures as drug targets: The prospect of the i-motif. *ChemMedChem, 9*(9), 2026–2030. https://doi.org/10.1002/cmdc.201402153.

Andrew, A. P., Isabel, N. G., Mario, R. B., Jimmy, K. G., & Mitchell, G. (2023). ChIP-DIP: A multiplexed method for mapping hundreds of proteins to DNA uncovers diverse regulatory elements controlling gene expression. *bioRxiv*, 2023.2012.2014.571730. https://doi.org/10.1101/2023.12.14.571730.

Asha, H., Green, J. A., Esposito, L., Santoro, F., & Improta, R. (2023). Computing the electronic circular dichroism spectrum of DNA quadruple helices of different topology: A critical test for a generalized excitonic model based on a fragment diabatization. *Chirality, 35*(5), 298–310. https://doi.org/10.1002/chir.23540.

Baek, M., McHugh, R., Anishchenko, I., Jiang, H., Baker, D., & DiMaio, F. (2024). Accurate prediction of protein–nucleic acid complexes using RoseTTAFoldNA. *Nature Methods, 21*(1), 117–121. https://doi.org/10.1038/s41592-023-02086-5.

Bandyopadhyay, D., & Mishra, P. P. (2021). Decoding the structural dynamics and conformational alternations of DNA secondary structures by single-molecule FRET microspectroscopy. *Frontiers in Molecular Biosciences, 8*, 725541. https://doi.org/10.3389/fmolb.2021.725541.

Banerjee, N., Panda, S., & Chatterjee, S. (2022). Frontiers in G-Quadruplex therapeutics in cancer: Selection of small molecules, peptides and aptamers. *Chemical Biology & Drug Design, 99*(1), 1–31. https://doi.org/10.1111/cbdd.13910.

Bansal, A., Kaushik, S., & Kukreti, S. (2022). Non-canonical DNA structures: Diversity and disease association. *Frontiers in Genetics, 13*, 959258. https://doi.org/10.3389/fgene.2022.959258.

Barnwal, R. P., Yang, F., & Varani, G. (2017). Applications of NMR to structure determination of RNAs large and small. *Archives of Biochemistry and Biophysics, 628*, 42–56. https://doi.org/10.1016/j.abb.2017.06.003.

Belmont, P., Constant, J.-F., & Demeunynck, M. (2001). Nucleic acid conformation diversity: from structure to function and regulation. *Chemical Society Reviews, 30*(1), 70–81. https://doi.org/10.1039/A904630E.

Belotserkovskii, B. P., Tornaletti, S., D'Souza, A. D., & Hanawalt, P. C. (2018). R-loop generation during transcription: Formation, processing and cellular outcomes. *DNA Repair (Amst), 71*, 69–81. https://doi.org/10.1016/j.dnarep.2018.08.009.

Berenson, A., Lane, R., Soto-Ugaldi, L. F., Patel, M., Ciausu, C., Li, Z., ... Fuxman Bass, J. I. (2023). Paired yeast one-hybrid assays to detect DNA-binding cooperativity and antagonism across transcription factors. *Nature Communications, 14*(1), 6570. https://doi. org/10.1038/s41467-023-42445-6.

Bhandari, Y. R., Fan, L., Fang, X., Zaki, G. F., Stahlberg, E. A., Jiang, W., ... Wang, Y.-X. (2017). Topological structure determination of RNA using small-angle X-ray scattering. *Journal of Molecular Biology, 429*(23), 3635–3649. https://doi.org/10.1016/j.jmb.2017.09.006.

Bhatia, V., Barroso, S. I., García-Rubio, M. L., Tumini, E., Herrera-Moyano, E., & Aguilera, A. (2014). BRCA2 prevents R-loop accumulation and associates with TREX-2 mRNA export factor PCID2. *Nature, 511*(7509), 362–365.

Binet, T., Padiolleau-Lefèvre, S., Octave, S., Avalle, B., & Maffucci, I. (2023). Comparative study of single-stranded oligonucleotides secondary structure prediction tools. *BMC Bioinformatics, 24*(1), 422. https://doi.org/10.1186/s12859-023-05532-5.

Binnig, G., Quate, C. F., & Gerber, C. (1986). Atomic force microscope. *Physical Review Letters, 56*(9), 930.

Bishop, G. R., & Chaires, J. B. (2003). Characterization of DNA structures by circular dichroism. *Chapter 7,* 7.11.11–17.11.18 *Current Protocols in Nucleic Acid Chemistry/Edited by Serge L. Beaucage ... [et al.].* https://doi.org/10.1002/0471142700.nc0711s11.

Biver, T. (2022). Discriminating between parallel, anti-parallel and hybrid G-quadruplexes: Mechanistic details on their binding to small molecules. *Molecules (Basel, Switzerland), 27*(13), https://doi.org/10.3390/molecules27134165.

Blake, R. D., Fresco, J. R., & Langridge, R. (1970). High-resolution X-ray diffraction by single crystals of mixtures of transfer ribonucleic acids. *Nature, 225*(5227), 32–35. https://doi.org/10.1038/225032a0.

Blue, S. M., Yee, B. A., Pratt, G. A., Mueller, J. R., Park, S. S., Shishkin, A. A., ... Yeo, G. W. (2022). Transcriptome-wide identification of RNA-binding protein binding sites using seCLIP-seq. *Nature Protocols, 17*(5), 1223–1265. https://doi.org/10.1038/s41596-022-00680-z.

Bochman, M. L., Paeschke, K., & Zakian, V. A. (2012). DNA secondary structures: Stability and function of G-quadruplex structures. *Nature Reviews. Genetics, 13*(11), 770–780. https://doi.org/10.1038/nrg3296.

Boguslawski, S. J., Smith, D. E., Michalak, M. A., Mickelson, K. E., Yehle, C. O., Patterson, W. L., & Carrico, R. J. (1986). Characterization of monoclonal antibody to DNA·RNA and its application to immunodetection of hybrids. *Journal of Immunological Methods, 89*(1), 123–130. https://doi.org/10.1016/0022-1759(86)90040-2.

Boguslawski, S. J., Smith, D. E., Michalak, M. A., Mickelson, K. E., Yehle, C. O., Patterson, W. L., & Carrico, R. J. (1986). Characterization of monoclonal antibody to DNA·RNA and its application to immunodetection of hybrids. *Journal of Immunological Methods, 89*(1), 123–130.

Bondeson, K., Frostellkarlsson, A., Fagerstam, L., & Magnusson, G. (1993). Lactose repressor-operator DNA interactions: Kinetic analysis by a surface plasmon resonance biosensor. *Analytical Biochemistry, 214*(1), 245–251. https://doi.org/10.1006/abio.1993.1484.

Boros-Oláh, B., Dobos, N., Hornyák, L., Szabó, Z., Karányi, Z., Halmos, G., ... Székvölgyi, L. (2019). Drugging the R-loop interactome: RNA-DNA hybrid binding proteins as targets for cancer therapy. *DNA Repair (Amst), 84*, 102642. https://doi.org/10.1016/j.dnarep.2019.102642.

Bou-Nader, C., Bothra, A., Garboczi, D. N., Leppla, S. H., & Zhang, J. (2022). Structural basis of R-loop recognition by the S9.6 monoclonal antibody. *Nature Communications, 13*(1), 1641. https://doi.org/10.1038/s41467-022-29187-7.

Bragg, W. (1962). *Fifty years of X-ray diffraction. International union of crystallography.* Oxford: Oxford University Press.

Brázda, V., Havlík, J., Kolomazník, J., Trenz, O., & Šťastný, J. (2021). R-loop tracker: Web access-based tool for R-loop detection and analysis in genomic DNA sequences. *International Journal of Molecular Sciences, 22*(23), https://doi.org/10.3390/ijms222312857.

Brünger, A. T. (1997). X-ray crystallography and NMR reveal complementary views of structure and dynamics. *Nature Structural Biology, 4(Suppl)*, 862–865.

Burke, J. E., & Butcher, S. E. (2012). Nucleic acid structure characterization by small angle X-ray scattering (SAXS). *Chapter 7,* Unit7.18 *Current Protocols in Nucleic Acid Chemistry / Edited by Serge L. Beaucage ... [et al.].* https://doi.org/10.1002/0471142700.nc0718s51.

Buske, F. A., Bauer, D. C., Mattick, J. S., & Bailey, T. L. (2012). Triplexator: Detecting nucleic acid triple helices in genomic and transcriptomic data. *Genome Research, 22*(7), 1372–1381. https://doi.org/10.1101/gr.130237.111.

Buske, F. A., Bauer, D. C., Mattick, J. S., & Bailey, T. L. (2013). Triplex-Inspector: An analysis tool for triplex-mediated targeting of genomic loci. *Bioinformatics (Oxford, England), 29*(15), 1895–1897. https://doi.org/10.1093/bioinformatics/btt315.

Butter, F., Scheibe, M., Mörl, M., & Mann, M. (2009). Unbiased RNA-protein interaction screen by quantitative proteomics. *Proceedings of the National Academy of Sciences of the United States of America, 106*(26), 10626–10631. https://doi.org/10.1073/pnas.0812099106.

Cai, Y.-H., & Huang, H. (2012). Advances in the study of protein–DNA interaction. *Amino Acids, 43*, 1141–1146.

Campagne, S., Gervais, V., & Milon, A. (2011). Nuclear magnetic resonance analysis of protein-DNA interactions. *Journal of the Royal Society, Interface / the Royal Society, 8*(61), 1065–1078. https://doi.org/10.1098/rsif.2010.0543.

Cappannini, A., Mosca, K., Mukherjee, S., Moafinejad, S. N., Sinden, R. R., Arluison, V., ... Wien, F. (2023). NACDDB: Nucleic acid circular dichroism database. *Nucleic Acids Research, 51*(D1), D226–D231. https://doi.org/10.1093/nar/gkac829.

Carlomagno, T. (2014). Present and future of NMR for RNA–protein complexes: A perspective of integrated structural biology. *Journal of Magnetic Resonance, 241*, 126–136.

Chan, Y. A., Aristizabal, M. J., Lu, P. Y., Luo, Z., Hamza, A., Kobor, M. S., ... Hieter, P. (2014). Genome-wide profiling of yeast DNA:RNA hybrid prone sites with DRIP-chip. *PLoS Genetics, 10*(4), e1004288. https://doi.org/10.1371/journal.pgen.1004288.

Chen, H., & Chang, G. D. (2001). Simultaneous immunoblotting analysis with activity gel electrophoresis in a single polyacrylamide gel. *Electrophoresis, 22*(10), 1894–1899.

Chen, J., Malone, B., Llewellyn, E., Grasso, M., Shelton, P. M., Olinares, P. D. B., ... Chait, B. T. (2020). Structural basis for helicase-polymerase coupling in the SARS-CoV-2 replication-transcription complex. *Cell, 182*(6), 1560–1573 e1513.

Chen, J. L., Bellaousov, S., & Turner, D. H. (2016). RNA secondary structure determination by NMR. *Methods in Molecular Biology, 1490*, 177–186. https://doi.org/10.1007/978-1-4939-6433-8_11.

Chen, J. Y., Zhang, X., Fu, X. D., & Chen, L. (2019). R-ChIP for genome-wide mapping of R-loops by using catalytically inactive RNASEH1. *Nature Protocols, 14*(5), 1661–1685. https://doi.org/10.1038/s41596-019-0154-6.

Chen, L., Chen, J.-Y., Zhang, X., Gu, Y., Xiao, R., Shao, C., ... Li, H. (2017). R-ChIP using inactive RNase H reveals dynamic coupling of R-loops with transcriptional pausing at gene promoters. *Molecular Cell, 68*(4), 745–757 e745.

Chen, L., Dickerhoff, J., Sakai, S., & Yang, D. (2022). DNA G-quadruplex in human telomeres and oncogene promoters: Structures, functions, and small molecule targeting. *Accounts of Chemical Research, 55*(18), 2628–2646. https://doi.org/10.1021/acs.accounts.2c00337.

Chen, Y., & Pollack, L. (2016). SAXS studies of RNA: Structures, dynamics, and interactions with partners. *Wiley Interdiscip Rev RNA, 7*(4), 512–526. https://doi.org/10.1002/wrna.1349.

Childs-Disney, J. L., Yang, X., Gibaut, Q. M. R., Tong, Y., Batey, R. T., & Disney, M. D. (2022). Targeting RNA structures with small molecules. *Nature Reviews. Drug Discovery, 21*(10), 736–762. https://doi.org/10.1038/s41573-022-00521-4.

Chu, B., Zhang, D., & Paukstelis, P. J. (2019). A DNA G-quadruplex/i-motif hybrid. *Nucleic Acids Research, 47*(22), 11921–11930. https://doi.org/10.1093/nar/gkz1008.

Chu, C., Zhang, Q. C., Da Rocha, S. T., Flynn, R. A., Bharadwaj, M., Calabrese, J. M., ... Chang, H. Y. (2015). Systematic discovery of Xist RNA binding proteins. *Cell, 161*(2), 404–416.

Chu, H.-P., Cifuentes-Rojas, C., Kesner, B., Aeby, E., Lee, H.-G., Wei, C., ... Lee, J. T. (2017). TERRA RNA antagonizes ATRX and protects telomeres. *Cell, 170*(1), 86–101 e116.

Coman, D., & Russu, I. M. (2004). Site-resolved stabilization of a DNA triple helix by magnesium ions. *Nucleic Acids Research, 32*(3), 878–883. https://doi.org/10.1093/nar/gkh228.

Corley, M., Burns, M. C., & Yeo, G. W. (2020). How RNA-binding proteins interact with RNA: Molecules and mechanisms. *Molecular Cell, 78*(1), 9–29. https://doi.org/10.1016/j.molcel.2020.03.011.

Cotton, A. (1895). *Comptes Rendus Paris, 120*(989–991), 1044–1046.

Cozzolino, F., Iacobucci, I., Monaco, V., & Monti, M. (2021). Protein-DNA/RNA interactions: An overview of investigation methods in the -omics era. *Journal of Proteome Research, 20*(6), 3018–3030. https://doi.org/10.1021/acs.jproteome.1c00074.

Crossley, M. P., Bocek, M. J., Hamperl, S., Swigut, T., & Cimprich, K. A. (2020). qDRIP: A method to quantitatively assess RNA–DNA hybrid formation genome-wide. e84–e84 *Nucleic Acids Research, 48*(14), https://doi.org/10.1093/nar/gkaa500.

Cunha, E. N., de Souza, M. F. B., Lanza, D. C. F., & Lima, J. (2020). A low-cost smart system for electrophoresis-based nucleic acids detection at the visible spectrum. *PLoS One, 15*(10), e0240536. https://doi.org/10.1371/journal.pone.0240536.

Dans, P. D., Gallego, D., Balaceanu, A., Darré, L., Gómez, H., & Orozco, M. (2019). Modeling, simulations, and bioinformatics at the service of RNA structure. *Chem, 5*(1), 51–73. https://doi.org/10.1016/j.chempr.2018.09.015.

Davey, C. A., Sargent, D. F., Luger, K., Maeder, A. W., & Richmond, T. J. (2002). Solvent mediated interactions in the structure of the nucleosome core particle at 1.9Å resolution††. We dedicate this paper to the memory of Max Perutz who was particularly inspirational and supportive to T.J.R. in the early stages of this study. *Journal of Molecular Biology, 319*(5), 1097–1113. https://doi.org/10.1016/S0022-2836(02)00386-8.

Davis, P. W., Adamiak, R. W., & Tinoco, I., Jr. (1990). Z-RNA: The solution NMR structure of r(CGCGCG). *Biopolymers, 29*(1), 109–122. https://doi.org/10.1002/bip.360290116.

Del Villar-Guerra, R., Trent, J. O., & Chaires, J. B. (2018). G-quadruplex secondary structure obtained from circular dichroism spectroscopy. *Angewandte Chemie (International Ed. in English), 57*(24), 7171–7175. https://doi.org/10.1002/anie.201709184.

Demczuk, S., Harbers, M., & Vennström, B. (1993). Identification and analysis of all components of a gel retardation assay by combination with immunoblotting. *Proceedings of the National Academy of Sciences, 90*(7), 2574–2578.

Dickerson, R. E., & Ng, H. L. (2001). DNA structure from A to B. *Proceedings of the National Academy of Sciences of the United States of America, 98*(13), 6986–6988. https://doi.org/10.1073/pnas.141238898.

Doerr, A. (2017). Cryo-electron tomography. *Nature Methods, 14*(1), 34. https://doi.org/10.1038/nmeth.4115.

Dufrêne, Y. F., Ando, T., Garcia, R., Alsteens, D., Martinez-Martin, D., Engel, A., ... Müller, D. J. (2017). Imaging modes of atomic force microscopy for application in molecular and cell biology. *Nature Nanotechnology, 12*(4), 295–307.

Dumelie, J. G., & Jaffrey, S. R. (2017). Defining the location of promoter-associated R-loops at near-nucleotide resolution using bisDRIP-seq. *Elife, 6*. https://doi.org/10.7554/eLife.28306.

Eisenstein, M., & Shakked, Z. (1995). Hydration patterns and intermolecular interactions inA-DNA crystal structures. Implications for DNA recognition. *Journal of Molecular Biology, 248*(3), 662–678. https://doi.org/10.1006/jmbi.1995.0250.

Electrophoretic mobility shift assays. (2005). *Nature Methods, 2*(7), 557–558. https://doi.org/10.1038/nmeth0705-557.

Esnault, C., Magat, T., Zine El Aabidine, A., Garcia-Oliver, E., Cucchiarini, A., Bouchouika, S., ... Andrau, J.-C. (2023). G4access identifies G-quadruplexes and their associations with open chromatin and imprinting control regions. *Nature Genetics, 55*(8), 1359–1369. https://doi.org/10.1038/s41588-023-01437-4.

Fang, X., Stagno, J. R., Bhandari, Y. R., Zuo, X., & Wang, Y.-X. (2015). Small-angle X-ray scattering: A bridge between RNA secondary structures and three-dimensional topological structures. *Current Opinion in Structural Biology, 30*, 147–160.

Farooq, S., Fijen, C., & Hohlbein, J. (2014). Studying DNA–protein interactions with single-molecule Förster resonance energy transfer. *Protoplasma, 251*(2), 317–332. https://doi.org/10.1007/s00709-013-0596-6.

Filius, M., Kim, S. H., Severins, I., & Joo, C. (2021). High-resolution single-molecule FRET via DNA eXchange (FRET X). *Nano Letters, 21*(7), 3295–3301. https://doi.org/10.1021/acs.nanolett.1c00725.

Fried, M., & Crothers, D. M. (1981). Equilibria and kinetics of lac repressor-operator interactions by polyacrylamide gel electrophoresis. *Nucleic Acids Research, 9*(23), 6505–6525.

Fusée, L., Salomao, N., Ponnuswamy, A., Wang, L., López, I., Chen, S., ... Fahraeus, R. (2023). The p53 endoplasmic reticulum stress-response pathway evolved in humans but not in mice via PERK-regulated p53 mRNA structures. *Cell Death & Differentiation, 30*(4), 1072–1081. https://doi.org/10.1038/s41418-023-01127-y.

Galas, D. J., & Schmitz, A. (1978). DNAase footprinting a simple method for the detection of protein-DNA binding specificity. *Nucleic Acids Research, 5*(9), 3157–3170. https://doi.org/10.1093/nar/5.9.3157.

Ganapathiraju, M. K., Subramanian, S., Chaparala, S., & Karunakaran, K. B. (2020). A reference catalog of DNA palindromes in the human genome and their variations in 1000 Genomes. *Human Genome Variation, 7*(1), 40. https://doi.org/10.1038/s41439-020-00127-5.

Garner, M. M., & Revzin, A. (1981). A gel electrophoresis method for quantifying the binding of proteins to specific DNA regions: Application to components of the Escherichia coli lactose operon regulatory system. *Nucleic Acids Research, 9*(13), 3047–3060.

Georgakopoulos-Soares, I., Chan, C. S. Y., Ahituv, N., & Hemberg, M. (2022). High-throughput techniques enable advances in the roles of DNA and RNA secondary structures in transcriptional and post-transcriptional gene regulation. *Genome Biology, 23*(1), 159. https://doi.org/10.1186/s13059-022-02727-6.

Georgakopoulos-Soares, I., Morganella, S., Jain, N., Hemberg, M., & Nik-Zainal, S. (2018). Noncanonical secondary structures arising from non-B DNA motifs are determinants of mutagenesis. *Genome Research, 28*(9), 1264–1271.

George, J., Mittal, S., Kadamberi, I. P., Pradeep, S., & Chaluvally-Raghavan, P. (2022). Optimized proximity ligation assay (PLA) for detection of RNA-protein complex interactions in cell lines. *STAR Protocol, 3*(2), 101340. https://doi.org/10.1016/j.xpro.2022.101340.

Ghersi, D., & Singh, M. (2013). Interaction-based discovery of functionally important genes in cancers. *Nucleic Acids Research, 42*(3), e18. https://doi.org/10.1093/nar/gkt1305.

Gilmour, D. S., Pflugfelder, G., Wang, J. C., & Lis, J. T. (1986). Topoisomerase I interacts with transcribed regions in Drosophila cells. *Cell, 44*(3), 401–407.

Granger-Schnarr, M., Lloubes, R., De Murcia, G., & Schnarr, M. (1988). Specific protein-DNA complexes: Immunodetection of the protein component after gel electrophoresis and Western blotting. *Analytical Biochemistry, 174*(1), 235–238.

Grosselin, K., Durand, A., Marsolier, J., Poitou, A., Marangoni, E., Nemati, F., ... Gérard, A. (2019). High-throughput single-cell ChIP-seq identifies heterogeneity of chromatin states in breast cancer. *Nature Genetics, 51*(6), 1060–1066. https://doi.org/10.1038/s41588-019-0424-9.

Guaita, M., Watters, S. C., & Loerch, S. (2022). Recent advances and current trends in cryo-electron microscopy. *Current Opinion in Structural Biology, 77*, 102484. https://doi.org/10.1016/j.sbi.2022.102484.

Gudanis, D., Zielińska, K., Baranowski, D., Kierzek, R., Kozlowski, P., & Gdaniec, Z. (2021). Impact of a single nucleotide change or non-nucleoside modifications in G-rich region on the quadruplex–duplex hybrid formation. *Biomolecules, 11*, 1236. https://doi.org/10.3390/biom11081236.

Guille, M. J., & Kneale, G. G. (1997). Methods for the analysis of DNA-protein interactions. *Molecular Biotechnology, 8*, 35–52.

Guo, A.-D., Yan, K.-N., Hu, H., Zhai, L., Hu, T.-F., Su, H., ... Chen, X.-H. (2023). Spatiotemporal and global profiling of DNA–protein interactions enables discovery of low-affinity transcription factors. *Nature Chemistry, 15*(6), 803–814. https://doi.org/10.1038/s41557-023-01196-z.

Haeusler, A. R., Donnelly, C. J., Periz, G., Simko, E. A., Shaw, P. G., Kim, M.-S., ... Sattler, R. (2014). C9orf72 nucleotide repeat structures initiate molecular cascades of disease. *Nature, 507*(7491), 195–200.

Hafner, M., Katsantoni, M., Köster, T., Marks, J., Mukherjee, J., Staiger, D., ... Zavolan, M. (2021). CLIP and complementary methods. *Nature Reviews Methods Primers, 1*(1), 20. https://doi.org/10.1038/s43586-021-00018-1.

Hafner, M., Landthaler, M., Burger, L., Khorshid, M., Hausser, J., Berninger, P., ... Tuschl, T. (2010). Transcriptome-wide identification of RNA-binding protein and microRNA target sites by PAR-CLIP. *Cell, 141*(1), 129–141. https://doi.org/10.1016/j.cell.2010.03.009.

Hartono, S. R., Malapert, A., Legros, P., Bernard, P., Chédin, F., & Vanoosthuyse, V. (2018). The affinity of the S9.6 antibody for double-stranded RNAs impacts the accurate mapping of R-loops in fission yeast. *Journal of Molecular Biology, 430*(3), 272–284. https://doi.org/10.1016/j.jmb.2017.12.016.

Heath, G. R., Kots, E., Robertson, J. L., Lansky, S., Khelashvili, G., Weinstein, H., & Scheuring, S. (2021). Localization atomic force microscopy. *Nature, 594*(7863), 385–390. https://doi.org/10.1038/s41586-021-03551-x.

Hellman, L. M., & Fried, M. G. (2007). Electrophoretic mobility shift assay (EMSA) for detecting protein–nucleic acid interactions. *Nature Protocols, 2*(8), 1849–1861. https://doi.org/10.1038/nprot.2007.249.

Hellman, L. M., & Fried, M. G. (2007). Electrophoretic mobility shift assay (EMSA) for detecting protein–nucleic acid interactions. *Nature Protocols, 2*(8), 1849–1861.

Hillen, H. S., Kokic, G., Farnung, L., Dienemann, C., Tegunov, D., & Cramer, P. (2020). Structure of replicating SARS-CoV-2 polymerase. *Nature, 584*(7819), 154–156. https://doi.org/10.1038/s41586-020-2368-8.

Ho, P. S., Ellison, M. J., Quigley, G. J., & Rich, A. (1986). A computer aided thermodynamic approach for predicting the formation of Z-DNA in naturally occurring sequences. *The EMBO Journal, 5*(10), 2737–2744. https://doi.org/10.1002/j.1460-2075.1986.tb04558.x.

Hu, J., Wu, M., Jiang, L., Zhong, Z., Zhou, Z., Rujiralai, T., & Ma, J. (2018). Combining gold nanoparticle antennas with single-molecule fluorescence resonance energy transfer (smFRET) to study DNA hairpin dynamics. *Nanoscale, 10*(14), 6611–6619.

Hwang, H., & Myong, S. (2014). Protein induced fluorescence enhancement (PIFE) for probing protein–nucleic acid interactions. *Chemical Society Reviews, 43*(4), 1221–1229. https://doi.org/10.1039/c3cs60201j.

Imeddourene, A. B., Xu, X., Zargarian, L., Oguey, C., Foloppe, N., Mauffret, O., & Hartmann, B. (2016). The intrinsic mechanics of B-DNA in solution characterized by NMR. *Nucleic Acids Research, 44*(7), 3432–3447. https://doi.org/10.1093/nar/gkw084.

Inagaki, H., Ohye, T., Kogo, H., Tsutsumi, M., Kato, T., Tong, M., ... Kurahashi, H. (2013). Two sequential cleavage reactions on cruciform DNA structures cause palindrome-mediated chromosomal translocations. *Nature Communications, 4*(1), 1592. https://doi.org/10.1038/ncomms2595.

Iwahara, J., Zweckstetter, M., & Clore, G. M. (2006). NMR structural and kinetic characterization of a homeodomain diffusing and hopping on nonspecific DNA. *Proceedings of the National Academy of Sciences of the United States of America, 103*(41), 15062–15067. https://doi.org/10.1073/pnas.0605868103.

Jana, J., Mondal, S., Bhattacharjee, P., Sengupta, P., Roychowdhury, T., Saha, P., ... Chatterjee, S. (2017). Chelerythrine down regulates expression of VEGFA, BCL2 and KRAS by arresting G-Quadruplex structures at their promoter regions. *Scientific Reports, 7*(1), 40706. https://doi.org/10.1038/srep40706.

Jeong, M., Lee, A.-R., Kim, H.-E., Choi, Y.-G., Choi, B.-S., & Lee, J.-H. (2014). NMR study of the Z-DNA binding mode and B–Z transition activity of the Zα domain of human ADAR1 when perturbed by mutation on the α3 helix and β-hairpin. *Archives of Biochemistry and Biophysics, 558*, 95–103.

Jerabek-Willemsen, M., André, T., Wanner, R., Roth, H. M., Duhr, S., Baaske, P., & Breitsprecher, D. (2014). MicroScale thermophoresis: Interaction analysis and beyond. *Journal of Molecular Structure, 1077*, 101–113. https://doi.org/10.1016/j.molstruc.2014.03.009.

Jie Xiao, P. D. (2008). Detecting Z-DNA forming regions in the human genome. *The International Conference on Genome Informatics.*

Josué Carvalho, J. A. Q., & Carla Cruz (2017). Circular dichroism of G-quadruplex: A laboratory experiment for the study of topology and ligand binding. *Journal of Chemical Education, 94*(10), 1547–1551. https://doi.org/10.1021/acs.jchemed.7b00160.

Kang, C. (2019). Applications of in-cell NMR in structural biology and drug discovery. *International Journal of Molecular Sciences, 20*(1), 139.

Kang, D., Lee, Y., & Lee, J. S. (2020). RNA-binding proteins in cancer: Functional and therapeutic perspectives. *Cancers (Basel), 12*(9), https://doi.org/10.3390/cancers12092699.

Kato, T., Goodman, R. P., Erben, C. M., Turberfield, A. J., & Namba, K. (2009). High-resolution structural analysis of a DNA nanostructure by cryoEM. *Nano Letters, 9*(7), 2747–2750. https://doi.org/10.1021/nl901265n.

Katsamba, P. S., Park, S., & Laird-Offringa, I. A. (2002). Kinetic studies of RNA-protein interactions using surface plasmon resonance. *Methods (San Diego, Calif.), 26*(2), 95–104. https://doi.org/10.1016/s1046-2023(02)00012-9.

Keene, J. D., Komisarow, J. M., & Friedersdorf, M. B. (2006). RIP-Chip: The isolation and identification of mRNAs, microRNAs and protein components of ribonucleoprotein complexes from cell extracts. *Nature Protocols, 1*(1), 302–307. https://doi.org/10.1038/nprot.2006.47.

Kennedy, S. D. (2016). NMR methods for characterization of RNA secondary structure. *Methods in Molecular Biology, 1490*, 253–264. https://doi.org/10.1007/978-1-4939-6433-8_16.

Khalil, A. M., Guttman, M., Huarte, M., Garber, M., Raj, A., Rivea Morales, D., ... Rinn, J. L. (2009). Many human large intergenic noncoding RNAs associate with chromatin-modifying complexes and affect gene expression. *Proceedings of the National Academy of Sciences of the United States of America, 106*(28), 11667–11672. https://doi.org/10.1073/pnas.0904715106.

Khan, E. S., & Danckwardt, S. (2022). Pathophysiological role and diagnostic potential of R-loops in cancer and beyond. *Genes, 13*(12), 2181.

Kosiol, N., Juranek, S., Brossart, P., Heine, A., & Paeschke, K. (2021). G-quadruplexes: A promising target for cancer therapy. *Molecular Cancer, 20*(1), 40. https://doi.org/10.1186/s12943-021-01328-4.

Kriel, N. L., Heunis, T., Sampson, S. L., Gey van Pittius, N. C., Williams, M. J., & Warren, R. M. (2020). Identifying nucleic acid-associated proteins in *Mycobacterium smegmatis* by mass spectrometry-based proteomics. *BMC Molecular and Cell Biology, 21*(1), 19. https://doi.org/10.1186/s12860-020-00261-6.

Kuo, C.-C., Hänzelmann, S., Sentürk Cetin, N., Frank, S., Zajzon, B., Derks, J.-P., ... Costa, I. G. (2019). Detection of RNA–DNA binding sites in long noncoding RNAs. *Nucleic Acids Research, 47*(6), e32. https://doi.org/10.1093/nar/gkz037.

Kypr, J., Kejnovska, I., Renciuk, D., & Vorlickova, M. (2009). Circular dichroism and conformational polymorphism of DNA. *Nucleic Acids Research, 37*(6), 1713–1725. https://doi.org/10.1093/nar/gkp026.

Lago, S., Nadai, M., Cernilogar, F. M., Kazerani, M., Domíniguez Moreno, H., Schotta, G., & Richter, S. N. (2021). Promoter G-quadruplexes and transcription factors cooperate to shape the cell type-specific transcriptome. *Nature Communications, 12*(1), 3885. https://doi.org/10.1038/s41467-021-24198-2.

Lahnsteiner, A., Craig, S. J. C., Kamali, K., Weissensteiner, B., McGrath, B., Risch, A., & Makova, K. D. (2023). In vivo detection of DNA secondary structures using Permanganate/S1 footprinting with direct adapter ligation and sequencing (PDAL-Seq). *bioRxiv*, 2023.2011.2028.569002. https://doi.org/10.1101/2023.11.28.569002.

Lane, D. P. (1992). Cancer. p53, guardian of the genome. *Nature, 358*(6381), 15–16. https://doi.org/10.1038/358015a0.

Lapinaite, A., Simon, B., Skjaerven, L., Rakwalska-Bange, M., Gabel, F., & Carlomagno, T. (2013). The structure of the box C/D enzyme reveals regulation of RNA methylation. *Nature, 502*(7472), 519–523. https://doi.org/10.1038/nature12581.

Lee, I. B., Lee, J. Y., Lee, N.-K., & Hong, S.-C. (2012). Direct observation of the formation of DNA triplexes by single-molecule FRET measurements. *Current Applied Physics, 12*(4), 1027–1032. https://doi.org/10.1016/j.cap.2011.12.026.

Lee, T. I., Johnstone, S. E., & Young, R. A. (2006). Chromatin immunoprecipitation and microarray-based analysis of protein location. *Nature Protocols, 1*(2), 729–748. https://doi.org/10.1038/nprot.2006.98.

Leisegang, M. S., Bains, J. K., Seredinski, S., Oo, J. A., Krause, N. M., Kuo, C.-C., ... Brandes, R. P. (2022). HIF1α-AS1 is a DNA:DNA:RNA triplex-forming lncRNA interacting with the HUSH complex. *Nature Communications, 13*(1), 6563. https://doi.org/10.1038/s41467-022-34252-2.

Lerner, E., Cordes, T., Ingargiola, A., Alhadid, Y., Chung, S., Michalet, X., & Weiss, S. (2018). Toward dynamic structural biology: Two decades of single-molecule Forster resonance energy transfer. *Science (New York, N. Y.), 359*(6373), https://doi.org/10.1126/science.aan1133.

Li, C., Zhou, Z., Ren, C., Deng, Y., Peng, F., Wang, Q., ... Jiang, Y. (2022). Triplex-forming oligonucleotides as an anti-gene technique for cancer therapy. *Frontiers in Pharmacology, 13*, 1007723. https://doi.org/10.3389/fphar.2022.1007723.

Li, J. J., & Herskowitz, I. (1993). Isolation of ORC6, a component of the yeast origin recognition complex by a one-hybrid system. *Science (New York, N. Y.), 262*(5141), 1870–1874.

Li, Q., Lin, C., Luo, Z., Li, H., Li, X., & Sun, Q. (2022). Cryo-EM structure of R-loop monoclonal antibody S9.6 in recognizing RNA:DNA hybrids. *Journal of Genetics and Genomics = Yi Chuan xue bao, 49*(7), 677–680. https://doi.org/10.1016/j.jgg.2022.04.011.

Li, X., Zhou, B., Chen, L., Gou, L.-T., Li, H., & Fu, X.-D. (2017). GRID-seq reveals the global RNA–chromatin interactome. *Nature Biotechnology, 35*(10), 940–950. https://doi.org/10.1038/nbt.3968.

Li, Y., Liu, S., Cao, L., Luo, Y., Du, H., Li, S., ... You, F. (2021). CBRPP: A new RNA-centric method to study RNA–protein interactions. *RNA Biology, 18*(11), 1608–1621. https://doi.org/10.1080/15476286.2021.1873620.

Li, Y., Syed, J., & Sugiyama, H. (2016). RNA-DNA triplex formation by long noncoding RNAs. *Cell Chemical Biology, 23*(11), 1325–1333. https://doi.org/10.1016/j.chembiol.2016.09.011.

Lin, C., Dickerhoff, J., & Yang, D. (2019). NMR studies of G-quadruplex structures and G-quadruplex-interactive compounds. *Methods in Molecular Biology, 2035*, 157–176. https://doi.org/10.1007/978-1-4939-9666-7_9.

Lin, M., & Guo, J.-T. (2019). New insights into protein–DNA binding specificity from hydrogen bond based comparative study. *Nucleic Acids Research, 47*(21), 11103–11113. https://doi.org/10.1093/nar/gkz963.

Lin, R., Zhong, X., Zhou, Y., Geng, H., Hu, Q., Huang, Z., ... Chen, J.-Y. (2021). R-loopBase: A knowledgebase for genome-wide R-loop formation and regulation. *Nucleic Acids Research, 50*(D1), D303–D315. https://doi.org/10.1093/nar/gkab1103.

Liu, J. J., Yu, C. S., Wu, H. W., Chang, Y. J., Lin, C. P., & Lu, C. H. (2021). The structure-based cancer-related single amino acid variation prediction. *Scientific Reports, 11*(1), 13599. https://doi.org/10.1038/s41598-021-92793-w.

Liu, Z., Wang, J., Cheng, H., Ke, X., Sun, L., Zhang, Q. C., & Wang, H. W. (2018). Cryo-EM structure of human dicer and its complexes with a pre-miRNA substrate. *Cell, 173*(5), 1191–1203.e1112. https://doi.org/10.1016/j.cell.2018.03.080.

Lorenz, D. A., Her, H.-L., Shen, K. A., Rothamel, K., Hutt, K. R., Nojadera, A. C., ... Yeo, G. W. (2023). Multiplexed transcriptome discovery of RNA-binding protein binding sites by antibody-barcode eCLIP. *Nature Methods, 20*(1), 65–69. https://doi.org/10.1038/s41592-022-01708-8.

Luger, K., Mäder, A. W., Richmond, R. K., Sargent, D. F., & Richmond, T. J. (1997). Crystal structure of the nucleosome core particle at 2.8 Å resolution. *Nature, 389*(6648), 251–260. https://doi.org/10.1038/38444.

Lunde, B. M., Moore, C., & Varani, G. (2007). RNA-binding proteins: Modular design for efficient function. *Nature Reviews. Molecular Cell Biology, 8*(6), 479–490. https://doi.org/10.1038/nrm2178.

Ma, H., Jia, X., Zhang, K., & Su, Z. (2022). Cryo-EM advances in RNA structure determination. *Signal Transduction and Targeted Therapy, 7*(1), 58. https://doi.org/10.1038/s41392-022-00916-0.

Majka, J., & Speck, C. (2007). Analysis of protein-DNA interactions using surface plasmon resonance. *Advances in Biochemical Engineering/Biotechnology, 104*, 13–36.

Makowski, M. M., Gräwe, C., Foster, B. M., Nguyen, N. V., Bartke, T., & Vermeulen, M. (2018). Global profiling of protein–DNA and protein–nucleosome binding affinities using quantitative mass spectrometry. *Nature Communications, 9*(1), 1653. https://doi.org/10.1038/s41467-018-04084-0.

Maldonado, R., Filarsky, M., Grummt, I., & Längst, G. (2018). Purine–and pyrimidine–triple-helix-forming oligonucleotides recognize qualitatively different target sites at the ribosomal DNA locus. *RNA (New York, N. Y.), 24*(3), 371–380.

Mariner, P. D., Walters, R. D., Espinoza, C. A., Drullinger, L. F., Wagner, S. D., Kugel, J. F., & Goodrich, J. A. (2008). Human Alu RNA is a modular transacting repressor of mRNA transcription during heat shock. *Molecular Cell, 29*(4), 499–509.

Marušič, M., Toplishek, M., & Plavec, J. (2023). NMR of RNA—Structure and interactions. *Current Opinion in Structural Biology, 79*, 102532. https://doi.org/10.1016/j.sbi.2023.102532.

Matos-Rodrigues, G., van Wietmarschen, N., Wu, W., Tripathi, V., Koussa, N. C., Pavani, R., ... Nussenzweig, A. (2022). S1-END-seq reveals DNA secondary structures in human cells. *Molecular Cell, 82*(19), 3538–3552.e3535. https://doi.org/10.1016/j.molcel.2022.08.007.

Mattay, J. (2023). Current technical approaches to study RNA–protein interactions in mRNAs and long non-coding RNAs. *BioChem, 3*(1), 1–14.

McHugh, C. A., & Guttman, M. (2018). RAP-MS: A method to identify proteins that interact directly with a specific RNA molecule in cells. *Methods in Molecular Biology, 1649*, 473–488. https://doi.org/10.1007/978-1-4939-7213-5_31.

McLean, M. J., Lee, J. W., & Wells, R. D. (1988). Characteristics of Z-DNA helices formed by imperfect (purine-pyrimidine) sequences in plasmids. *The Journal of Biological Chemistry, 263*(15), 7378–7385.

Mei, Y., Deng, Z., Vladimirova, O., Gulve, N., Johnson, F. B., Drosopoulos, W. C., ... Lieberman, P. M. (2021). TERRA G-quadruplex RNA interaction with TRF2 GAR domain is required for telomere integrity. *Scientific Reports, 11*(1), 3509. https://doi.org/10.1038/s41598-021-82406-x.

Mellacheruvu, D., Wright, Z., Couzens, A. L., Lambert, J.-P., St-Denis, N. A., Li, T., ... Nesvizhskii, A. I. (2013). The CRAPome: A contaminant repository for affinity purification–mass spectrometry data. *Nature Methods, 10*(8), 730–736. https://doi.org/10.1038/nmeth.2557.

Minajigi, A., Froberg, J., Wei, C., Sunwoo, H., Kesner, B., Colognori, D., ... Lee, J. T. (2015). Chromosomes. A comprehensive Xist interactome reveals cohesin repulsion and an RNA-directed chromosome conformation. *Science (New York, N. Y.)*, *349*(6245), https://doi.org/10.1126/science.aab2276.

Mitsui, Y., Langridge, R., Shortle, B. E., Cantor, C. R., Grant, R. C., Kodama, M., & Wells, R. D. (1970). Physical and enzymatic studies on poly d (I–C). Poly d (I–C), an unusual double-helical DNA. *Nature*, *228*(5277), 1166–1169.

Mittler, G., Butter, F., & Mann, M. (2009). A SILAC-based DNA protein interaction screen that identifies candidate binding proteins to functional DNA elements. *Genome Research*, *19*(2), 284–293. https://doi.org/10.1101/gr.081711.108.

Mondal, T., Subhash, S., Vaid, R., Enroth, S., Uday, S., Reinius, B., ... Kanduri, C. (2015). MEG3 long noncoding RNA regulates the TGF-β pathway genes through formation of RNA-DNA triplex structures. *Nature Communications*, *6*, 7743. https://doi.org/10. 1038/ncomms8743.

Monsen, R. C., Chua, E. Y. D., Hopkins, J. B., Chaires, J. B., & Trent, J. O. (2023). Structure of a 28.5 kDa duplex-embedded G-quadruplex system resolved to 7.4 Å resolution with cryo-EM. *Nucleic Acids Research*, *51*(4), 1943–1959. https://doi.org/10. 1093/nar/gkad014.

Mueller, A. M., Breitsprecher, D., Duhr, S., Baaske, P., Schubert, T., & Längst, G. (2017). MicroScale thermophoresis: A rapid and precise method to quantify protein–nucleic acid interactions in solution. *Methods in Molecular Biology*, *1654*, 151–164. https://doi.org/10. 1007/978-1-4939-7231-9_10.

Müller, G. A., & Engeland, K. (2021). DNA affinity purification: A pulldown assay for identifying and analyzing proteins binding to nucleic acids. *Methods in Molecular Biology*, *2267*, 81–90. https://doi.org/10.1007/978-1-0716-1217-0_6.

Nadel, J., Athanasiadou, R., Lemetre, C., Wijetunga, N. A., Ó Broin, P., Sato, H., ... Greally, J. M. (2015). RNA:DNA hybrids in the human genome have distinctive nucleotide characteristics, chromatin composition, and transcriptional relationships. *Epigenetics Chromatin*, *8*(46), https://doi.org/10.1186/s13072-015-0040-6 Nov 16, PMID: 26579211; PMCID: PMC4647656.

Nasrullah, U., Haeussler, K., Biyanee, A., Wittig, I., Pfeilschifter, J., & Eberhardt, W. (2019). Identification of TRIM25 as a negative regulator of caspase-2 expression reveals a novel target for sensitizing colon carcinoma cells to intrinsic apoptosis. *Cells*, *8*(12), https://doi.org/10.3390/cells8121622.

Nichols, P. J., Krall, J. B., Henen, M. A., Vogeli, B., & Vicens, Q. (2023). Z-RNA biology: A central role in the innate immune response? *RNA (New York, N. Y.)*, *29*(3), 273–281. https://doi.org/10.1261/rna.079429.122.

Niranjanakumari, S., Lasda, E., Brazas, R., & Garcia-Blanco, M. A. (2002). Reversible cross-linking combined with immunoprecipitation to study RNA-protein interactions in vivo. *Methods (San Diego, Calif.)*, *26*(2), 182–190. https://doi.org/10.1016/s1046-2023(02)00021-x.

O'Leary, V. B., Ovsepian, S. V., Carrascosa, L. G., Buske, F. A., Radulovic, V., Niyazi, M., ... Anastasov, N. (2015). PARTICLE, a triplex-forming long ncRNA, regulates locus-specific methylation in response to low-dose irradiation. *Cell Reports*, *11*(3), 474–485.

Ohlendorf, D. H., & Matthew, J. B. (1985). Electrostatics and flexibility in protein-DNA interactions. *Advances in Biophysics*, *20*, 137–151. https://doi.org/10.1016/0065-227x (85)90034-6.

Onel, B., Wu, G., Sun, D., Lin, C., & Yang, D. (2019). Electrophoretic mobility shift assay and dimethyl sulfate footprinting for characterization of G-quadruplexes and G-quadruplex-protein complexes. *Methods in Molecular Biology*, *2035*, 201–222. https://doi.org/ 10.1007/978-1-4939-9666-7_11.

Paramasivan, S., Rujan, I., & Bolton, P. H. (2007). Circular dichroism of quadruplex DNAs: Applications to structure, cation effects and ligand binding. *Methods (San Diego, Calif.), 43*(4), 324–331. https://doi.org/10.1016/j.ymeth.2007.02.009.

Park, P. J. (2009). ChIP–seq: Advantages and challenges of a maturing technology. *Nature Reviews. Genetics, 10*(10), 669–680. https://doi.org/10.1038/nrg2641.

Parvathy, V. R., Bhaumik, S. R., Chary, K. V., Govil, G., Liu, K., Howard, F. B., & Miles, H. T. (2002). NMR structure of a parallel-stranded DNA duplex at atomic resolution. *Nucleic Acids Research, 30*(7), 1500–1511. https://doi.org/10.1093/nar/30.7.1500.

Patel, M. J., Yilmaz, G., Bhatia, L., Biswas-Fiss, E. E., & Biswas, S. B. (2018). Site-specific fluorescence double-labeling of proteins and analysis of structural changes in solution by Fluorescence Resonance Energy Transfer (FRET). *MethodsX, 5,* 419–430. https://doi.org/10.1016/j.mex.2018.03.006.

Peng, Y., Soper, T. J., & Woodson, S. A. (2012). RNase footprinting of protein binding sites on an mRNA target of small RNAs. *Methods in Molecular Biology, 905,* 213–224. https://doi.org/10.1007/978-1-61779-949-5_13.

Pilch, D. S., Levenson, C., & Shafer, R. H. (1990). Structural analysis of the (dA)10.2(dT) 10 triple helix. *Proceedings of the National Academy of Sciences of the United States of America, 87*(5), 1942–1946. https://doi.org/10.1073/pnas.87.5.1942.

Plavec, J. (2022). NMR study on nucleic acids. In N. Sugimoto (Ed.). *Handbook of chemical biology of nucleic acids* (pp. 1–44) Singapore: Springer Nature Singapore.

Porter, D. F., Miao, W., Yang, X., Goda, G. A., Ji, A. L., Donohue, L. K. H., ... Khavari, P. A. (2021). easyCLIP analysis of RNA-protein interactions incorporating absolute quantification. *Nature Communications, 12*(1), 1569. https://doi.org/10.1038/s41467-021-21623-4.

Postepska-Igielska, A., Giwojna, A., Gasri-Plotnitsky, L., Schmitt, N., Dold, A., Ginsberg, D., & Grummt, I. (2015). LncRNA Khps1 regulates expression of the proto-oncogene SPHK1 via triplex-mediated changes in chromatin structure. *Molecular Cell, 60*(4), 626–636.

Qiao, Y., Luo, Y., Long, N., Xing, Y., & Tu, J. (2021). Single-molecular Förster resonance energy transfer measurement on structures and interactions of biomolecules. *Micromachines, 12*(5), 492.

Raghavan, S. C., Chastain, P., Lee, J. S., Hegde, B. G., Houston, S., Langen, R., ... Lieber, M. R. (2005). Evidence for a triplex DNA conformation at the bcl-2 major breakpoint region of the t (14; 18) translocation. *Journal of Biological Chemistry, 280*(24), 22749–22760.

Raiber, E.-A., Kranaster, R., Lam, E., Nikan, M., & Balasubramanian, S. (2012). A non-canonical DNA structure is a binding motif for the transcription factor SP1 in vitro. *Nucleic Acids Research, 40*(4), 1499–1508.

Rashid, F., Raducanu, V.-S., Zaher, M. S., Tehseen, M., Habuchi, S., & Hamdan, S. M. (2019). Initial state of DNA-Dye complex sets the stage for protein induced fluorescence modulation. *Nature Communications, 10*(1), 2104. https://doi.org/10.1038/s41467-019-10137-9.

Rau, D. C., & Parsegian, V. A. (1992). Direct measurement of the intermolecular forces between counterion-condensed DNA double helices. Evidence for long range attractive hydration forces. *Biophysical Journal, 61*(1), 246–259. https://doi.org/10.1016/s0006-3495(92)81831-3.

Ravichandran, S., Razzaq, M., Parveen, N., Ghosh, A., & Kim, K. K. (2021). The effect of hairpin loop on the structure and gene expression activity of the long-loop G-quadruplex. *Nucleic Acids Research, 49*(18), 10689–10706. https://doi.org/10.1093/nar/gkab739.

Ravichandran, S., Subramani, V. K., & Kim, K. K. (2019). Z-DNA in the genome: From structure to disease. *Biophysical Reviews, 11*(3), 383–387. https://doi.org/10.1007/s12551-019-00534-1.

Ray, J., Kruse, A., Ozer, A., Kajitani, T., Johnson, R., MacCoss, M., ... Lis, J. T. (2020). RNA aptamer capture of macromolecular complexes for mass spectrometry analysis. *Nucleic Acids Research, 48*(15), e90. https://doi.org/10.1093/nar/gkaa542.

Reddy, K., Tam, M., Bowater, R. P., Barber, M., Tomlinson, M., Nichol Edamura, K., ... Pearson, C. E. (2010). Determinants of R-loop formation at convergent bidirectionally transcribed trinucleotide repeats. *Nucleic Acids Research, 39*(5), 1749–1762. https://doi.org/10.1093/nar/gkq935.

Reece-Hoyes, J. S., & Marian Walhout, A. J. (2012). Yeast one-hybrid assays: A historical and technical perspective. *Methods (San Diego, Calif.), 57*(4), 441–447. https://doi.org/10.1016/j.ymeth.2012.07.027.

Richard, P., & Manley, J. L. (2017). R loops and links to human disease. *Journal of Molecular Biology, 429*(21), 3168–3180.

Roberts, R. W., & Crothers, D. M. (1992). Stability and properties of double and triple helices: Dramatic effects of RNA or DNA backbone composition. *Science (New York, N. Y.), 258*(5087), 1463–1466.

Romero-López, C., Roda-Herreros, M., Berzal-Herranz, B., Ramos-Lorente, S. E., & Berzal-Herranz, A. (2023). Inter- and intramolecular RNA–RNA interactions modulate the regulation of translation mediated by the 3′ UTR in west nile virus. *International Journal of Molecular Sciences, 24*(6), 5337.

Sabarinathan, R., Tafer, H., Seemann, S. E., Hofacker, I. L., Stadler, P. F., & Gorodkin, J. (2013). The RNAsnp web server: Predicting SNP effects on local RNA secondary structure. *Nucleic Acids Research, 41*(Web Server issue), W475–W479. https://doi.org/10.1093/nar/gkt291.

Sanz, L. A., Castillo-Guzman, D., & Chédin, F. (2021). Mapping R-loops and RNA:DNA hybrids with S9.6-Based immunoprecipitation methods. *Journal of Visualized Experiments* (174), https://doi.org/10.3791/62455.

Sanz, L. A., & Chédin, F. (2019). High-resolution, strand-specific R-loop mapping via S9.6-based DNA–RNA immunoprecipitation and high-throughput sequencing. *Nature Protocols, 14*(6), 1734–1755.

Sanz, L. A., Hartono, S. R., Lim, Y. W., Steyaert, S., Rajpurkar, A., Ginno, P. A., ... Chédin, F. (2016). Prevalent, dynamic, and conserved R-loop structures associate with specific epigenomic signatures in mammals. *Molecular Cell, 63*(1), 167–178.

Sasmal, D. K., Pulido, L. E., Kasal, S., & Huang, J. (2016). Single-molecule fluorescence resonance energy transfer in molecular biology. *Nanoscale, 8*(48), 19928–19944. https://doi.org/10.1039/c6nr06794h.

Sathyamoorthy, B., Sannapureddi, R. K. R., Negi, D., & Singh, P. (2022). Conformational characterization of duplex DNA with solution-state NMR spectroscopy. *Journal of Magnetic Resonance Open, 10-11*, 100035. https://doi.org/10.1016/j.jmro.2022.100035.

Sato, K., Akiyama, M., & Sakakibara, Y. (2021). RNA secondary structure prediction using deep learning with thermodynamic integration. *Nature Communications, 12*(1), 941. https://doi.org/10.1038/s41467-021-21194-4.

Schmidtmann, E., Anton, T., Rombaut, P., Herzog, F., & Leonhardt, H. (2016). Determination of local chromatin composition by CasID. *Nucleus, 7*(5), 476–484. https://doi.org/10.1080/19491034.2016.1239000.

Schmitz, K.-M., Mayer, C., Postepska, A., & Grummt, I. (2010). Interaction of noncoding RNA with the rDNA promoter mediates recruitment of DNMT3b and silencing of rRNA genes. *Genes & Development (Cambridge, England), 24*(20), 2264–2269.

Shiroma, Y., Takahashi, R. U., Yamamoto, Y., & Tahara, H. (2020). Targeting DNA binding proteins for cancer therapy. *Cancer Science, 111*(4), 1058–1064. https://doi.org/10.1111/cas.14355.

Simon, M. D., Pinter, S. F., Fang, R., Sarma, K., Rutenberg-Schoenberg, M., Bowman, S. K., ... Lee, J. T. (2013). High-resolution Xist binding maps reveal two-step spreading during X-chromosome inactivation. *Nature, 504*(7480), 465–469. https://doi.org/10.1038/nature12719.

Singh, H., LeBowitz, J. H., Baldwin, A. S., & Sharp, P. A. (1988). Molecular cloning of an enhancer binding protein: isolation by screening of an expression library with a recognition site DNA. *Cell, 52*(3), 415–423.

Skourti-Stathaki, K. (2022). Detection of R-loop structures by immunofluorescence using the S9.6 monoclonal antibody. *Methods in Molecular Biology, 2528*, 21–29. https://doi.org/10.1007/978-1-0716-2477-7_2.

Song, S., Lee, J. U., Kang, J., Park, K. H., & Sim, S. J. (2020). Real-time monitoring of distinct binding kinetics of hot-spot mutant p53 protein in human cancer cells using an individual nanorod-based plasmonic biosensor. *Sensors and Actuators B: Chemical, 322*, 128584. https://doi.org/10.1016/j.snb.2020.128584.

Sosnick, T. R. (2001). Characterization of tertiary folding of RNA by circular dichroism and urea. *Chapter 11*, Unit 11.15 *Current Protocols in Nucleic Acid Chemistry/Edited by Serge L. Beaucage ... [et al.].* https://doi.org/10.1002/0471142700.nc1105s04.

Spitale, R. C., & Incarnato, D. (2023). Probing the dynamic RNA structurome and its functions. *Nature Reviews. Genetics, 24*(3), 178–196. https://doi.org/10.1038/s41576-022-00546-w.

Sridhar, B., Rivas-Astroza, M., Nguyen, T. C., Chen, W., Yan, Z., Cao, X., ... Zhong, S. (2017). Systematic mapping of RNA-chromatin interactions in vivo. *Current Biology: CB, 27*(4), 602–609. https://doi.org/10.1016/j.cub.2017.01.011.

Stahelin, R. V. (2013). Surface plasmon resonance: A useful technique for cell biologists to characterize biomolecular interactions. *Molecular Biology of the Cell, 24*(7), 883–886. https://doi.org/10.1091/mbc.E12-10-0713.

Stead, J. A., Keen, J. N., & McDowall, K. J. (2006). The identification of nucleic acid-interacting proteins using a simple proteomics-based approach that directly incorporates the electrophoretic mobility shift assay. *Molecular & Cellular Proteomics, 5*(9), 1697–1702.

Stiewe, T., & Haran, T. E. (2018). How mutations shape p53 interactions with the genome to promote tumorigenesis and drug resistance. *Drug Resistance Updates: Reviews and Commentaries in Antimicrobial and Anticancer Chemotherapy, 38*, 27–43. https://doi.org/10.1016/j.drup.2018.05.001.

Stockley, P. G., & Persson, B. (2009). Surface plasmon resonance assays of DNA-protein interactions. *Methods in Molecular Biology, 543*, 653–669. https://doi.org/10.1007/978-1-60327-015-1_38.

Stork, C. T., Bocek, M., Crossley, M. P., Sollier, J., Sanz, L. A., Chedin, F., ... Cimprich, K. A. (2016). Co-transcriptional R-loops are the main cause of estrogen-induced DNA damage. *Elife, 5*, e17548.

Subramani, V. K., & Kim, K. K. (2023). Characterization of Z-DNA using circular dichroism. *Methods in Molecular Biology, 2651*, 33–51. https://doi.org/10.1007/978-1-0716-3084-6_2.

Suzuki, Y., Endo, M., & Sugiyama, H. (2015). Studying RNAP–promoter interactions using atomic force microscopy. *Methods (San Diego, Calif.), 86*, 4–9.

Szlachta, K., Thys, R. G., Atkin, N. D., Pierce, L. C., Bekiranov, S., & Wang, Y.-H. (2018). Alternative DNA secondary structure formation affects RNA polymerase II promoter-proximal pausing in human. *Genome Biology, 19*, 1–19.

Takizawa, Y., & Kurumizaka, H. (2022). Chromatin structure meets cryo-EM: Dynamic building blocks of the functional architecture. *Biochimica et Biophysica Acta (BBA)—Gene Regulatory Mechanisms, 1865*(7), 194851. https://doi.org/10.1016/j.bbagrm.2022.194851.

Tan, J., & Lan, L. (2020). The DNA secondary structures at telomeres and genome instability. *Cell Bioscience, 10*, 47. https://doi.org/10.1186/s13578-020-00409-z.

Tang, L. (2020). Mapping RNA–RNA interactions. 760–760 *Nature Methods, 17*(8), https://doi.org/10.1038/s41592-020-0922-9.

Tateishi-Karimata, H., & Sugimoto, N. (2021). Roles of non-canonical structures of nucleic acids in cancer and neurodegenerative diseases. *Nucleic Acids Research, 49*(14), 7839–7855. https://doi.org/10.1093/nar/gkab580.

Teh, H. F., Peh, W. Y., Su, X., & Thomsen, J. S. (2007). Characterization of protein—DNA interactions using surface plasmon resonance spectroscopy with various assay schemes. *Biochemistry, 46*(8), 2127–2135. https://doi.org/10.1021/bi061903t.

Thomas, J. M. (2012). The birth of X-ray crystallography. *Nature, 491*(7423), 186–187. https://doi.org/10.1038/491186a.

Thomas, M., White, R. L., & Davis, R. W. (1976). Hybridization of RNA to double-stranded DNA: Formation of R-loops. *Proceedings of the National Academy of Sciences, 73*(7), 2294–2298.

Tijerina, P., Mohr, S., & Russell, R. (2007). DMS footprinting of structured RNAs and RNA-protein complexes. *Nature Protocols, 2*(10), 2608–2623. https://doi.org/10.1038/nprot.2007.380.

Tijerina, P., Mohr, S., & Russell, R. (2007). DMS footprinting of structured RNAs and RNA–protein complexes. *Nature Protocols, 2*(10), 2608–2623. https://doi.org/10.1038/nprot.2007.380.

Trembinski, D. J., Bink, D. I., Theodorou, K., Sommer, J., Fischer, A., van Bergen, A., ... Boon, R. A. (2020). Aging-regulated anti-apoptotic long non-coding RNA Sarrah augments recovery from acute myocardial infarction. *Nature Communications, 11*(1), 2039. https://doi.org/10.1038/s41467-020-15995-2.

Trendel, J., Schwarzl, T., Horos, R., Prakash, A., Bateman, A., Hentze, M. W., & Krijgsveld, J. (2019). The human RNA-binding proteome and its dynamics during translational arrest. *Cell, 176*(1), 391–403.e319.

Tresini, M., Warmerdam, D. O., Kolovos, P., Snijder, L., Vrouwe, M. G., Demmers, J. A., ... Hoeijmakers, J. H. (2015). The core spliceosome as target and effector of non-canonical ATM signalling. *Nature, 523*(7558), 53–58.

Tsai, N.-C., Hsu, T.-S., Kuo, S.-C., Kao, C.-T., Hung, T.-H., Lin, D.-G., ... Lin, Y.-C. J. (2021). Large-scale data analysis for robotic yeast one-hybrid platforms and multi-disciplinary studies using GateMultiplex. *BMC Biology, 19*(1), 214. https://doi.org/10.1186/s12915-021-01140-y.

Tsuji, Y. (2023). Optimization of biotinylated RNA or DNA pull-down assays for detection of binding proteins: Examples of IRP1, IRP2, HuR, AUF1, and Nrf2. *International Journal of Molecular Science, 24*(4), https://doi.org/10.3390/ijms24043604.

Uhm, H., Kang, W., Ha, K. S., Kang, C., & Hohng, S. (2018). Single-molecule FRET studies on the cotranscriptional folding of a thiamine pyrophosphate riboswitch. *Proceedings of the National Academy of Sciences of the United States of America, 115*(2), 331–336. https://doi.org/10.1073/pnas.1712983115.

Ule, J., Jensen, K. B., Ruggiu, M., Mele, A., Ule, A., & Darnell, R. B. (2003). CLIP identifies Nova-regulated RNA networks in the brain. *Science (New York, N. Y.), 302*(5648), 1212–1215. https://doi.org/10.1126/science.1090095.

Umeyama, T., & Ito, T. (2017). DMS-Seq for in vivo genome-wide mapping of protein-DNA interactions and nucleosome centers. *Cell Reports, 21*(1), 289–300. https://doi.org/10.1016/j.celrep.2017.09.035.

Underwood, K. F., Mochin, M. T., Brusgard, J. L., Choe, M., Gnatt, A., & Passaniti, A. (2013). A quantitative assay to study protein:DNA interactions, discover transcriptional regulators of gene expression, and identify novel anti-tumor agents. *Journal of Visualized Experiments*(78), https://doi.org/10.3791/50512.

Van Nostrand, E. L., Pratt, G. A., Shishkin, A. A., Gelboin-Burkhart, C., Fang, M. Y., Sundararaman, B., ... Yeo, G. W. (2016). Robust transcriptome-wide discovery of RNA-binding protein binding sites with enhanced CLIP (eCLIP). *Nature Methods, 13*(6), 508–514. https://doi.org/10.1038/nmeth.3810.

Van Oeffelen, L., Peeters, E., Nguyen Le Minh, P., & Charlier, D. (2014). The 'Densitometric Image Analysis Software' and its application to determine stepwise equilibrium constants from electrophoretic mobility shift assays. *PLoS One, 9*(1), e85146. https://doi.org/10.1371/journal.pone.0085146.

Van Treeck, B., & Parker, R. (2018). Emerging roles for intermolecular RNA-RNA interactions in RNP assemblies. *Cell, 174*(4), 791–802. https://doi.org/10.1016/j.cell.2018.07.023.

Varshney, D., Spiegel, J., Zyner, K., Tannahill, D., & Balasubramanian, S. (2020). The regulation and functions of DNA and RNA G-quadruplexes. *Nature Reviews. Molecular Cell Biology, 21*(8), 459–474. https://doi.org/10.1038/s41580-020-0236-x.

Velema, W. A., & Lu, Z. (2023). Chemical RNA cross-linking: Mechanisms, computational analysis, and biological applications. *JACS Au, 3*(2), 316–332. https://doi.org/10.1021/jacsau.2c00625.

Vierstra, J., Lazar, J., Sandstrom, R., Halow, J., Lee, K., Bates, D., ... Stamatoyannopoulos, J. A. (2020). Global reference mapping of human transcription factor footprints. *Nature, 583*(7818), 729–736. https://doi.org/10.1038/s41586-020-2528-x.

Volpato, A., Ollech, D., Alvelid, J., Damenti, M., Müller, B., York, A. G., ... Testa, I. (2023). Extending fluorescence anisotropy to large complexes using reversibly switchable proteins. *Nature Biotechnology, 41*(4), 552–559. https://doi.org/10.1038/s41587-022-01489-7.

Waldrip, Z. J., Byrum, S. D., Storey, A. J., Gao, J., Byrd, A. K., Mackintosh, S. G., ... Tackett, A. J. (2014). A CRISPR-based approach for proteomic analysis of a single genomic locus. *Epigenetics: Official Journal of the DNA Methylation Society, 9*(9), 1207–1211. https://doi.org/10.4161/epi.29919.

Wan, Y., Kertesz, M., Spitale, R. C., Segal, E., & Chang, H. Y. (2011). Understanding the transcriptome through RNA structure. *Nature Reviews. Genetics, 12*(9), 641–655.

Wan, Y., Qu, K., Zhang, Q. C., Flynn, R. A., Manor, O., Ouyang, Z., ... Chang, H. Y. (2014). Landscape and variation of RNA secondary structure across the human transcriptome. *Nature, 505*(7485), 706–709. https://doi.org/10.1038/nature12946.

Wang, A. H. J., Quigley, G. J., Kolpak, F. J., Crawford, J. L., van Boom, J. H., van der Marel, G., & Rich, A. (1979). Molecular structure of a left-handed double helical DNA fragment at atomic resolution. *Nature, 282*(5740), 680–686. https://doi.org/10.1038/282680a0.

Wang, M. M., & Reed, R. R. (1993). Molecular cloning of the olfactory neuronal transcription factor Olf-1 by genetic selection in yeast. *Nature, 364*(6433), 121–126.

Wang, Q., Wu, J., Wang, H., Gao, Y., Liu, Q., Mu, A., ... Zhu, C. (2020). Structural basis for RNA replication by the SARS-CoV-2 polymerase. *Cell, 182*(2), 417–428 e413.

Wang, S., Poon, G. M., & Wilson, W. D. (2015). Quantitative investigation of protein-nucleic acid interactions by biosensor surface plasmon resonance. *Methods in Molecular Biology, 1334*, 313–332. https://doi.org/10.1007/978-1-4939-2877-4_20.

Wang, X., Terashi, G., & Kihara, D. (2023). CryoREAD: De novo structure modeling for nucleic acids in cryo-EM maps using deep learning. *Nature Methods, 20*(11), 1739–1747. https://doi.org/10.1038/s41592-023-02032-5.

Warwick, T., Brandes, R. P., & Leisegang, M. S. (2023). Computational methods to study DNA:DNA:RNA triplex formation by lncRNAs. *Non-Coding RNA (New York, N. Y.), 9*(1), 10.

Warwick, T., Seredinski, S., Krause, N. M., Bains, J. K., Althaus, L., Oo, J. A., ... Brandes, R. P. (2022). A universal model of RNA.DNA:DNA triplex formation accurately predicts genome-wide RNA-DNA interactions. *Briefings in Bioinformatics, 23*(6), https://doi.org/10.1093/bib/bbac445.

Watson, J. D., & Crick, F. H. (1953). Molecular structure of nucleic acids; a structure for deoxyribose nucleic acid. *Nature, 171*(4356), 737–738. https://doi.org/10.1038/171737a0.

Webb, R. (2008). New resonance. S10–S10 *Nature Physics, 4*(1), https://doi.org/10.1038/nphys863.

Wells, S. E., Hughes, J. M., Igel, A. H., & Ares, M., Jr. (2000). Use of dimethyl sulfate to probe RNA structure in vivo. *Methods Enzymol, 318*, 479–493. https://doi.org/10.1016/s0076-6879(00)18071-1.

West, J. A., Davis, C. P., Sunwoo, H., Simon, M. D., Sadreyev, R. I., Wang, P. I., ... Kingston, R. E. (2014). The long noncoding RNAs NEAT1 and MALAT1 bind active chromatin sites. *Molecular Cell, 55*(5), 791–802. https://doi.org/10.1016/j.molcel.2014.07.012.

Westin, L., Blomquist, P., Milligan, J. F., & Wrange, Ö. (1995). Triple helix DNA alters nucleosomal histone-DNA interactions and acts as a nucleosome barrier. *Nucleic Acids Research, 23*(12), 2184–2191.

Winnerdy, F. R., Bakalar, B., Maity, A., Vandana, J. J., Mechulam, Y., Schmitt, E., & Phan, A. T. (2019). NMR solution and X-ray crystal structures of a DNA molecule containing both right- and left-handed parallel-stranded G-quadruplexes. *Nucleic Acids Research, 47*(15), 8272–8281. https://doi.org/10.1093/nar/gkz349.

Woo, A. J., Dods, J. S., Susanto, E., Ulgiati, D., & Abraham, L. J. (2002). A proteomics approach for the identification of DNA binding activities observed in the electrophoretic mobility shift assay. *Molecular & Cellular Proteomics, 1*(6), 472–478.

Wulfridge, P., & Sarma, K. (2021). A nuclease- and bisulfite-based strategy captures strand-specific R-loops genome-wide. *Elife, 10*, e65146. https://doi.org/10.7554/eLife.65146.

Xu, W., Li, K., Li, Q., Li, S., Zhou, J., & Sun, Q. (2022). Quantitative, convenient, and efficient genome-wide R-loop profiling by ssDRIP-Seq in multiple organisms. *Methods in Molecular Biology, 2528*, 445–464. https://doi.org/10.1007/978-1-0716-2477-7_29.

Xu, W., Xu, H., Li, K., Fan, Y., Liu, Y., Yang, X., & Sun, Q. (2017). The R-loop is a common chromatin feature of the Arabidopsis genome. *Nature Plants, 3*(9), 704–714. https://doi.org/10.1038/s41477-017-0004-x.

Yan, Q., & Sarma, K. (2020). MapR: A method for identifying native R-loops genome wide. *130*(1), e113. https://doi.org/10.1002/cpmb.113.

Yan, Z., Huang, N., Wu, W., Chen, W., Jiang, Y., Chen, J., ... Zhong, S. (2019). Genome-wide colocalization of RNA–DNA interactions and fusion RNA pairs. *Proceedings of the National Academy of Sciences, 116*(8), 3328–3337. https://doi.org/10.1073/pnas.1819788116.

Yang, H.-Y., & Moerner, W. (2018). Resolving mixtures in solution by single-molecule rotational diffusivity. *Nano Letters, 18*(8), 5279–5287.

Yeh, S. Y., & Rhee, H. S. (2023). The ChIP-Exo method to identify genomic locations of DNA-binding proteins at near single base-pair resolution. *Methods in Molecular Biology, 2599*, 33–48. https://doi.org/10.1007/978-1-0716-2847-8_4.

Yesudhas, D., Batool, M., Anwar, M. A., Panneerselvam, S., & Choi, S. (2017). Proteins recognizing DNA: Structural uniqueness and versatility of DNA-binding domains in stem cell transcription factors. *Genes (Basel), 8*(8), https://doi.org/10.3390/genes8080192.

Yoo, J., & Aksimentiev, A. (2016). The structure and intermolecular forces of DNA condensates. *Nucleic Acids Research, 44*(5), 2036–2046. https://doi.org/10.1093/nar/gkw081.

Yu, K., Chedin, F., Hsieh, C.-L., Wilson, T. E., & Lieber, M. R. (2003). R-loops at immunoglobulin class switch regions in the chromosomes of stimulated B cells. *Nature Immunology, 4*(5), 442–451.

Yuan, W., Zhou, J., Tong, J., Zhuo, W., Wang, L., Li, Y., ... Qian, W. (2019). ALBA protein complex reads genic R-loops to maintain genome stability in Arabidopsis. *Science Advances, 5*(5), eaav9040. https://doi.org/10.1126/sciadv.aav9040.

Zhang, J.-y, Zheng, K.-w, Xiao, S., Hao, Y.-h, & Tan, Z. (2014). Mechanism and manipulation of DNA: RNA hybrid G-quadruplex formation in transcription of G-rich DNA. *Journal of the American Chemical Society, 136*(4), 1381–1390.

Zhang, S., Liu, Y., Sun, Y., Liu, Q., Gu, Y., Huang, Y., ... Ouyang, Y. (2024). Aberrant R-loop–mediated immune evasion, cellular communication, and metabolic reprogramming affect cancer progression: A single-cell analysis. *Molecular Cancer, 23*(1), 11. https://doi.org/10.1186/s12943-023-01924-6.

Zhang, W., Xie, M., Shu, M. D., Steitz, J. A., & DiMaio, D. (2016). A proximity-dependent assay for specific RNA-protein interactions in intact cells. *RNA (New York, N. Y.)*, *22*(11), 1785–1792. https://doi.org/10.1261/rna.058248.116.

Zhang, Y., El Omari, K., Duman, R., Liu, S., Haider, S., Wagner, A., ... Wei, D. (2020). Native de novo structural determinations of non-canonical nucleic acid motifs by X-ray crystallography at long wavelengths. *Nucleic Acids Research*, *48*(17), 9886–9898. https://doi.org/10.1093/nar/gkaa439.

Zhao, J., Ohsumi, T. K., Kung, J. T., Ogawa, Y., Grau, D. J., Sarma, K., ... Lee, J. T. (2010). Genome-wide identification of polycomb-associated RNAs by RIP-seq. *Molecular Cell*, *40*(6), 939–953. https://doi.org/10.1016/j.molcel.2010.12.011.

Zhou, J. C., Feller, B., Hinsberg, B., Sethi, G., Feldstein, P., Hihath, J., ... Miller, R. (2015). Immobilization-mediated reduction in melting temperatures of DNA–DNA and DNA–RNA hybrids: Immobilized DNA probe hybridization studied by SPR. *Colloids and Surfaces A: Physicochemical and Engineering Aspects*, *481*, 72–79. https://doi.org/10.1016/j.colsurfa.2015.04.046.

Znosko, B. M., Barnes, T. W., 3rd, Krugh, T. R., & Turner, D. H. (2003). NMR studies of DNA single strands and DNA:RNA hybrids with and without 1-propynylation at C5 of oligopyrimidines. *Journal of the American Chemical Society*, *125*(20), 6090–6097. https://doi.org/10.1021/ja021285d.

Zyner, K. G., Simeone, A., Flynn, S. M., Doyle, C., Marsico, G., Adhikari, S., ... Balasubramanian, S. (2022). G-quadruplex DNA structures in human stem cells and differentiation. *Nature Communications*, *13*(1), 142. https://doi.org/10.1038/s41467-021-27719-1.

> CHAPTER NINE

Multi-omics based artificial intelligence for cancer research

Lusheng Li[a,1], Mengtao Sun[a,1], Jieqiong Wang[b], and Shibiao Wan[a,*]
[a]Department of Genetics, Cell Biology and Anatomy, University of Nebraska Medical Center, Omaha, NE, United States
[b]Department of Neurological Sciences, University of Nebraska Medical Center, Omaha, NE, United States
*Corresponding author. e-mail address: swan@unmc.edu

Contents

1. Introduction	304
2. Artificial intelligence and machine learning	305
2.1 Supervised learning	306
2.2 Unsupervised learning	309
2.3 Semi-supervised learning	312
2.4 Deep learning	313
3. Omics techniques	314
3.1 Genomics	316
3.2 Epigenomics	317
3.3 Transcriptomics	318
3.4 Proteomics	318
3.5 Metabolomics	319
4. Application of AI and multi-omics in cancer research	320
4.1 Early cancer detection	320
4.2 Diagnosis	323
4.3 Prognosis	325
4.4 Treatment response prediction	328
4.5 Biomarker identification	329
4.6 Pathology identification	332
5. Multi-omics integration	334
5.1 Early integration	335
5.2 Late integration	340
5.3 Intermediate integration	341
6. Challenges	343
6.1 Data harmonization	343
6.2 Interpretability	343
6.3 Ethical considerations	344
Acknowledgments	345
Funding	345
Authors' contributions	346

[1] These authors contributed equally to this work, and they should be regarded as co-first authors.

Advances in Cancer Research, Volume 163
ISSN 0065-230X, https://doi.org/10.1016/bs.acr.2024.06.005
Copyright © 2024 Elsevier Inc. All rights are reserved, including those for text and data mining, AI training, and similar technologies.

Competing interests	346
References	346

Abstract

With significant advancements of next generation sequencing technologies, large amounts of multi-omics data, including genomics, epigenomics, transcriptomics, proteomics, and metabolomics, have been accumulated, offering an unprecedented opportunity to explore the heterogeneity and complexity of cancer across various molecular levels and scales. One of the promising aspects of multi-omics lies in its capacity to offer a holistic view of the biological networks and pathways underpinning cancer, facilitating a deeper understanding of its development, progression, and response to treatment. However, the exponential growth of data generated by multi-omics studies present significant analytical challenges. Processing, analyzing, integrating, and interpreting these multi-omics datasets to extract meaningful insights is an ambitious task that stands at the forefront of current cancer research. The application of artificial intelligence (AI) has emerged as a powerful solution to these challenges, demonstrating exceptional capabilities in deciphering complex patterns and extracting valuable information from large-scale, intricate omics datasets. This review delves into the synergy of AI and multi-omics, highlighting its revolutionary impact on oncology. We dissect how this confluence is reshaping the landscape of cancer research and clinical practice, particularly in the realms of early detection, diagnosis, prognosis, treatment and pathology. Additionally, we elaborate the latest AI methods for multi-omics integration to provide a comprehensive insight of the complex biological mechanisms and inherent heterogeneity of cancer. Finally, we discuss the current challenges of data harmonization, algorithm interpretability, and ethical considerations. Addressing these challenges necessitates a multidisciplinary collaboration, paving the promising way for more precise, personalized, and effective treatments for cancer patients.

1. Introduction

Cancer, as a complex and multifactorial disease, involves changes in the process of occurrence and development across multiple levels, including genes (Wu, Zhu, Thompson, & Hannun, 2018), proteins (Yang et al., 2021), metabolites (Seyfried & Shelton, 2010), and even environmental factors (Mbemi, Khanna, Njiki, Yedjou, & Tchounwou, 2020). It remains one of the most formidable challenges in modern medicine, characterized by large heterogeneity and intricate molecular dynamics (Dagogo-Jack & Shaw, 2018; Meacham & Morrison, 2013). This complexity not only impedes our understanding of cancer's mechanisms but also presents significant challenges in developing effective, personalized treatments. With remarkable advancements in omics technologies, large amounts of multi-omics data such as epigenomes, genomes, transcriptomes, proteomes, and metabolomes, can be obtained through high-throughput next-generation sequencing techniques. These multi-omics approaches provide unprecedented opportunities for a deeper understanding of cancer's complexities (Bock, Farlik, & Sheffield, 2016; Hasin, Seldin, & Lusis, 2017; Menyhárt & Győrffy 2021). In the field of cancer research, the application of multi-omics approaches has proven instrumental in unravelling the complex mechanisms underlying tumor occurrence, development, progression, and metastasis (Akhoundova & Rubin, 2022; Ghaffari et al., 2021; Heo, Hwa, Lee, Park & An, 2021; Lu & Zhan, 2018). Additionally, multi-omics approaches are pivotal in identifying new therapeutic targets (Ivanisevic & Sewduth, 2023) and predicting patients' responses to treatments (Lapuente-Santana, van Genderen, Hilbers, Finotello, & Eduati, 2021). However, how to effectively process, analyze, integrate, and interpret these large-scale multi-omics data to extract valuable information remains a major challenge for current cancer research. With the continuous development of artificial intelligence (AI) technologies, an increasing number of researchers are beginning to integrate AI into cancer research to address these challenges (Acosta, Falcone, Rajpurkar, & Topol, 2022; Cai, Poulos, Liu, & Zhong, 2022; He, Xiaowei Liu, Shi, & Jing, 2023; Swanson, Wu, Zhang, Alizadeh, & Zou, 2023; Wu, Li, & Tu, 2023). AI has demonstrated remarkable capabilities in deciphering complex patterns and extracting meaningful insights from large and intricate datasets (Sarker, 2022; Xu et al., 2021). The confluence of AI with multi-omics approaches in cancer research offers novel insights and powerful tools to deeply understand cancer's complexities. This review aims to explore the latest applications of AI and multi-omics within the context of cancer research, discussing the current challenges and future directions in this dynamic field.

2. Artificial intelligence and machine learning

Artificial intelligence (AI) is centered on the development of computer systems capable of simulating human intelligence and the cognitive processes of the human brain. These computer systems are expected to encompass a broad spectrum of capabilities (Ertel, 2018), including learning, reasoning, problem-solving, pattern recognition, understanding natural language, and decision-making. The AI application in cancer research represents a burgeoning field, characterized by the deployment of machine learning and deep learning algorithms, which is instrumental in processing and analyzing large-scale biomedical datasets to reveal the complexity of cancer at a molecular and cellular level and drive innovative approaches in cancer treatments (Lapuente-Santana et al., 2021; Seyhan & Carini, 2019). Machine learning (ML), a critical subset of artificial intelligence, involves the automated discovery of patterns and rules within extensive datasets (Langley, 2011; Sarker, 2021). ML is capable of making correct predictions and decisions by identifying and extracting meaningful features from the data. The basic principle of ML lies in its capacity to learn patterns and rules from data based on extracted features and then to apply the patterns and rules learned to new, unseen data to make predictions or classifications (Janiesch, Zschech, & Heinrich, 2021). The process of ML includes data collection, data pre-processing, feature extraction, model training, model evaluation, model optimization and model application. The model training process involves adjusting or optimizing the model's parameters to minimize the difference between the predicted output and the actual label. This difference, commonly referred to as 'error' or 'loss', can be quantified using specific loss functions, which provides guidance for the learning and optimization of the model towards improved accuracy. Based on the model training approaches, ML can be primarily categorized into three distinct types: supervised learning, unsupervised learning, semi-supervised learning, and reinforcement learning. Given that reinforcement learning is not extensively applied in cancer research, this review will focus on the first three types (Fig. 1).

2.1 Supervised learning

Supervised learning plays an important role for a range of tasks in cancer research, such as classifying cancer types, cancer early detection, cancer prognosis, and predicting patient's response to treatment. Supervised learning train models on a labeled dataset, where the training dataset consists of input data and corresponding output labels, allowing the model to be able to make accurate predictions on classification and regression tasks. The cross-entropy loss is a commonly used loss function for classification tasks and the mean square error (MSE) is a commonly employed loss function for regression tasks. A variety of standard methods are employed for classification tasks (Fig. 1), such as logistic regression (LR) (Cramer, 2002), decision tree (DT) based method (James, Witten, Hastie, Tibshirani, & Taylor, 2023), naïve bayes (NB) (Rish, 2001), support vector machine (SVM) (Hearst, Susan, Osuna, Platt, & Scholkopf, 1998),

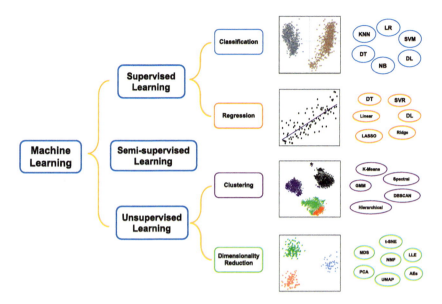

Fig. 1 **The overview of machine learning paradigms.** Machine learning can be categorized into supervised learning, semi-supervised learning, and unsupervised learning. In supervised learning, machine learning tasks include classification and regression. For unsupervised learning, it can be categorized into clustering methods and dimensionality reduction techniques. See texts for details about each ML algorithm.

K-nearest neighbor (KNN) (Cunningham & Delany, 2021). Among these algorithms, logistic regression algorithm is linear model, which constructs the relationship between independent variables (features) and a binary dependent variable (outcome) by estimating probabilities using a logistic function (sigmoid function) that transform the linear combination to the probability of 0 and 1. Logistic regression is straightforward to implement and interpret, but its performance can be deteriorating when dealing with datasets with a large number of features. Another reason that limits the effectiveness of logistic regression is the complex and non-linear of data. In addition, decision tree (DT) based methods, SVM (especially with non-linear kernels) and KNN can be categorized as non-linear models. Non-linear models offer powerful capabilities for capturing the complex and non-linear relationship within data, allowing them effective for a wide range of real-world problems. However, it presents challenges in computational requirements, susceptibility to overfitting, and interpretability when contrasted with linear models. Decision Tree based methods can identify the importance of various features during decision-making process, which facilitates the model interpretability. This is especially important for cancer research because medical researchers and doctors can directly understand how models make predictions to the outcomes of patients based on a variety of features, such as gene expression, genetic markers, proteins, lifestyle habits, medical history, etc. It is important to note that if the decision tree is too complex, it may result in overfitting, which reduces the model's generalizability to new, unseen data. For very large datasets, building decision trees may require long computation time and large computation memory. The fundamental principle of SVM involves identifying an optimal decision boundary that most effectively separates different categories of data points. It applies kernel function to transform original feature space to higher-dimensional space where a linear separation is possible, to achieve efficient classification. SVM can provide good performance on small and medium-sized datasets, especially in high-dimensional spaces. With proper regularization, it can avoid overfitting risk and has good generalization ability. However, the selection of kernel functions and the adjustment of parameters require specialized knowledge, which may impact the model performance. The KNN is to find the K nearest neighbors of a sample, and then using the information of these K neighbors to make predictions. For classification tasks, the majority voting method is typically adopted to predict the categories. For regression tasks, the prediction is made by calculating the average of the values from the neighboring points. KNN algorithm is computationally intensive and storage demanding due to it computes distances for all training samples. In addition, distance metrics may no longer be effective in high-dimensional spaces, resulting in performance degradation. Based on the characteristics of KNN, it can perform well on small, low-dimensional, class-balanced datasets. Naïve Bayes' algorithm is probabilistic model, that based on Bayes' theorem and conditional independence assumption between the features. It classifies samples by calculating the probability of each class given the sample, based on the feature probabilities. While Naïve Bayes demonstrates good performance on the small datasets, its efficacy may diminish when dealing with highly correlated features due to its strong assumptions regarding the independence of features. Common algorithms employed for regression tasks include linear regression (Maulud & Abdulazeez, 2020), Ridge Regression (Hoerl & Kennard, 1970; McDonald, 2009) and least absolute shrinkage and selection operator (LASSO) (Tibshirani, 1996). Linear regression involves identifying a line or hyperplane that optimally fits the training data, thereby minimizing the discrepancy between the model's predictions and the actual observations. The validity of the linear regression model is based on two critical hypotheses: the linear relationship hypothesis and the independence hypothesis. Specifically, linear regression assumes a linear correlation between the dependent and independent variables and that each observation is independent of the others. Ridge regression is a regularization method of linear regression used to deal with collinearity problems, that refers to situations where there is a high correlation or linear dependence between independent variables. It limits the coefficient size of the model by introducing $L2$ regularization term in the objective function to reduce the risk of overfitting. LASSO regression is a regularization method of linear regression for dealing collinearity problems and feature selection. It implements automatic feature selection by introducing $L1$ regularization term into the objective function to constrain the coefficients of the model and force some coefficients to become zero.

2.2 Unsupervised learning

Compared to supervised learning, unsupervised learning discovers hidden patterns, structures, or subgroups in the unlabeled data through clustering and dimensionality reduction. Clustering aims to divide a dataset into distinct categories based on inherent similarities and dissimilarities of internal data structures, which is to ensure that data points within the same category exhibit high degrees of similarity and data points belonging to different categories are separated as much as possible. The similarity can be measured through various methods, such as Euclidean distance, cosine similarity, probability distribution, density, spectral similarity. Common clustering algorithms (Fig. 1) include hierarchical clustering (Murtagh & Contreras, 2012), K-means clustering (Sinaga & Yang, 2020), spectral clustering (Ng, Jordan, & Weiss, 2001), Gaussian mixture model (GMM) (Maugis, Celeux, & Magniette, 2009), Density-Based Spatial Clustering of Applications with Noise (DBSCAN) clustering (Khan, Rehman, Aziz, Fong, & Sarasvady, 2014). Hierarchical clustering is a connectivity-based clustering method. It divides data points into different clusters based on distance or similarity measures to form a dendrogram with hierarchical structure. It does not require the pre-definition of the number of clusters, making it suitable for scenarios where the optimal number of clusters is unclear. Hierarchical clustering is effective for datasets that exhibit hierarchical structure, although it typically has a high computational complexity. K-means, a centroid-based clustering method, is to identify K optimal centroids based on similarity according to the predefined number of clusters,

K, and then assigns each data point to the cluster represented by the nearest centroid. K-means is straightforward and efficient across numerous scenarios. However, its effectiveness may diminish for data characterized by complex geometries, including non-spherical clusters or clusters that vary in size and density. Spectral clustering is a graph-based clustering method, which is especially suitable for discovering clusters with complex shapes and non-spherical clusters. Spectral clustering constructs a graph by connecting all data points with edges where points that are closer together are connected by edges with higher weights. The clustering is then achieved by partitioning this graph into subgraphs in such a way that the weights of the edges between different subgraphs are minimized, while the weights of the edges within each subgraph are maximized. Spectral clustering only requires a similarity matrix between the data, making it particularly efficient for handling sparse datasets. However, it is very important to select appropriate similarity measures and parameters to obtain effective clustering results. The GMM is a probabilistic model used to represent that a dataset is generated from a combination of multiple Gaussian distributions. GMM assigns data points to clusters by fitting the data's Gaussian distributions, effectively classifying each point based on the likelihood of its belonging to a particular distribution. However, GMM is very sensitive to the initial parameter settings, requiring the number of clusters and the initial parameters of each Gaussian distribution, such as the mean and covariance matrix, to be specified in advance. The improper selection of these initial parameters can lead to issues with model convergence or result in inaccurate clustering outcomes. DBSCAN, a density-based spatial clustering algorithm, divides regions of adequate density into distinct clusters. It can automatically discover the number of clusters according to the characteristics of the dataset. Moreover, DBSCAN can identify clusters of arbitrary shape while identifying points that do not belong to any cluster as noise, which is especially important for noisy and nonlinear distributed datasets in the real world. The choice of density parameters and neighborhood radius in DBSCAN significantly impacts the clustering outcomes. Additionally, density-based clustering algorithms, like DBSCAN, may encounter challenges with high-dimensional datasets due to the curse of dimensionality. In high-dimensional spaces, density distribution of data points tends to become more uniform, making it challenging to identify efficient clustering structures. The performance of the clustering methods that calculate the distance, such as hierarchical clustering, K-means clustering, spectral clustering and DBSCAN, can be compromised in high-dimensional spaces where distance measurements become increasingly less reliable. Furthermore, dimensionality reduction algorithm aims to map high-dimensional data to low-dimensional space, which is to effectively compress data from complex high-dimensional space to simple low-dimensional representation through linear or nonlinear transformation. The basic principle of dimensionality reduction algorithm is to compress the dimensionality of data as much as possible to reduce the interference of noise while preserving the most significant information and features inherent in the data. Common dimensionality reduction algorithms (Fig. 1) include principal component analysis (PCA) (Dunteman, 1989), multidimensional scaling (MDS) (Hout, Papesh, & Goldinger, 2013), non-negative matrix factorization (NMF) (Lee & Seung, 2000), locally linear embedding (LLE) (Chen & Liu, 2011), t-distributed neighborhood embedding algorithm (t-SNE) (Van Der Maaten & Hinton, 2008) uniform manifold approximation and projection (UMAP) (McInnes, Healy, & Melville, 2018) and autoencoders (AEs) (Rumelhart, Hinton, & Williams, 1985). Among these methods, PCA, MDS, and NMF are categorized as linear dimensionality reduction techniques, which rely on linear transformations of the data to reduce its dimensionality. PCA is the process of converting high-dimensional data into low-dimensional data through linear transformations to discover the inherent structure and patterns in the data. PCA achieves dimensionality reduction and feature extraction by searching for principal components (i.e., maximum variance direction) in the dataset. The MDS algorithm can transform high-dimensional data into low-dimensional space, which applies a distance matrix to represent the similarity or correlation between data points. The NMF is to decompose a non-negative data matrix into the product of two or more non-negative matrices, thereby revealing its underlying structure and patterns while preserving the non-negative properties of the data. The nonlinear dimensionality reduction techniques include LLE, t-SNE, UMAP and AEs. LLE is designed to project high-dimensional data into lower-dimensional spaces while preserving the local geometric structure of the data. The fundamental principle of LLE involves capturing the local geometry of the data through the maintenance of linear relationships between each data point and its nearest neighbors. In addition, t-SNE is to project the high-dimensional data into lower-dimensional spaces while preserving the similarity relationship between data points as much as possible. It exhibits strong performance in capturing the nonlinear structure of data effectively and producing intuitive visualizations of high-dimensional datasets. However, t-SNE is computationally intensive and may require considerable time to process large datasets. Additionally, the outcomes of t-SNE are sensitive to the choice of initialization parameters. UMAP is a novel manifold learning technique based on Riemannian geometry and algebraic topology for dimensionality reduction. It aims to present the global structure of the data as much as possible while preserving the local structure of the data. UMAP demonstrated competitive performance in term of visualization quality, the ability to preserve the global structure, and run time. However, the nonlinear dimensionality reduction methods face challenges with interpretability. For instance, the dimension of embedding space generated by UMAP may not represent any meanings.

2.3 Semi-supervised learning

Semi-supervised learning algorithms leverage both a large amount of unlabeled data and a small amount of labelled data for model training. In many real scenarios, the collection of labelled data is very difficult, and the annotation of unlabeled data is time-consuming, laborious and resource intensive. Semi-supervised algorithms train the model with a small amount of labelled data and apply the model trained to annotate unlabeled data. Semi-supervised learning approaches (Reddy, Viswanath, & Eswara Reddy, 2018; van Engelen & Hoos, 2020) include self-training, co-training, and label propagation. Self-training (Amini, Feofanov, Pauletto, Devijver, & Maximov, 2022) train the initial model on a small, labelled dataset and apply the trained model to predict the "pseudo-labels" of unlabeled data. These newly pseudo-labelled data are then combined with the original labelled dataset to retrain the model. For co-training (Qiao, Shen, Zhang, Wang, & Yuille, 2018; Zhou & Li, 2005), two models are trained separately on different views of the labelled data. Each model then predicts the "pseudo-labels" of unlabeled data and the algorithm selects newly pseudo-labelled data with high confidence false labels for trained models in another view. This process is iterated until both models no longer change, or a pre-set number of iterations is reached. Label propagation (Iscen, Tolias, Avrithis, & Chum, 2019) involves constructing a similarity graph where nodes represent data points (both labelled and unlabeled) and edges represent similarities between them. The pseudo-label is then propagated from the labelled data to the unlabeled data through the graph, where the propagation weight is proportional to the similarity of the pairs in the graph. Semi-supervised learning significantly contributes to the model training on labeled and unlabeled dataset. LNP-Chlo (Wan, Mak, & Kung, 2016), a multi-label predictor based on ensemble linear neighborhood propagation (LNP) (Wang & Zhang, 2008), was proposed to predict localization of chloroplast proteins. The semi-supervised learning method was applied to extract hybrid sequence-based feature information from both labeled and unlabeled proteins. In addition, Wan, Mak, and Kung (2017) leveraged transductive learning, a subset of semi-supervised learning, to develop a model based on the least squares and nearest neighbor algorithms to process the features for multi-label protein subchloroplast localization prediction.

2.4 Deep learning

Deep learning, a subfield of machine learning, employs artificial neural networks (Hassoun, 1995) for representation learning from massive amounts of data. A deep neural network consists of an input layer, multiple hidden layers, and an output layer, each of which receives the output of the previous layer of neurons as input and performs nonlinear transformation processing, thereby gradually transforming the raw data into meaningful feature representations. Popular deep neural network architectures (Fig. 2A) include multilayer perceptron (MLP), convolutional neural networks (CNNs) (Alzubaidi et al., 2021), recurrent neural networks (RNNs) (Schuster & Paliwal, 1997), autoencoders (AEs) (Rumelhart et al., 1985), generative adversarial networks

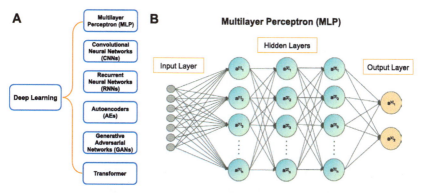

Fig. 2 **The overview of deep learning paradigms.** (A) The common architectures within deep learning, including multilayer perceptron (MLP), convolutional neural networks (CNNs), recurrent neural networks (RNNs), autoencoders (AEs), generative adversarial networks (GANs), and transformer. (B) An example for a deep learning architecture, e.g., multilayer perceptron (MLP). MLP is a basic architecture of the neural network, which consists of an input layer, hidden layers, and an output layer. The input layer receives data, where each neuron corresponds to a feature of the input data. The hidden layers perform computations and transformations on the input data through weighted sums and non-linear activation functions. These processed signals are then conveyed to the output layer, which generates the final output of the network.

(GANs) (Goodfellow et al., 2020) and transformer (Vaswani et al., 2017). These deep neural network architectures can be employed according to specific tasks and data types. Fig. 2B illustrates the architecture of MLP. CNN architecture is composed of multiple convolution layers, ReLU layer and pooling layers, which are applied to representation learning of data. Among these layers, the convolution layer serves the purpose of extracting features from various locations within the image. The ReLU layer is responsible for transforming numerical features into nonlinear representations, and the pooling layer plays a pivotal role in diminishing the feature count while preserving the essential characteristics of the features. The CNNs have shown a wide range of applications, including image classification, object detection and natural language processing (Gu et al., 2018). For multi-omics data, several studies have devised a strategy where graphs are constructed based on omics data. Subsequently, CNNs are employed to extract features from these graphs, which are then utilized for downstream analytical tasks. RNNs simulate the memory capacity in neural networks and are able to process data with time series properties, which makes RNNs advantageous for tasks such as time series analysis and prediction. In 2017, Google proposed the Transformer model (Vaswani et al., 2017), which employed a Self-Attention structure to replace the RNN network structure. Self-attention mechanisms and multi-head self-attention mechanisms are core components of the Transformer model, allowing the model to attend to information from different representation subspaces at different positions and capture contextual information in input sequences. Autoencoder compresses data into a lower-dimensional representation (latent space) using an encoder and then employs a decoder to reconstruct the original input from this compressed form. Autoencoders are effective in reducing the dimensionality, extracting the features and denoising the data. Generative Adversarial Networks (GAN) architecture consists of two main components: Generator and Discriminator. During training, these components work against each other, with the aim of refining the Generator's ability to produce new data that is indistinguishable from real data by the Discriminator.

3. Omics techniques

Omics encompasses the application of high-throughput sequencing technologies, with the aim of detecting and analyzing the overall structure and function of biological systems at a specific level, including genomics,

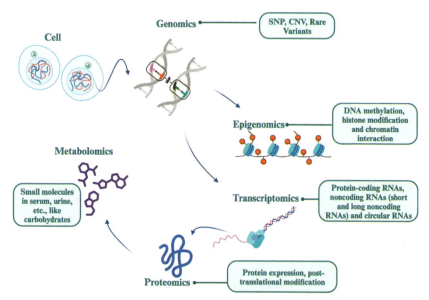

Fig. 3 **The relationship among different omics data, including genomics, metabolomics, epigenomics, transcriptomics, and proteomics.** Each type of data lists the key elements. Genomics contains the whole genome sequence and can be used to identify small variants such as single nucleotide polymorphisms (SNP) and copy number variations (CNV). Epigenomics includes changes in chromosomes (DNA and histones) without altering DNA sequence. Transcriptomics sequences whole RNA components in the organism. Proteomics explores the entire set of proteins, including their modifications, interactions, and functions. Metabolomics investigates small molecules in the body environment, such as serum and urine.

transcriptomics, proteomics, metabolomics and epigenomic. These omics datasets predominantly comprise small genetic variants, the methylation of DNA/histone, coding RNA sequences, protein sequences and various small molecules (Fig. 3). It offers potential biological insight to identify the causes of disease, advancing the understanding of biological systems functions and facilitating the research into human diseases by employing large-scale data exploration and bioinformatics methods. The vital characteristics of omics data have been summarized in Table 1. Investigations of biological systems or disease-specific studies through omics techniques fundamentally relies on two principal techniques: sequencing and mass spectrometry (MS). Based on sequencing technology, genomics, transcriptomics and epigenomics data could be obtained, whereas MS processes the details of proteomics and metabolomics.

Table 1 The summary of general characteristics and its applications for each type of omics data.

Omics	Characteristics	Best use scenarios
Genomics	Complete sequence of DNA in an organism	Subtype identification, diagnosis/prognosis, drug response prediction
Epigenomics	Reversible chemical changes to the DNA or histones	Detect molecular patterns, subtype identification, diagnosis/prognosis
Transcriptomics	Complete set of RNA transcripts in an organism	Detect molecular patterns, subtype identification, diagnosis/prognosis
Proteomics	The entire set of proteins	Detect molecular patterns, subtype identification, diagnosis/prognosis
Metabolomics	Small molecule metabolites in specimen	Detect molecular patterns, diagnosis/prognosis, subtype identification

3.1 Genomics

Genomics focuses on the entire genome, investigating the complex inter-relationship and impacts of genes on organisms, and quantifying all genes within a biological entity. This discipline has seen widespread applications in the medical field, such as the Human Genome Project (Collins, Morgan, & Patrinos, 2003), where genomics technologies are utilized for sequencing, assembling, and annotating of the entire human genome (Horner et al., 2010). In addition, Genome-Wide Association Studies (GWAS) (Tam et al., 2019) have been pivotal in identifying sequence variations, like single nucleotide polymorphisms (SNPs), throughout the human genome, explaining genetic variations associated with diseases. Over the past decade, genomics has revolutionized disease prediction, diagnosis, and treatment, providing impartial and precise insights. Current sequencing technologies can be categorized into the first, second, and third generations. The first-generation Sanger sequencing (Sanger, 1981), invented by Fred Sanger in the 1970s, employs the chain termination method and is characterized by its long read length of 700–900 base pairs with high accuracy. The second generation (Behjati & Tarpey, 2013; Van Dijk, Auger, Jaszczyszyn, &

Thermes, 2014), also known as next-generation sequencing (NGS), comprises platforms such as Roche's 454 sequencer, Illumina's Genome Analyzer, and ABI's SOLiD sequencer, which provide improved throughput but produce shorter read lengths. The third generation (Schadt, Turner, & Kasarskis, 2010) introduces single-molecule real-time sequencing technologies, enabling long-read sequencing with lengths reaching megabases (Eid et al., 2009), though it comes with a trade-off of higher error rates. It can be roughly divided into two categories: nanopore electrical signal sequencing and single-molecule fluorescence signal sequencing. The former mainly refers to Nanopore sequencing platform, which relies on nanopores to decode genetic information (Levene et al., 2003), while the latter includes tSMS™ and PacBio sequencing platform, which identify bases based on fluorescent labels on nucleotides (Harris et al., 2008).

3.2 Epigenomics

Epigenomics, genomic scale analysis of epigenetic mechanisms, investigates reversible modifications that regulate gene expression without altering the DNA structure, including three main processes, DNA methylation, histone modification and chromatin interaction (Stricker, Köferle, & Beck, 2017). DNA methylation is the foundational mechanism of epigenetic regulation, affecting gene expression and cellular metabolic processes by adding methyl groups to cytosine residues (Zhu et al., 2013). The epigenome is characterized by tissue-specific and cell-specific patterns, exhibiting dynamic and responsive to environmental and genetic factors due to these modifications (Taudt, Colomé-Tatché, & Johannes, 2016). With the advent of NGS, the complexities of epigenome have been simplified, allowing for extensive analysis of DNA methylation, chromatin organization, and histone modification. This technology has been served as a powerful tool for identifying tissue-specific epigenetic signatures and associations with diseases (Dawson & Kouzarides, 2012; Ling & Rönn 2019). Epigenomic modifications provide biological insights into the biological nuances of tissue/cell-specific changes across conditions from cancer to metabolic syndrome. In essence, epigenomics offers a comprehensive understanding of how these modifications influence gene expression, cellular function, and their implications for health and disease. The exploration of epigenomics advances the process for non-invasive molecular phenotyping and rapid diagnosis of diseases, uncovering the interrelationship between genetics, environment, and the epigenome.

3.3 Transcriptomics

Transcriptomics is defined as complete set of RNAs transcripts produced by the genome within a given cell population, comprehensively exploring RNA expression and its dynamic responses to environmental factors or pathogenic agents. The transcriptome encompasses various RNA categories, including protein coding RNAs, noncoding RNAs (both short and long noncoding RNAs) and circular RNAs. Notably, noncoding RNAs show associations with diseases (Schmitz, Grote, & Herrmann, 2016), highlighting the importance of microRNAs, small-interfering RNAs and small nuclear RNAs. To quantify the expression levels of RNAs at different stages of disease or metabolism, a range of platforms have been established with similar protocols but differ in RNA-capturing techniques (Hrdlickova, Toloue, & Tian, 2017). The widely employed technology for transcriptomics is RNA-seq (Hrdlickova et al., 2017), enabling both qualitative and quantitative analysis of RNA transcripts, even in limited samples. Furthermore, single-cell transcriptomics (scRNA-seq) (Aldridge & Teichmann, 2020) has emerged as a powerful method that facilitates the detection of cell-type-specific transcripts associated with diseases. Long-read RNA-seq, utilizing third-generation sequencing technologies (Schadt et al., 2010), such as PacBio and Oxford Nanopore, obtains longer RNA sequences and processes the complex transcript structures. Whole-transcriptome analysis (Jiang et al., 2015) captures both known and novel features, allowing researchers to identify biomarkers across the broadest range of transcripts and to enable links between genotypes and phenotypes with a more comprehensive understanding of biological processes.

3.4 Proteomics

Proteomics studies the entire set of proteins expressed by a cell, tissue or organism, revealing the functional relevance of proteins as a dynamic snapshot of the complex protein environment at any given time (Horgan & Kenny, 2011). The proteome plays a crucial role in understanding cellular processes and disease states, characterized by its complexity due to post-translational modifications, spatial configurations, and interactions. It also facilitates the detection and quantification of proteins, providing insights directly associated with environmental changes or disease progression. Post-translational modifications, such as phosphorylation and glycosylation (Mann & Jensen, 2003), play pivotal roles in signal transduction, protein trafficking, and enzyme activities (Wu et al., 2011). This underlines the importance of studying modified proteins to gain a deeper understanding of disease mechanisms. With the development of mass spectrometry (MS) technologies, like the high-resolution LTQ™Orbitrap™ (Wilhelm et al., 2014), combined with powerful analytical tools demonstrates the strong ability of proteomics to identify and characterize proteins. Integration of chromatography-based separation modules with liquid chromatography-mass spectrometry (LC-MS) (Vogeser & Parhofer, 2007) and gas chromatography-mass spectrometry (GC-MS) (Vogeser & Parhofer, 2007), mitigate matrix effects and ion suppression, enhancing the analytical precision. Proteomics proves invaluable in diverse applications, ranging from the discovery of novel ceramide-binding proteins and the analysis of microbial cultures to the early disease diagnosis, therapeutic target identification, and unraveling the complex functions of genes.

3.5 Metabolomics

Metabolomics is dedicated to exploring the set of small molecule metabolites, which are fundamental for energy provision, signaling, and protein modulation within complex biological systems. These metabolites, including carbohydrates, fatty acids, and amino acids, arise from cellular metabolism and provide precise insights into disease-correlated dynamic changes, as well as fluctuations in metabolite levels or ratios. Therefore, the development of metabolomics aids in searching the underlying mechanisms of disease progression (Cui, Lu, & Lee, 2018; Newgard, 2017; Wang, Li, Lam, & Shui, 2020) and exhibits quantifiable correlations with other omics domains, such as genomics and proteomics (Cambiaghi, Ferrario, & Masseroli, 2017). The metabolome is regarded as dynamic, subject to variations effected by diet, stress, physical activity, pharmacological effects, and disease. Cutting-edge analytical technologies, such as fourier transform-infrared (FT-IR) spectroscopy (Ferraro, 2012), Raman spectroscopy (Lin et al., 2020), NMR spectroscopy (Rule & Hitchens, 2006), and MS-based approaches including MS (Want, Cravatt, & Siuzdak, 2005), MS/MS (Ceglarek et al., 2009), liquid chromatography (LC)-MS (Ardrey, 2003) and gas chromatography (GC)-MS (Hubschmann, 2015), have provided stable basis to ensure accurate and comprehensive profiling analysis. Metabolomics studies branches into targeted which quantify specific metabolites in predefined pathways, and non-targeted approaches which explore a broader spectrum for *de novo* target discovery (Zhang, Zhu, Wang, Zhang, & Cai, 2016). Understanding of the metabolome facilitates the candidate genes identification and sheds light on the dynamic interplay between metabolites and various physiological factors.

4. Application of AI and multi-omics in cancer research

This section will explore the integration of AI and multi-omics in cancer research and clinical practice, specifically focus on the realms of early detection, diagnosis, prognosis, and treatment. Fig. 4 shows the overview of multi-omics based artificial intelligence for cancer research. The machine learning and deep learning methods were employed on the multi-omics data for facilitating cancer research, from early detection to treatment, which will be elaborated below.

4.1 Early cancer detection

Early cancer detection is one of the important measures for cancer prevention and treatment. By facilitating early screening, detection, and treatment, the prognosis for cancer patients can be significantly improved,

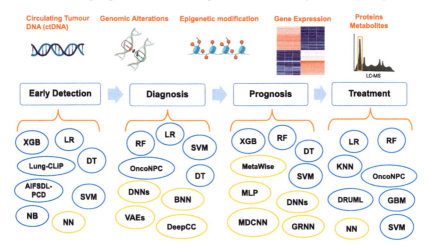

Fig. 4 The overview of multi-omics based AI approaches for cancer research. Multi-omics techniques are capable of capturing comprehensive molecular insights from multiple views, including circulating tumor DNA (ctDNA), genomic alteration, epigenetic modification, gene expression and metabolites. This comprehensive molecular information is extensively utilized in various aspects of cancer research, from early detection to treatment. The figure illustrates the deployment of algorithms and models to corresponding tasks, with the blue circle representing conventional machine learning methods and the yellow circle denoting deep learning methods.

prolonging survival time. Cancer screening aims to identify cancer at an asymptomatic or precancerous stage, allowing for early interventions that can substantially reduce cancer mortality. In recent years, cell-free DNA (cfDNA) have shown promising potential in enhancing early cancer detection. For instance, Wan et al. (2019) utilized LR and SVM to detect early-stage colorectal cancer through plasma cell-free DNA analysis via whole-genome sequencing. Klein et al. (2021) demonstrated the utility of a targeted methylation-based multi-cancer early detection (MCED) test, leveraging cfDNA sequencing and machine learning to detect signals indicative of multiple cancer types and to predict the origin of these cancer signals. Similarly, Chabon et al. (2020) introduced the Lung Cancer Likelihood in Plasma (Lung-CLiP) method, designed to differentiate early-stage lung cancer patients from risk-matched controls. This Lung-CLiP model integrates the outputs of two constituent models targeting single-nucleotide variants (SNVs) and copy number variations (CNVs), employing a variety of machine learning techniques including 5-nearest neighbor (5NN), 3NN, Naive Bayes, LR, and DT. The SNV model leveraged semi-supervised learning in the elastic net logistic regression model to distinguish mutations in cfDNA of patients from controls. The CNV model was designed to enumerate somatic copy number alterations (CNAs) present in circulating cfDNA and in leukocytes. Microarray gene expression data was used to train a deep learning model (AIFSDL-PCD) for the early detection of prostate cancer and RMSprop optimizer is employed to adjust the hyperparameters which help to improve model performance in effectively classifying prostate cancer (Alshareef et al., 2022a). Plasma proteomics data was used to train four machine learning model, including Elastic Net, Random Forest, SVM, and XGBoost, for the early prediction of lung cancer. Valuable biomarkers provided by plasma proteome can identify individuals at high risk of lung cancer, with the potential for early detection (Davies et al., 2023). Metabolites also can serve as critical bio-markers in cancer early detection. Xie et al. (2021) combined plasma metabolites and Naïve Bayes algorithm to discover potential screening biomarkers for early detection of lung cancer. Early cancer detection plays a pivotal role in accurately identifying and classifying individuals into high-risk groups for cancer, potentially leading to the development of tailored risk assessment models. The clinical data encompass a wide range of information, including genetic markers, lifestyle factors, and medical histories, which are pivotal in understanding individual health risks and conditions. Stark, Hart, Nartowt, and Deng (2019) constructed a model by integrating LR, linear discriminant analysis, and neural network models based on personal health data of patients with breast cancer. This model significantly enhances the five-year breast cancer risk prediction, serving as a non-invasive, cost–effective tool for risk stratification, thereby promoting early breast cancer detection and preventive measures. Furthermore, Feng et al. (2022) developed a machine learning-based model to predict lymph node metastasis (LNM) in kidney cancer patients. After filtering clinical features through LASSO, and univariate and multivariate logistic regression analyses, statistically significant risk factors were used to train

Table 2 The summary of AI application in early cancer detection.

Omics	Algorithms	Topics of interest	Sources
Clinical data	Logistic regression, linear discriminant analysis and neural network models	Breast cancer risk prediction	Stark et al. (2019)
	XGBoost	Probability prediction of LNM in patients with kidney cancer	Feng et al. (2022)
Genomics	LR and SVM	Early-stage colorectal cancer detection	Wan et al. 2019
Genomics	Lung-CLiP	Early-stage lung cancer identification	Chabon et al. (2020)
Epigenomics	Ensemble logistic regressions	Cancer signals detection	Klein et al. (2021)
Transcriptomic	AIFSDL-PCD	Prostate cancer classification	Alshareef et al. (2022b)
Metabolomics	Naïve Bayes	Early detection of lung cancer	Xie et al. (2021)
Proteomics	EN, RF, SVM, XGBoost	Early detection of lung cancer	Davies et al. (2023)

XGBoost, eXtreme gradient boosting; LR, logistic regression; SVM, support vector machine; NN, neural network; LNM, lymph node metastasis; AIFSDL-PCD, artificial intelligence based feature selection with deep learning model for prostate cancer detection.

XGBoost algorithms. They suggested a threshold probability with 54.6% that could accurately distinguish approximately 89% of LNM patients, showcasing the model's promising clinical application prospects. The above methods and applications are summarized in the Table 2.

4.2 Diagnosis

Cancer diagnosis involves the identification and classification of cancer types and subtypes, which is critical for prognosis assessment and effective tailored treatment. Traditional diagnostic approaches, relying on biopsies, blood tests and medical imaging technology, face limitations and encounter misdiagnosis rates in certain instances. The emergence of multi-omics techniques and artificial intelligence (AI) has revolutionized cancer diagnosis, offering new avenues for improved accuracy and precision. Recent studies (Table 3) have demonstrated the potential of AI-driven multi-omics analysis in cancer diagnosis. For instance, Herrgott et al. (2023) developed a machine learning model, d-MeLB, leveraging DNA methylation signatures in serum to distinguish meningioma from other central nervous system (CNS) diseases and predict recurrence risk. This model, utilizing random forest to generate a binary classifier, effectively differentiated meningioma (MNG) specimens from non–MNG samples and stratified patients into low and high recurrence risk groups based on serum-derived cfDNA and/or tissue-derived DNA methylation signatures. The effectiveness of the d-MeLB classifier was validated through model selection and independent validation, demonstrating its potential as a non-invasive and reliable diagnostic tool for diagnosing and predicting outcomes in meningioma patients. Similarly, (Wang et al., 2022) employed CNVs analysis and found widespread dysregulation of lipid metabolism in early-stage lung cancer, based on the scRNA-seq results from patients with early-stage lung cancer and healthy lung tissues. A significant influencing metabolite compose was filtered by analyzing the expression of abnormal lipid metabolism in early-stage lung cancer based on high-performance liquid chromatography-mass spectrometry (HPLC-MS) technology. The metabolite compose was then used to train a SVM classification model, LCAID, for early lung cancer diagnosis, which achieved an impressive 98.90% accuracy in early lung cancer diagnosis within the validation cohort. Li, Kugeratski, and Kalluri (2023) developed a Random Forest based model to identify cancer-specific exosomes employing proteomics data. The established model identifies protein markers to differentiate cancer exosomes from non-cancerous ones and classify cancer subtype with high accuracy. Moon et al. (2023) utilized a gradient tree boosting

Table 3 The summary of AI application in cancer diagnosis.

Omics	Algorithms	Topics of interest	Sources
Epigenomics	VAEs	Subtypes of LUAD and LUSC classification	Wang and Wang (2019)
Epigenomics	Random forest	Meningioma identification and recurrence risk prediction	Herrgott et al. (2023)
Genomics	OncoNPC (XGBoost)	CUP classification	Moon et al. (2023)
	Random forest and Deep neural network	Cancer types of classification	Lee et al. (2019)
	Neural network	Cancer types of classification	Jiao et al. (2020)
	Deep neural network	12 cancer types of classification	Sun et al. (2019)
Transcriptomics and lipidomics	LCAID (SVM)	Early detection of lung cancer	Wang et al. (2022)
Transcriptomics	DeepCC (MXNet)	Cancer molecular subtype classification	Gao et al. (2019)
	BNN	Cancer types of classification	Joshi and Dhar (2022)
Proteomics	Random forest	Non-invasive cancer diagnosis	Li et al. (2023)

VAE, Variational autoencoder; BNN, Bayesian neural network; LUAD, lung adenocarcinoma; LUSC, lung squamous cell carcinoma; CNS, central nervous system; CUP, cancer of unknown primary.

framework (XGBoost) to develop Oncology NGS-based primary cancer-type classifier (OncoNPC) for predicting cancer types on targeted NGS data from 36,445 tumors across 22 cancer types. The somatic alterations, including mutations (SNVs and indels), mutational signatures, CNAs were used for model training. For each tumor sample, a cancer type with the highest probability was chosen as the predicted primary site. The OncoNPC suggested that many CUP tumors can be classified into meaningful sub-groups with the potential to support clinical decision-making for patients with CUP. Deep learning has revolutionized the field of cancer research by enabling the analysis of complex multi-omics data at unprecedented depths. For example, the variational autoencoders (VAEs) were employed to extract a biologically relevant latent space of DNA methylation data to classify the lung adenocarcinoma (LUAD) and lung squamous cell carcinoma (LUSC) subtypes accurately (Wang & Wang, 2019). The Cancer Predictor using an Ensemble Model (CPEM) introduced by Lee, Hyoung-oh Jeong, and Jeong (2019) leveraged an ensemble of a random forest and a deep neural network to classify 31 different cancer types based on somatic alterations. Similarly, Jiao et al. (2020) developed a neural network–based classifier capable of classifying 24 cancer types solely based on SNVs type and mutational distribution. The constructed model, named Genome Deep Learning, employed a deep neural network architecture to differentiate between 12 types of cancer and healthy tissues based on genomic variations (Sun et al., 2019). Furthermore, Gao et al. (2019) implemented Deep cancer subtype classification (DeepCC) based on MXNet (Chen et al., 2015) to classify cancer molecular subtypes through functional spectra using gene set enrichment analysis (GSEA). Additionally, the Epistemic Invariance in Cancer Classification (EpICC) employed a Bayesian neural network (BNN) for classifying 31 cancer types, using gene expression data of feature genes (Joshi & Dhar, 2022). These studies (Table 3) highlight the growing utility of AI-driven multi-omics analysis in cancer diagnosis, offering improved accuracy, precision, and potential for clinical translation.

4.3 Prognosis

Prognosis refers to predicting the development trend and survival time of cancer patients. Understanding the prognosis of patients is critical for developing appropriate treatment. In traditional prognostic prediction, oncologists usually make assessments with statistical models and previous experience based on the clinical information of the patient and the pathological features of the tumor, which include age, health background, cancer subtype, cancer stage and tumor grade. However, the predictive capacities of these features and statistical models are inherently limited due to variability in individual patients, reliability of data, and the experience of oncologists. These statistical data and models only can provide general trends and probabilities rather than accurate prediction for individual patients. Leveraging the comprehensive information from multi-omics techniques, AI has shown great potential for discovering prognostic biomarkers to improve the accuracy of prognosis (Table 4). Kalafi et al. (2019)

Table 4 The summary of AI application in cancer prognosis.

Omics	Algorithm	Topics of interest	Sources
Clinical data	MLP, SVM, DT and RF	Survival prediction of breast cancer	Kalafi et al. (2019)
	XGBoost	Patients at high risk of distant rectal metastasis screening	Qiu et al. (2023)
Genomics	MetaWise (DNN)	Classification of metastatic tumors from primary tumors	Zheng et al. (2023)
Transcriptomics	Elastic net penalized Cox proportional hazards regression	Survival and locoregional recurrence prediction of patients with advanced oral cancer	Tseng et al. (2020)
	Graph Recurrent Neural networks	Clinical outcomes prediction of patient with neuroblastoma	Tranchevent et al. (2019)
	Neural networks	Identification of patients with the greatest risk of relapse	Katipally et al. (2023)
	MDCNN	Occurrence prediction of metastasis	Cai et al. (2024)
	A multitask deep learning model	Classification of disease state, tissue origin and neoplastic subclass	Hong et al. (2022)
	Semi-supervised learning algorithm	Cancer recurrence prediction	Park et al. (2014)
Metabolomics	Neural networks	ER status classification	Alakwaa et al. (2018)

RF, Random forest; MDA, mixture discriminant analysis; DNN, deep neural networks; MDCNN, multiple dimension convolutional neural network; ER, estrogen receptor.

applied random forests to assess the importance of features and ensembled decision trees to determine the most relevant variables collected from clinical and demographic data. They trained various models, including a MLP with two hidden layers, SVM, DT, and RF, using processed features to predict breast cancer survival. Similarly, Qiu et al. (2023) applied XGBoost model for identifying patients at high risk of distant metastasis from rectal cancer, utilizing common clinicopathological factors. Tseng, Hsin-Yao Wang, Jang-Jih Lu, and Liao (2020) constructed a model to differentiate the high-risk group from the low-risk group in cancer-specific survival and locoregional recurrence–free survival for postoperative patients with advanced oral cancer. Utilizing cancer-related gene variant profiles, an elastic net penalized Cox proportional hazards regression model was constructed, providing survival index of each patient treated with different curative therapeutics. Innovative approaches such as graph-based semi-supervised learning models, feed-forward neural networks, and graph recurrent neural networks have been applied to predict clinical outcomes and prognostic states using omics data, showcasing the model's superiority over traditional classification methods. For instance, Katipally et al. (2023) utilized mRNA and miRNA expression profiles from a discovery cohort of patients to train a neural network molecular classifier for classifying colorectal liver metastases into three molecular subtypes: canonical, immune, and stromal. Park, Ahn, Kim, and Park (2014) identified informative gene pairs from integrated gene expression data with Protein-Protein Interaction (PPI). These gene pairs were then utilized to construct a graph-based semi-supervised learning model aimed at distinguishing between recurrence and non-recurrence in cancer patients. This innovative approach involved regularizing the graph and predicting the labels (indicating non-recurrence or recurrence) for the previously unlabeled samples. Alakwaa, Chaudhary, and Garmire (2018) applied feed-forward neural networks to predict the estrogen receptor status in breast cancer based on metabolomics data, which highlights the potential of metabolomics in enhancing the predictive capabilities in cancer prognosis. In another innovative approach, Tranchevent, Azuaje, and Rajapakse (2019) employed a graph recurrent neural network to predict clinical outcomes in neuroblastoma patients based on microarray and RNA-seq data. The transcriptomics data were structured into graphs, with each node representing a patient and the edges signifying the correlations between patients' omics profiles. Cai et al. (2024) applied LASSO regression and Pearson correlation coefficients to identify genes associated with cancer metastasis to different tissues. They proposed a Multi-Dimensional Convolutional Neural Network (MDCNN) designed to predict metastasis based on the expression of these genes. Furthermore, Hong, Hachem, and Fehlings (2022) introduced a multitask deep learning model designed to classify disease state, tissue origin, and neoplastic subclass using RNA-seq datasets of non-neoplastic, neoplastic, and peri-neoplastic tissue. Zheng et al. (2023) developed MetaWise, a Deep Neural Network (DNN) model, leveraging mutational signatures obtained from Whole-Exome Sequencing (WES) data to classify metastatic tumors from primary ones.

4.4 Treatment response prediction

The prediction of patients' response to specific treatment is critical for personalizing treatment and improving outcomes. Multiple factors, including patient's health background, cancer subtype and stage, genetic mutations, contribute to the variability in treatment response and outcomes among cancer patients. The complexity of these multiple factors presents a significant challenge for traditional prediction methods to accurately predict the response of treatment. The limitations of these traditional methods underscore the necessity for innovative approaches, potentially harnessing the power of AI and multi-omics (Table 5). Based on genomic sequencing data, including somatic alterations and demographic factors, XGBoost was trained to classify cancers of unknown primary origin and predict treatment responses (Moon et al., 2023). Another study employed LR and RF to investigate the treatment outcomes of TACE in patients with HCC. Cirrhosis and tumor signal intensity are strong predictors and were used to train the predicting models which present high accuracy (Abajian et al., 2018a). Kong et al. (2022) developed a network-based method to effectively identify immunotherapy-response-associated biomarkers to construct a robust NetBio-based logistic regression model for precision oncology. More than 700 ICI-treated patients with three different cancer types—melanoma, gastric cancer, and bladder cancer were applied to accurately predict ICI treatment response. Expression levels of genes/pathways against drug responses (classified as responders and non-responders) were used to train the model. In another study, Huang et al. (2018a) leveraged Linear SVM to classify patients into drug sensitive/resistant groups based on gene expression data. Clinical data, including drug response information, from TCGA for 32 types of cancer were analyzed to select the corresponding viable candidate drugs. Cross validation evidenced that current model could predict response of cancer patients to current standard-of-care drug

Table 5 The summary of AI application in cancer treatment response prediction.

Omics technique	Algorithm	Topics of interest	Sources
Clinical data	LR, RF	Outcome prediction of TACE in patients with HCC	Abajian et al. (2018b)
Genomics	OncoNPC (XGBoost)	Treatment response prediction	Moon et al. (2023)
Transcriptomics and clinical data	SVM	Patient response to drugs prediction	Huang et al. (2018b)
Transcriptomics	Logistic regression	ICI treatment response prediction	Kong et al. (2022)
	KNN, SVM, GBM, DT, RF, NN	FOLFOX treatment response in CRC prediction	Lu et al. (2020)
Proteomic	DRUML	Efficacy of anti-cancer drugs prediction	Gerdes et al. (2021)

ICI, Immune checkpoint inhibitor; CRC, metastatic or recurrent colorectal cancer; GBM, gradient boosting machine; HCC, hepatocellular carcinoma; DRUML, drug ranking using machine learning.

therapies with high accuracy. Similarly, significant genes were selected through differential expressed gene analysis based on microarray data collected from CRC patients and employed as input for six ML algorithms, including KNN, SVM, GBM, decision tree, random forest, and neural network, to predict FOLFOX treatment response (Lu et al., 2020). To predict the efficacy of anti-cancer drugs, a machine learning model named DRUML was constructed based on proteomic and phosphoproteomic data collected from cancer cell lines. The model presents could rank anti-cancer drugs with valuable performance according with the area above the curve values (Gerdes et al., 2021).

4.5 Biomarker identification

Biomarkers are one type of measurable indicators that reflect biological states or conditions of an organism, usually used for disease diagnosis, progression and treatment response prediction. In the context of cancer, proteins, metabolites, gene mutations or other molecules can be regarded as biomarkers to differentiate between the cancerous tissue and normal one. Detecting reliable biomarkers is critical for cancer diagnosis and treatment. ML and DL are powerful tools for biomarker discovery due to their ability to learn extreme complex patterns in large, high-dimensional dataset, with potential to advance precision medicine (Table 6). For transcriptomics, Xu, Zhang, Zhu, and Xu (2017) used microarray data to train a SVM model and identify 15 gene signatures to predict the recurrence and prognosis of colon cancer. Ghanat Bari, Ung, Zhang, Zhu, and Li (2017) utilized gene expression data collected from 9 cancer types to develop a Machine Learning-Assisted Network Inference (MALANI) approach, more than 200 million SVM models were created to identify a new class of cancer-related genes. Approximately 3% of non-differentially expressed genes were discovered and termed Class II cancer genes, providing potential therapeutic targets. Jha et al. (2022) utilized three feed-forward neural networks to find commonly shared features of transcriptome from 18 cancer types and identified transcriptome signatures across all cancers. In addition, based on images and transcriptomic data, Zadeh Shirazi et al. (2021) devised a deep convolutional neural network and found gene signatures correlated strongly with survival in genetically defined glioblastoma patient groups. For genomics, Shi, Zhang, and Wang (2023) constructed a DL model based on the VGG19 architecture and transfer learning strategy. The tumor microenvironment signature was developed from the genomic data collected from colorectal cancer patients for prognosis prediction. For epigenomics, Queirós et al. (2015) utilized DNA methylation data to train a SVM model with five epigenetic biomarkers. Chronic lymphocytic leukemia (CLL) patients were accurately classified into three biologically distinct subgroups: naive B-cell-like, intermediate, and memory B-cell-like CLL. By analyzing DNA methylation array data with a machine-learning-based approach involving support vector machines (SVM), Aref-Eshghi et al. (2018) identified specific DNA methylation signatures associated with 14 Mendelian neurodevelopmental disorders. Heiss and Brenner (2017) developed a model to diagnose colorectal cancer (CRC) based on differentially methylated CpG sites detected in an epigenome-wide association study (EWAS). A logistic regression model was trained to differentiate between CRC cases and controls with methylation markers and demographic risk factors. Based on the analysis of proteomics data, Vizza et al. (2024) developed a ML based pipeline involving RF to identify proteomic

Table 6 The summary of AI application in cancer biomarker identification.

Omics	Algorithm	Topics of interest	Sources
Transcriptomics	SVM	Colon cancer recurrence and prognosis prediction	Xu et al. (2017)
	Neural Network	Transcriptomic features identification between cancer types	Jha et al. (2022)
	SVM	New cancer–related genes identification	Ghanat Bari et al. (2017)
Transcriptomics and pathology	Neural Network	Glioblastoma survival prediction	Zadeh Shirazi et al. (2021)
Genomics and pathology	CNN	Colorectal cancer prognostic prediction	Shi et al. (2023)
Epigenomics	SVM	Chronic lymphocytic leukemia classification	Queirós et al. (2015)
	SVM	Neurodevelopmental disorder diagnosis	Aref-Eshghi et al. (2018)
	LR	Colorectal cancer diagnosis	Heiss and Brenner (2017)
Proteomics	RF	Identification of proteomic biomarkers between prostate cancer and benign prostatic hyperplasia	Vizza et al. (2024)
	RF	Liquid biopsy-based cancer diagnosis	Halner et al. (2023)
	Neural Network	Benign and malignant thyroid nodule detection	Sun et al. (2022)
Metabolomics	LASSO, RF	Gastric cancer diagnosis and prognosis prediction	Chen et al. (2024)

SVM, Support vector machine; CNN, convolutional neural network; LR, logistic regression; RF, random forest; LASSO, least absolute shrinkage and selection operator.

biomarkers to differentiate between prostate cancer (PCa) and benign prostatic hyperplasia (BPH). Identified markers promote the accuracy of RF model performance in PCa and BPH classification, aiding in early diagnosis. Halner et al. (2023) also established a RF based architecture, DEcancer, for the diagnosis of liquid biopsy-based cancer and the identification of biomarker. Recursive feature elimination was performed using the RF Gini-index based variable importance score to select the most relevant proteins which can be used to distinguish between cancerous and non-cancerous samples. Sun et al. (2022) used machine learning techniques, including an artificial neural network (ANN), to identify 14 protein biomarkers for distinguishing between benign and malignant thyroid nodules. Furthermore, Chen et al. (2024) applied machine learning algorithms to identify metabolomic biomarkers for the diagnosis and prognosis prediction of gastric cancer. Using LASSO regression to select features and RF for model training during the process of marker filtering, and it was demonstrated better performance compared to traditional clinical methods. Therefore, ML and DL have made it more efficient to detect potential promising biomarkers from large scale complex dataset, enhancing significantly the accuracy of cancer diagnosis, prognosis, and treatment prediction.

4.6 Pathology identification

Cancer pathology, which involves the study of the changes in cells, tissues, and organs caused by cancer, is essential for diagnosis, prognosis, and treatment management. One of the primary challenges in cancer pathology is the heterogeneity of tumors. The intratumor heterogeneity significantly complicates the accurate diagnosis and elucidation of cancer progression (Ramón y Cajal et al., 2020; Marusyk, Janiszewska, & Polyak, 2020), which requires pathologists with extensive expertise and clinical experience. Additionally, the rising prevalence of cancer, combined with resource constraints, has significantly increased the workload for pathologists, further intensifying the challenges in delivering timely and precise diagnostic services (Maung, 2016). AI has revolutionized the field of cancer pathology by enabling more accurate diagnosis and efficient decision-making processes through advanced histology image analysis (Echle et al., 2021; Shmatko, Ghaffari Laleh, Gerstung, & Kather, 2022). Deep learning techniques, especially convolutional neural networks (CNNs), have demonstrated remarkable efficacy in pathological image analysis tasks (Table 7). For example, Lu et al. (2021) proposed a deep-learning-based algorithm—Tumor Origin Assessment via Deep Learning (TOAD) to

Table 7 The summary of AI application in cancer pathology.

Algorithm	Topics of interest	Sources
TOAD	Prediction of the primary tumor origin	Lu et al. (2021)
CNN	Classification of diffuse gliomas	Wang et al. (2023)
GNN	Prediction of diagnosis and prognosis	Lee et al. (2022)
MuRCL	Cancer subtype classification	Zhu et al. (2023)
TransMIL	Pathology diagnosis	Shao et al. (2021)
Weakly supervised regression method	Biomarker prediction	El Nahhas et al. (2024)
MOMA	Prediction of the multi-omics aberrations and prognoses	Tsai et al. (2023)
Weakly supervised learning multiple instance learning approach	Expression prediction of the biomarker PDL1	Jin et al. (2024)
Optimal Transport-based Co-Attention Transformer framework	Survival prediction	Xu and Chen (2023)

TOAD, Tumor origin assessment via deep learning; CNN, convolutional neural network; GNN, graph neural network; MuRCL, multi-instance reinforcement contrastive learning framework; TransMIL, transformer based MIL; MIL, multiple instance learning; MOMA, multi-omics multi-cohort assessment.

predict the origin of the primary tumor based on histology slides. The whole-slide images (WSI) of tumors with known primary origins were used to train a CNN model to simultaneously identifies the tumor as primary or metastatic and predicts its site of origin. Similarly, the CNN based model that learned imaging features containing both pathological morphology and underlying biological clues was developed to automatic classification of diffuse gliomas from annotation-free standard WSIs (Wang et al., 2023). In addition, Lee et al. (2022) presented a graph neural-network model in a semi-supervised manner that leverages attention techniques to learn contextual features of the heterogeneous tumor microenvironment from aggregates presentations of highly correlated image patches. This model was trained with WSI of kidney, breast, lung and uterine cancers to predict the diagnosis and prognosis. Recent advances include the Multi-instance Reinforcement Contrastive Learning framework (MuRCL), proposed by Zhu et al. (2023), which effectively explored the inherent semantic relationships of different patches of WSI for cancer subtype classification. Similarly, Shao et al. (2021) developed a Transformer based MIL (TransMIL) framework to explored both morphological and spatial information between different instances for pathology diagnosis. Moreover, a self-supervised attention-based weakly supervised regression method was developed by El Nahhas et al. (2024) to predict continuous biomarkers from WSIs of cancer patients. Further innovations also include the Multi-omics Multi-cohort Assessment (MOMA) platform, which was developed by Tsai, Tsung-Hua Lee, Fang-Yi Su, and Eliana Marostica (2023) to systematically identify and interpret the relationship between patients' histologic patterns, multi-omics, and clinical profiles. The capability of MOMA is to predict the multi-omics aberrations and prognoses based on histopathology images. Furthermore, a weakly supervised learning multiple instance learning approach with the teacher-student framework was proposed to predict the biomarker PDL1 expression from hematoxylin and eosin (H&E) slides (Jin et al., 2024). Additionally, the Multimodal Optimal Transport-based Co-Attention Transformer framework was developed to identify the interactions within WSIs and co-expression within genomics for survival prediction (Xu & Chen, 2023). In summary, the application of AI into cancer pathology is transforming the field, offering more accurate, efficient diagnostic capabilities and personalized treatment strategies. These cutting-edge technologies not only enhance the efficiency and precision of tumor identification and classification but also enable the prediction of treatment responses and patient outcomes.

5. Multi-omics integration

Multi-Omics Integration can provide a profoundly comprehensive insight of the complex biological mechanisms and heterogeneity of cancer. It holds great promise in revolutionizing several key aspects of oncology, including the classification of cancer subtypes, identification of biomarkers, development of precision treatment strategies, and discovery of novel drug targets. Due to the heterogeneous nature of multi-omics data and intricate regulation mechanisms across different molecular levels, advanced models and algorithms need to be developed to overcome these challenges. The integration of multi-omics data (Table 8) can be approached through three primary strategies: early, intermediate, and late integration (Fig. 5) (Rappoport & Shamir, 2018).

5.1 Early integration

Early integration involves the direct concatenation of raw data or extracted features from different omics at the initial stage before model training (Fig. 5A). However, the early integration strategy also faces some challenges, such as the different scale between different omics data or the extremely high dimension of the integrated dataset. A bimodal deep neural network (DNN) model was developed to predict the overall survival of non-small cell lung cancer (NSCLC) patients based on gene expression and clinical data (Lai, Wei-Ning Chen, Che Lin, & Wu, 2020). The output layers of two DNNs, that separately encoded microarray data and clinical data, were concatenated for final prediction. DualGCN (Ma et al., 2022) used graph convolutional networks to encode the chemical structures of drugs and omics data to predict cancer drug response. The DualGCN framework was composed of drug-GCN module and bio-GCN module, where the drug-GCN module encoded the chemical structure information of a drug and the bio-GCN module encodes biological features of cancer samples such as gene expression and CNV. The two embeddings generated by drug-GCN and bio-GCN module were concatenated and subsequently fed into a MLP to predict the drug response on the cancer sample. In addition, network-based integration method of multi-omics data was proposed to predict clinical outcome prediction in neuroblastoma (Wang et al., 2022). The Patient Similarity Networks (PSN) was obtained by calculating distances between patients from single omics features and the integration of different omics was achieved by Similarity Network Fusion (SNF) algorithm or fusing features of patient similarity networks. Explainable multi omics graph integration (EMOGI), based on GCNs, was introduced to identify new cancer genes by integration of mutations, copy number changes, DNA methylation, gene expression and protein–protein interaction (PPI) networks (Schulte-Sasse, Budach, Hnisz, & Marsico, 2021). The different omics were computed across 16 TCGA tumor types for high-confidence cancer/non-cancer genes. These high-confidence genes were concatenated by an early integration strategy and the combined with the PPI networks to construct a network for subsequent model

Table 8 The summary of AI application in multi-omics integration.

Integration type	Omics	Algorithm	Result	Sources
Early integration	Genomics, transcriptomics, and pathology	AMIL and SNN	Outcomes prediction and prognostic features discovery	Chen et al. (2022)
	Genomics, transcriptomics, and chemical information	Graph convolutional networks	Cancer drug response prediction	Ma et al. (2022)
	Genomics, transcriptomics, and chemical information	DNN	Drug sensitivity prediction	Tang and Gottlieb (2021)
	Transcriptomics and clinical data	DNN	Survival prediction of NSCLC patients	Lai et al. (2020)
	Genomics, transcriptomics, and clinical data	Gated attentive convolutional neural network, random forest	Prognosis prediction of breast cancer	Arya and Saha (2021)
	Transcriptomics and epigenomics	Similarity Network Fusion algorithm	Clinical outcome prediction in neuroblastoma	Wang et al. (2022)
	Transcriptomics, epigenomics and proteomics	EMOGI (GCN)	Biomarker prediction for complex diseases	Schulte-Sasse et al. (2021)

Late integration	Radiology, pathology, and genomics	Deep attention-based multiple-instance learning model	Prediction of immunotherapy response in patients with NSCLC	Vanguri et al. (2022)
	Genomics and transcriptomics	MOGONET	Integration of omics data	Wang et al. (2021)
	Clinical data, pathology, genomics, and transcriptomics	LR, SVM, RF	Breast cancer therapy response prediction	Sammut et al. (2022)
	Genomics and transcriptomics	MDNNMD	Prognosis prediction of breast cancer	Sun et al. (2018)
	Radiology and transcriptomics	Multimodal fusion with uncertainty	survival prediction for NSCLC patients	Wang et al. (2021)
Intermediate integration	Genomics and transcriptomics	P-NET	Molecular drivers of therapeutic targeting resistance identification	Elmarakeby et al. (2021)
	Genomics and transcriptomics	MOLI	Drug response prediction	Sharifi-Noghabi et al. (2019)
	Transcriptomics and proteomics	DBN	Disease–gene associations prediction	Luo et al. (2019)

(continued)

Table 8 The summary of AI application in multi-omics integration. (*cont'd*)

Integration type	Omics	Algorithm	Result	Sources
	Transcriptomics and epigenomics	Autoencoder	Prediction of prognosis in patients with esophageal squamous cell carcinoma	Yu et al. (2020)
	Clinical data, genomics, epigenomics, transcriptomics, and pathology	MultiSurv	Pan-cancer survival prediction	Vale-Silva and Rohr (2021)
	Transcriptomics, epigenomics and proteomics	Multi-view Factorization AutoEncoder	Prediction of target clinical variables	Ma and Zhang (2019)
	Transcriptomics, epigenomics and proteomics	VAEs	Multi-omics data integration from incomplete multi-view observations	Lee and Van der Schaar (2021)

AMIL, Attention-based multiple-instance learning; SNN, self-normalizing network; GCN, graph convolutional networks; DNN, deep neural network; P-NET, pathway-aware multi-layered hierarchical network; DBN, deep belief net; NSCLC, non-small cell lung cancer.

training through consecutive layers of graph convolutions. This study, Arya and Saha (2021) introduced a two-phase model that combines gated attentive convolutional neural networks on multi-modal data (clinical, gene expression and CNA profile) stacked with random forest classifiers for

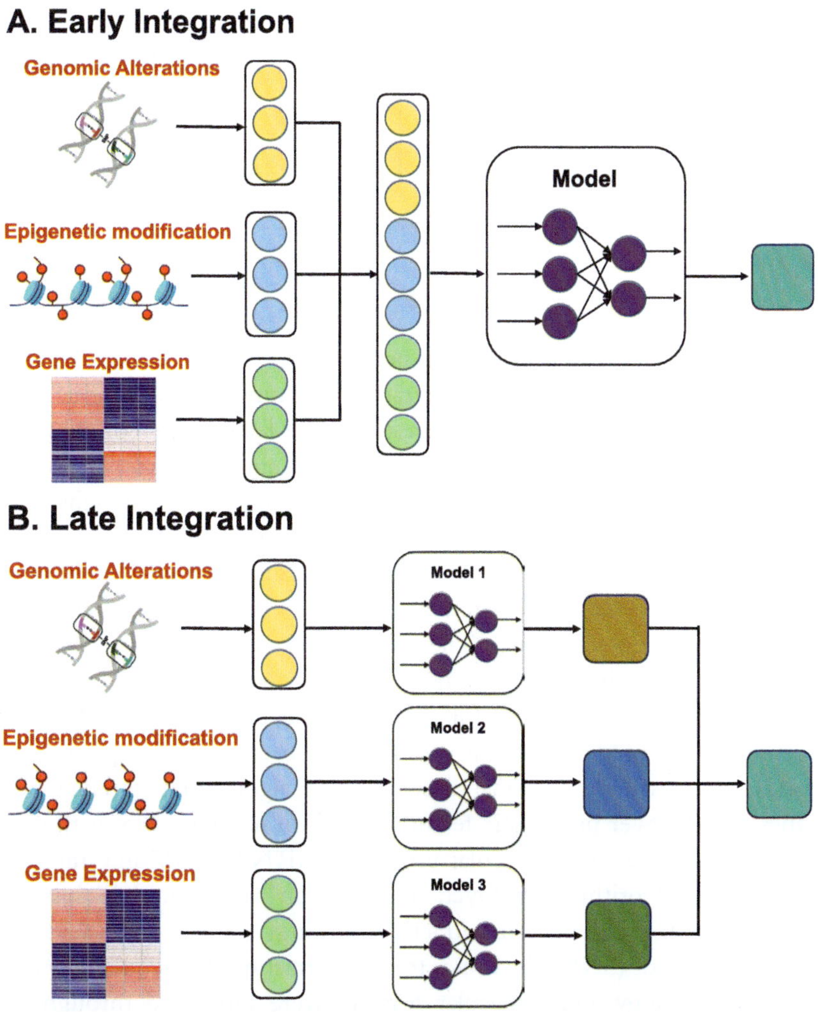

Fig. 5 **Different types of multi-omics integration strategies.** (A) Early Integration. The raw data or extracted features from different omics is concatenated at the initial stage for following model training. (B) Late Integration. The predictions from individual models trained on different omics data are integrated at the final decision stage. (C) Intermediate Integration. The joint representation is learned from the different representation of omics for following model training.

C. Intermediate Integration

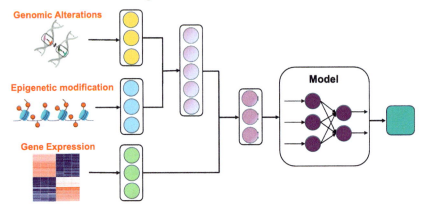

Fig. 5 (*Continued*)

improving breast cancer prognosis predictions. At phase one, separate SiGaAtCNN were leveraged to extract the features from clinical, gene expression and CNA profile. And these separate features are concatenated for the final survival classification task with random forest classifiers. PathDSP (Tang & Gottlieb, 2021) was proposed to predict drug sensitivity based on a pathway-based deep neural network model, which integrated chemical structure information and cancer signaling pathway enrichment analysis involving drug-associated genes, gene expression, mutation and copy number variation data. The drug features and cell line features extracted from different omics using pathway enrichment analysis are concatenated and fed into the fully connected neural network (FNN) for drug sensitivity prediction. The research, Chen et al. (2022) proposed a weakly supervised, multimodal deep learning algorithm to predict outcomes and discover prognostic features based on integration of pathology whole-slide images and molecular profile data (RNA, CNV and mutation status). This algorithm employed an attention-based multiple instance learning (AMIL) network for extracting histology features and a self-normalizing network (SNN) for omics feature extraction. Subsequently, these extracted histology and molecular features were integrated through the Kronecker Product.

5.2 Late integration

In late integration, models are trained separately for different omics, and then the predictions from individual models are integrated at the final decision stage (Fig. 5B). The character of late integration is beneficial for highly heterogeneous omics data. The common integrated methods include averaging, majority voting and weighted fusion. A Multimodal Deep Neural Network by integrating Multi-dimensional Data (MDNNMD) was developed to predict the prognosis of breast cancer (Sun, Wang, & Li, 2018). This model leverages a weighted linear method to aggregate the initial outputs from each DNN model on gene expression profile, CNA profile and clinical data. DyAM (Vanguri et al., 2022), a dynamic deep attention-based multiple-instance learning model with masking, was proposed to predict immunotherapy response in patients with NSCLC treated with PD-(L)1 blockade. This multimodal model has the capacity to integrate the features from radiology, pathology and genomics and the final output result is a weighted sum of individually optimized LR models corresponding to each specific modality. The experiment results showed that the multimodal model achieved better performance in predicting immunotherapy response than unimodal and bimodal model. The clinical, digital pathology, genomic and transcriptomic features were integrated by a machine learning framework to predict the breast cancer therapy response (Sammut et al., 2022). The machine learning framework was proposed by unweighted ensemble classifier of logistic regression, support vector machine and random forest. Multi-Omics Graph cOnvolutional NETworks (MOGONET) was introduced for patient classification and biomarker identification by omics-specific learning and cross-omics correlation learning (Wang, Subramanian, & Syeda-Mahmood, 2021). The initial predictions generated by each omics-specific GCN are integrated via the cross-omics discovery tensor using cross-omics correlations. Subsequently, the cross-omics discovery tensor is fed into View Correlation Discovery Network (VCDN) to obtain the final prediction. Wang et al. (2021) proposed a fusion method based on model uncertainty in prediction. This method reduces the uncertainty in the fused prediction by weighing the prediction from each modality with its uncertainty.

5.3 Intermediate integration

Intermediate integration is to learn the joint representation from the different representation of omics, which can effectively capture the complementary information from different omics layers (Fig. 5C). A multi-omics late integration method, MOLI (Sharifi-Noghabi, Zolotareva,

Collins, & Ester, 2019), was proposed to predict drug response based on deep neural networks. The omics-specific features from gene mutation, CNA and gene expression data were concatenated into one representation for drug response prediction. Multi-view Factorization AutoEncoder (MAE) with network constraints was developed to integrate multi-omics data (gene expression, miRNA expression, protein expression, and DNA methylation) and domain knowledge such as molecular interaction networks for predicting target clinical variables (Ma & Zhang, 2019). Luo, Ping, Li, Tian, and Wu (2019) developed multimodal deep belief net (DBN) model to predict the disease–gene associations. The latent representations of protein-protein interaction networks and gene ontology terms are initially learned through two independent Deep Belief Networks (DBNs), and then cross-modality representation of two latent representations was learned by joint DBN. Based on joint multimodal representation strategy, Yu et al. (2020) proposed a joint representation learning-based classification model for prognosis prediction in patients of esophageal squamous cell carcinoma with mRNA and methylation data. The omics-specific features learned from autoencoder were fused based on joint multimodal representation strategy. Different views from different omics are integrated in the latent space and then used for supervised learning if labeled data is available or unsupervised learning. In addition, a pathway-aware multi-layered hierarchical network (P-NET) was introduced to stratify prostate cancer patients based on their resistance to treatment and to identify molecular drivers of this resistance for therapeutic targeting (Elmarakeby et al., 2021). P-NET integrates biologically informed architecture that encodes different biological entities (patient molecular profile, genes, pathways, biological processes, and outcome) into a neural network to inform generation of biological hypotheses. Vale-Silva and Rohr (2021) introduced a multimodal deep learning method, MultiSurv, to predict long-term pan-cancer survival. The MultiSurv model integrated the modality-specific feature representations from six different data modalities into a single fused representation according to the row-wise maximum over the different data modalities. A deep variational information bottleneck (IB) approach was introduced for multi-omics data integration through learning from incomplete multi-view observations (Lee & Van Der Schaar, 2021). The joint representation was modelled as a product-of-experts (PoE) over the marginal representations, which combine both common and complementary information across the observed views.

6. Challenges

This review underscores the profound advancements achieved in cancer research through the integration of artificial intelligence (AI) and multi-omics. The confluence of AI and multi-omics has catalyzed a paradigm shift in oncology, revolutionizing various aspects of cancer including cancer screening, diagnosis, prognosis, treatment, biomarker identification and pathology. Despite these significant advancements, this field continues to face many challenges.

6.1 Data harmonization

The analysis of multi-omics data requires strict standardization and quality control to ensure reliability of results. Due to variations in experimental methods, sample handling, and pre-processing, bias may be introduced that can affect the accuracy and reliability of the multi-omics data. Therefore, a series of standardized experimental protocols and processing methods are needed to address these problems. In addition, different omics techniques produce data that vary widely in type, scale, and format. This may lead to some useful and valuable information lost to some extent in the standardization process. For multi-omics data integration, advanced computational methods are required to improve data consistency and protecting data integrity. Handling missing data across different omics is a significant issue. Data imputation and augmentation techniques such as GANs (Goodfellow et al., 2020) and diffusion model (Yang et al., 2023) can be leveraged to mitigate this issue. The data harmonization is significant for data sharing across different research institutions. In a real situation, the accessibility of data emerges as a critical concern for many researchers (Dash, Shakyawar, Sharma, & Kaushik, 2019), which can significantly impede the effective utilization of data and the progress of research. While a variety of publicly accessible multi-omics databases currently exist such as The Cancer Genome Atlas (TCGA) and Genotype-Tissue Expression (GTEx), there is a pressing need for more efforts in promoting data access through collaborative research networks and data management regulations.

6.2 Interpretability

Due to the heterogeneity of multi-omics data, it's challenging to explicitly stimulate and model the complex interaction and regulatory mechanism across different omics layer. Many advanced machine learning algorithms, particularly deep neural networks, lack the interpretability due to their sophisticated learning and decision-making process. The interpretability of algorithms is particularly important for practical clinical applications, as clinicians need to understand the decision-making processes of algorithms in order to trust and ultimately leverage those decisions (Amann et al., 2020). In addition, there may be a trade-off between interpretability and the performance of the algorithm, as some algorithms such as decision tree-based algorithms that are easy to interpret may not be as good at predicting performance as some complex 'black-box' algorithms. Grinsztajn, Oyallon, and Varoquaux (2022) showed that tree-based models outperform on medium-sized data (~10 K samples) than Neural Networks. Several strategies have been leveraged to improve the interpretability of algorithms. For example, some algorithms such as tree-based models and SHapley Additive exPlanations (SHAP) (Lundberg & Lee, 2017) can generate feature importance scores or contribution scores that can help in interpreting the decisions of algorithms. Furthermore, incorporating domain knowledge (Muralidhar, Islam, Marwah, Karpatne, & Ramakrishnan, 2018; Dash, Chitlangia, Ahuja, & Srinivasan, 2022), such as biological and clinical knowledge, into algorithms can guide and constrain the learning process, making the models' outputs more interpretable and relevant.

6.3 Ethical considerations

The rapid progression of AI in cancer research raises important ethical considerations (Carter et al., 2020; Murdoch, 2021). These ethical issues relate to data privacy and AI model fairness. Multi-omics data contains a lot of personal information, such as gene sequence, living habits, health status, etc., which is highly sensitive and private. How to achieve data sharing and cooperation cross multiple institutes while ensuring data privacy is also an important challenge. Ensuring the confidentiality and security of this data is crucial to protect patient privacy. Advanced data encryption and anonymization techniques are needed to protect the confidentiality and integrity of data. Several advanced computational methods have been used to preserve privacy, such as federated learning (McMahan, Moore, Ramage, Hampson, & Aguera Y Arcas, 2017), swarm learning (Warnat-Herresthal et al., 2021) and blockchain technique (Zheng, Xie, Dai, Chen, & Wang, 2017). The basic principle of federated learning and swarm learning is to train the model independently across multiple decentralized devices and then transmitted the results from independent models to other participants for aggregation. Cao et al. (2022) developed a computational framework, dsMTL, to preserve privacy based on distributed multi-task machine learning. This approach avoids the privacy and security risks associated with the centralized storage and transfer of data. Furthermore, the establishment of rigorous regulatory frameworks is necessary to review and monitor the usage of data. For example, the Health Insurance Portability and Accountability Act of 1996 (HIPAA) was enacted by the United States Act of Congress to protect patient health information. Fairness of the AI model is also an important ethical consideration. Recent statistics (Martin et al., 2019; Peterson et al., 2019) showed that a substantial proportion of bio-medical data used in AI research, including genomics and clinical datasets, primarily originates from individuals of European descent. The issue of data inequality among ethnic groups has emerged as a global health concern, as the AI model trained on inequality data may introduce bias. It would directly impact the accuracy and robustness of AI models, which can exacerbate healthcare disparities among minority groups. To reduce health care disparities among ethnic groups, a transfer learning model based on domain adaptation was employed (Gao & Cui, 2020). This method pre-train the model on the majority group and then fine-tune the model using the minority group. In addition, there is a need to ensure that all patients have equitable access to AI-based diagnostics and treatment methods. The open sourcing of models, while ensuring data privacy and security, is crucial to enabling AI models to benefit a larger patient population.

Effectively addressing these challenges necessitates a multidisciplinary colla-boration such as biology, medicine, mathematics, computer science and policy management. The integration of AI and multi-omics, harnesses the advanced information processing and analytical capabilities of AI and the comprehensive biological insights provided by multi-omics, presents vast application prospects and potential in cancer research. This progression is expected to bridge the gap between cutting-edge research and practical healthcare applications, enabling more precise, personalized, and effective treatments for cancer patients.

Acknowledgments

Some figures of this manuscript were generated based on the software BioRender.

Funding

Research reported in this publication was supported by the National Cancer Institute of the National Institutes of Health under Award Number P30CA036727. This work was supported by the American Cancer Society under award number IRG-22-146-07-IRG, and by the Buffett Cancer Center, which is supported by the National Cancer Institute under award number CA036727. This work was supported by the Buffet Cancer Center, which is sup-ported by the National Cancer Institute under award number CA036727, in collaboration with the UNMC/Children's Hospital & Medical Center Child Health Research Institute Pediatric Cancer Research Group. This study was supported, in part, by the National Institute on Alcohol Abuse and Alcoholism (P50AA030407-5126, Pilot Core grant). This study was also supported by the Nebraska EPSCoR FIRST Award (OIA-2044049). This work was also partially supported by the National Institute of General Medical Sciences under Award Numbers P20GM103427 and P20GM130447. This study was in part financially supported by the Child Health Research Institute at UNMC/Children's Nebraska. This work was also partially supported by the University of Nebraska Collaboration Initiative Grant from the Nebraska Research Initiative (NRI). The content is solely the responsibility of the authors and does not necessarily represent the official views from the funding organizations.

Authors' contributions

LL, MS, JW, and SW wrote the manuscript. SW supervised the manuscript. The manuscript was approved by all authors.

Competing interests

The authors declare no competing interests.

References

Abajian, A., Murali, N., Savic, L. J., Laage-Gaupp, F. M., Nezami, N., Duncan, J. S., ... Chapiro, J. (2018a). Predicting treatment response to intra-arterial therapies for hepatocellular carcinoma with the use of supervised machine learning—An artificial intelligence concept. *Journal of Vascular and Interventional Radiology, 29*(6), 850–857.

Abajian, A., Murali, N., Savic, L. J., Laage-Gaupp, F. M., Nezami, N., Duncan, J. S., ... Chapiro, J. (2018b). Predicting treatment response to intra-arterial therapies for hepatocellular carcinoma with the use of supervised machine learning—An artificial intelligence concept. *Journal of Vascular and Interventional Radiology, 29*(6), 850–857.e1. https://doi.org/10.1016/j.jvir.2018.01.769.

Acosta, J. N., Falcone, G. J., Rajpurkar, P., & Topol, E. J. (2022). Multimodal biomedical AI. *Nature Medicine, 28*(9), 1773–1784.

Akhoundova, D., & Rubin, M. A. (2022). Clinical application of advanced multi-omics tumor profiling: Shaping precision oncology of the future. *Cancer Cell, 40*(9), 920–938. https://doi.org/10.1016/j.ccell.2022.08.011.

Alakwaa, F. M., Chaudhary, K., & Garmire, L. X. (2018). Deep learning accurately predicts estrogen receptor status in breast cancer metabolomics data. *Journal of Proteome Research, 17*(1), 337–347.

Aldridge, S., & Teichmann, S. A. (2020). Single cell transcriptomics comes of age. *Nature Communications, 11*(1), 4307.

Alshareef, A. M., Alsini, R., Alsieni, M., Alrowais, F., Marzouk, R., Abunadi, I., & Nemri, N. (2022a). Optimal deep learning enabled prostate cancer detection using microarray gene expression. *Journal of Healthcare Engineering, 2022.*

Alshareef, A. M., Alsini, R., Alsieni, M., Alrowais, F., Marzouk, R., ... Nemri, N. (2022b). Optimal deep learning enabled prostate cancer detection using microarray gene expression (Edited by K. Shankar) *Journal of Healthcare Engineering, 2022*, 1–12. https://doi.org/10.1155/2022/7364704.

Alzubaidi, L., Zhang, J., Humaidi, A. J., Al-Dujaili, A., Duan, Y., Al-Shamma, O., ... Farhan, L. (2021). Review of deep learning: Concepts, CNN architectures, challenges, applications, future directions. *Journal of Big Data, 8*, 1–74.

Amann, J., Blasimme, A., Vayena, E., Frey, D., & Madai, V. I. Precise4Q Consortium. (2020). Explainability for Artificial Intelligence in healthcare: A multidisciplinary perspective. *BMC Medical Informatics and Decision Making, 20,* 1–9.

Amini, M.-R., Feofanov, V., Pauletto, L., Devijver, E., & Maximov, Y. (2022). Self-training: A survey. arXiv Preprint arXiv:2202.12040.

Ardrey, R. E. (2003). *Liquid chromatography-mass spectrometry: An introduction, Vol. 2.* John Wiley & Sons.

Aref-Eshghi, E., Rodenhiser, D. I., Schenkel, L. C., Lin, H., Skinner, C., Ainsworth, P., Paré, G., et al. (2018). Genomic DNA methylation signatures enable concurrent diagnosis and clinical genetic variant classification in neurodevelopmental syndromes. *The American Journal of Human Genetics, 102*(1), 156–174. https://doi.org/10.1016/j.ajhg.2017.12.008.

Arya, N., & Saha, S. (2021). Multi-modal advanced deep learning architectures for breast cancer survival prediction. *Knowledge-Based Systems, 221,* 106965.

Behjati, S., & Tarpey, P. S. (2013). What is next generation sequencing? *Archives of Disease in Childhood-Education and Practice.*

Bock, C., Farlik, M., & Sheffield, N. C. (2016). Multi-omics of single cells: Strategies and applications. *Trends in Biotechnology, 34*(8), 605–608.

Cai, W.-L., Cheng, M., Wang, Y., Xu, P.-H., Yang, X., Sun, Z.-W., & Yan, W.-J. (2024). Prediction and related genes of cancer distant metastasis based on deep learning. *Computers in Biology and Medicine, 168,* 107664.

Cai, Z., Poulos, R. C., Liu, J., & Zhong, Q. (2022). Machine learning for multi-omics data integration in cancer. *Iscience.*

Cambiaghi, A., Ferrario, M., & Masseroli, M. (2017). Analysis of metabolomic data: Tools, current strategies and future challenges for omics data integration. *Briefings in Bioinformatics, 18*(3), 498–510.

Cao, H., Zhang, Y., Baumbach, J., Burton, P. R., Dwyer, D., Koutsouleris, N., ... Rieg, T. (2022). dsMTL: A computational framework for privacy-preserving, distributed multi-task machine learning. *Bioinformatics (Oxford, England), 38*(21), 4919–4926.

Carter, S. M., Rogers, W., Win, K. T., Frazer, H., Richards, B., & Houssami, N. (2020). The ethical, legal and social implications of using artificial intelligence systems in breast cancer care. *The Breast, 49,* 25–32.

Ceglarek, U., Leichtle, A., Bruegel, M., Kortz, L., Brauer, R., Bresler, K., ... Fiedler, G. M. (2009). Challenges and developments in tandem mass spectrometry based clinical metabolomics. *Molecular and Cellular Endocrinology, 301*(1–2), 266–271.

Chabon, J. J., Hamilton, E. G., Kurtz, D. M., Esfahani, M. S., Moding, E. J., Stehr, H., ... Chaudhuri, A. A. (2020). Integrating genomic features for non-invasive early lung cancer detection. *Nature, 580*(7802), 245–251.

Chen, J., & Liu, Y. (2011). Locally linear embedding: A survey. *Artificial Intelligence Review, 36,* 29–48.

Chen, R. J., Lu, M. Y., Williamson, D. F. K., Chen, T. Y., Lipkova, J., Noor, Z., ... Joo, B. (2022). Pan-cancer integrative histology-genomic analysis via multimodal deep learning. *Cancer Cell, 40*(8), 865–878.

Chen, T., Li, M., Li, Y., Lin, M., Wang, N., Wang, M., ... Zhang, Z. (2015). Mxnet: A flexible and efficient machine learning library for heterogeneous distributed systems. arXiv Preprint arXiv:1512.01274.

Chen, Y., Wang, B., Zhao, Y., Shao, X., Wang, M., Ma, F., Yang, L., et al. (2024). Metabolomic machine learning predictor for diagnosis and prognosis of gastric cancer. *Nature Communications, 15*(1), 1657. https://doi.org/10.1038/s41467-024-46043-y.

Collins, F. S., Morgan, M., & Patrinos, A. (2003). The human genome project: Lessons from large-scale biology. *Science (New York, N. Y.), 300*(5617), 286–290.

Cramer, J. S., The Origins of Logistic Regression (December 2002). Tinbergen Institute Working Paper No. 2002-119/4, Available at SSRN: https://ssrn.com/abstract=360300 or http://dx.doi.org/10.2139/ssrn.360300.

Cui, L., Lu, H., & Lee, Y. H. (2018). Challenges and emergent solutions for LC-MS/MS based untargeted metabolomics in diseases. *Mass Spectrometry Reviews, 37*(6), 772–792.

Cunningham, P., & Delany, S. J. (2021). K-nearest neighbour classifiers—A tutorial. *ACM Computing Surveys (CSUR), 54*(6), 1–25.

Dagogo-Jack, I., & Shaw, A. T. (2018). Tumour heterogeneity and resistance to cancer therapies. *Nature Reviews Clinical Oncology, 15*(2), 81–94.

Dash, S., Shakyawar, S. K., Sharma, M., & Kaushik, S. (2019). Big data in healthcare: Management, analysis and future prospects. *Journal of Big Data, 6*(1), 1–25.

Dash, T., Chitlangia, S., Ahuja, A., & Srinivasan, A. (2022). A review of some techniques for inclusion of domain-knowledge into deep neural networks. *Scientific Reports, 12*(1), 1040.

Davies, M. P. A., Sato, T., Ashoor, H., Hou, L., Liloglou, T., Yang, R., & Field, J. K. (2023). Plasma protein biomarkers for early prediction of lung cancer. *eBioMedicine, 93*, 104686.

Davies, M. P. A., Sato, T., Ashoor, H., Hou, L., Liloglou, T., Yang, R., & Field, J. K. (2023). Plasma protein biomarkers for early prediction of lung cancer. *eBioMedicine, 93*, 104686. https://doi.org/10.1016/j.ebiom.2023.104686.

Dawson, M. A., & Kouzarides, T. (2012). Cancer epigenetics: From mechanism to therapy. *Cell, 150*(1), 12–27.

Dunteman, G. H. (1989). *Principal components analysis, Vol. 69*. Sage.

Echle, A., Rindtorff, N. T., Brinker, T. J., Luedde, T., Pearson, A. T., & Kather, J. N. (2021). Deep learning in cancer pathology: A new generation of clinical biomarkers. *British Journal of Cancer, 124*(4), 686–696. https://doi.org/10.1038/s41416-020-01122-x.

Eid, J., Fehr, A., Gray, J., Luong, K., Lyle, J., Otto, G., ... Bettman, B. (2009). Real-time DNA sequencing from single polymerase molecules. *Science (New York, N. Y.), 323*(5910), 133–138.

El Nahhas, O. S. M., Loeffler, C. M. L., Carrero, Z. I., van Treeck, M., Kolbinger, F. R., Hewitt, K. J., Muti, H. S., et al. (2024). Regression-based deep-learning predicts molecular biomarkers from pathology slides. *Nature Communications, 15*(1), 1253. https://doi.org/10.1038/s41467-024-45589-1.

Elmarakeby, H. A., Hwang, J., Arafeh, R., Crowdis, J., Gang, S., Liu, D., ... Richter, C. (2021). Biologically informed deep neural network for prostate cancer discovery. *Nature, 598*(7880), 348–352.

Ertel, W. (2018). *Introduction to Artificial Intelligence*. Springer.

Feng, X., Hong, T., Liu, W., Xu, C., Li, W., Yang, B., ... Zhou, H. (2022). Development and validation of a machine learning model to predict the risk of lymph node metastasis in renal carcinoma. *Frontiers in Endocrinology, 13*, 1054358.

Ferraro, J. R. (2012). *Practical Fourier transform infrared spectroscopy: Industrial and laboratory chemical analysis*. Elsevier.

Gao, F., Wang, W., Tan, M., Zhu, L., Zhang, Y., Fessler, E., ... Wang, X. (2019). DeepCC: A novel deep learning-based framework for cancer molecular subtype classification. *Oncogenesis, 8*(9), 44.

Gao, Y., & Cui, Y. (2020). Deep transfer learning for reducing health care disparities arising from biomedical data inequality. *Nature Communications, 11*(1), 5131.

Gerdes, H., Casado, P., Dokal, A., Hijazi, M., Akhtar, N., Osuntola, R., ... Britton, D. (2021). Drug ranking using machine learning systematically predicts the efficacy of anti-cancer drugs. *Nature Communications, 12*(1), 1850.

Ghaffari, S., Hanson, C., Schmidt, R. E., Bouchonville, K. J., Offer, S. M., & Sinha, S. (2021). An integrated multi-omics approach to identify regulatory mechanisms in cancer metastatic processes. *Genome Biology, 22*(1), 19. https://doi.org/10.1186/s13059-020-02213-x.

Ghanat Bari, M., Ung, C. Y., Zhang, C., Zhu, S., & Li, H. (2017). Machine learning-assisted network inference approach to identify a new class of genes that coordinate the functionality of cancer networks. *Scientific Reports, 7*(1), 6993. https://doi.org/10.1038/s41598-017-07481-5.

Goodfellow, I., Pouget-Abadie, J., Mirza, M., Xu, B., Warde-Farley, D., Ozair, S., ... Bengio, Y. (2020). Generative adversarial networks. *Communications of the ACM, 63*(11), 139–144.

Grinsztajn, L., Oyallon, E., & Varoquaux, G. (2022). Why do tree-based models still outperform deep learning on typical tabular data? *Advances in Neural Information Processing Systems, 35*, 507–520.

Gu, J., Wang, Z., Kuen, J., Ma, L., Shahroudy, A., Shuai, B., ... Cai, J. (2018). Recent advances in convolutional neural networks. *Pattern Recognition, 77*, 354–377.

Halner, A., Hankey, L., Liang, Z., Pozzetti, F., Szulc, D. A., Mi, E., ... Liu, P. J. (2023). DEcancer: Machine learning framework tailored to liquid biopsy based cancer detection and biomarker signature selection. *iScience, 26*(5), 106610. https://doi.org/10.1016/j.isci.2023.106610.

Harris, T. D., Buzby, P. R., Babcock, H., Beer, E., Bowers, J., Braslavsky, I., ... Efcavitch, J. W. (2008). Single-molecule DNA sequencing of a viral genome. *Science (New York, N. Y.), 320*(5872), 106–109.

Hasin, Y., Seldin, M., & Lusis, A. (2017). Multi-omics approaches to disease. *Genome Biology, 18*(1), 1–15.

Hassoun, M. H. (1995). *Fundamentals of artificial neural networks*. MIT Press.

He, X., Xiaowei Liu, F. Z., Shi, H., & Jing, J. (2023). Artificial Intelligence-based multi-omics analysis fuels cancer precision medicine. *Semin. Cancer Biol. 88*, 187–200. Elsevier.

Hearst, M. A., Susan, T. D., Osuna, E., Platt, J., & Scholkopf, B. (1998). Support vector machines. *IEEE Intelligent Systems and Their Applications, 13*(4), 18–28.

Heiss, J. A., & Brenner, H. (2017). Epigenome-wide discovery and evaluation of leukocyte DNA methylation markers for the detection of colorectal cancer in a screening setting. *Clinical Epigenetics, 9*(1), 24. https://doi.org/10.1186/s13148-017-0322-x.

Heo, Y. J., Hwa, C., Park, J.-M., Lee, G.-H., & An, J.-Y. (2021). Integrative multi-omics approaches in cancer research: From biological networks to clinical subtypes. *Molecules and Cells, 44*(7), 433–443. https://doi.org/10.14348/molcells.2021.0042.

Herrgott, G. A., Snyder, J. M., She, R., Malta, T. M., Sabedot, T. S., Lee, I. Y., ... Zhang, J. (2023). Detection of diagnostic and prognostic methylation-based signatures in liquid biopsy specimens from patients with meningiomas. *Nature Communications, 14*(1), 5669.

Hoerl, A. E., & Kennard, R. W. (1970). Ridge regression: Biased estimation for non-orthogonal problems. *Technometrics, 12*(1), 55–67.

Hong, J., Hachem, L. D., & Fehlings, M. G. (2022). A deep learning model to classify neoplastic state and tissue origin from transcriptomic data. *Scientific Reports, 12*(1), 9669.

Horgan, R. P., & Kenny, L. C. (2011). Omic'Technologies: Genomics, transcriptomics, proteomics and metabolomics. *The Obstetrician & Gynaecologist, 13*(3), 189–195.

Horner, D., Stephen, G., Pavesi, T., Castrignano, P. D.'O. D., Meo, S., Liuni, M., ... Pesole, G. (2010). Bioinformatics approaches for genomics and post genomics applications of next-generation sequencing. *Briefings in Bioinformatics, 11*(2), 181–197.

Hout, M. C., Papesh, M. H., & Goldinger, S. D. (2013). Multidimensional scaling. *Wiley Interdisciplinary Reviews: Cognitive Science, 4*(1), 93–103.

Hrdlickova, R., Toloue, M., & Tian, B. (2017). RNA-Seq methods for transcriptome analysis. *Wiley Interdisciplinary Reviews: RNA, 8*(1) e1364.

Huang, C., Clayton, E. A., Matyunina, L. V., McDonald, L. D. E., Benigno, B. B., Vannberg, F., & McDonald, J. F. (2018a). Machine learning predicts individual cancer patient responses to therapeutic drugs with high accuracy. *Scientific Reports, 8*(1), 16444.

Huang, C., Clayton, E. A., Matyunina, L. V., McDonald, L. D. E., Benigno, B. B., Vannberg, F., & McDonald, J. F. (2018b). Machine learning predicts individual cancer patient responses to therapeutic drugs with high accuracy. *Scientific Reports, 8*(1), 16444. https://doi.org/10.1038/s41598-018-34753-5.

Hubschmann, H.-J. (2015). *Handbook of GC-MS: Fundamentals and applications.* John Wiley & Sons.

Iscen, A. Tolias, G., Avrithis, Y., & Chum, O. (2019). *Label propagation for deep semi-supervised learning,* In Proceedings of the IEEE Conference on Computer Vision and Pattern Recognition, (pp. 5070–5079).

Ivanisevic, T., & Sewduth, R. N. (2023). Multi-omics integration for the design of novel therapies and the identification of novel biomarkers. *Proteomes, 11*(4), 34.

James, G., Witten, D., Hastie, T., Tibshirani, R., & Taylor, J. (2023). *Tree-based methods. An introduction to statistical learning: With applications in python.* Springer, 331–366.

Janiesch, C., Zschech, P., & Heinrich, K. (2021). Machine learning and deep learning. *Electronic Markets, 31*(3), 685–695.

Jha, A., Quesnel-Vallières, M., Wang, D., Thomas-Tikhonenko, A., Lynch, K. W., & Barash, Y. (2022). Identifying common transcriptome signatures of cancer by interpreting deep learning models. *Genome Biology, 23*(1), 117. https://doi.org/10.1186/s13059-022-02681-3.

Jiang, Z., Zhou, X., Li, R., Michal, J. J., Zhang, S., Dodson, M. V., ... Harland, R. M. (2015). Whole transcriptome analysis with sequencing: Methods, challenges and potential solutions. *Cellular and Molecular Life Sciences, 72*, 3425–3439.

Jiao, W., Atwal, G., Polak, P., Karlic, R., Cuppen, E., Al-Shahrour, F., Atwal, G., et al. (2020). A deep learning system accurately classifies primary and metastatic cancers using passenger mutation patterns. *Nature Communications, 11*(1), 728. https://doi.org/10.1038/s41467-019-13825-8.

Jin, D., Liang, S., Shmatko, A., Arnold, A., Horst, D., Grünewald, T. G. P., ... Bai, X. (2024). Teacher-student collaborated multiple instance learning for pan-cancer PDL1 expression prediction from histopathology slides. *Nature Communications, 15*(1), 3063. https://doi.org/10.1038/s41467-024-46764-0.

Joshi, P., & Dhar, R. (2022). EpICC: A Bayesian neural network model with uncertainty correction for a more accurate classification of cancer. *Scientific Reports, 12*(1), 14628.

Kalafi, E. Y., Nor, N. A. M., Taib, N. A., Ganggayah, M. D., Town, C., & Dhillon, S. K. (2019). Machine learning and deep learning approaches in breast cancer survival prediction using clinical data. *Folia Biologica, 65*(5/6), 212–220.

Katipally, R. R., Martinez, C. A., Pugh, S. A., Bridgewater, J. A., Primrose, J. N., Domingo, E., ... Weichselbaum, R. R. (2023). Integrated clinical-molecular classification of colorectal liver metastases: A biomarker analysis of the phase 3 new EPOC randomized clinical trial. *JAMA Oncology, 9*(9), 1245–1254.

Khan, K., Rehman, S. U., Aziz, K., Fong, S., & Sarasvady, S. (2014). *DBSCAN: Past, present and future* (pp. 232–238). IEEE.

Klein, E. A., Richards, D., Cohn, A., Tummala, M., Lapham, R., Cosgrove, D., ... Hunkapiller, N. (2021). Clinical validation of a targeted methylation-based multi-cancer early detection test using an independent validation set. *Annals of Oncology, 32*(9), 1167–1177.

Kong, J. H., Ha, D., Lee, J., Kim, I., Park, M., Im, S.-H., ... Kim, S. (2022). Network-based machine learning approach to predict immunotherapy response in cancer patients. *Nature Communications, 13*(1), 3703.

Lai, Y.-H., Wei-Ning Chen, T.-C. H., Che Lin, Y. T., & Wu, S. (2020). Overall survival prediction of non-small cell lung cancer by integrating microarray and clinical data with deep learning. *Scientific Reports, 10*(1), 4679.

Langley, P. (2011). The changing science of machine learning. *Machine Learning, 82*(3), 275–279.

Lapuente-Santana, Ó., van Genderen, M., Hilbers, P. A. J., Finotello, F., & Eduati, F. (2021). Interpretable systems biomarkers predict response to immune-checkpoint inhibitors. *Patterns, 2*(8).

Lee, C., & Van Der Schaar, M. (2021). *A variational information bottleneck approach to multi-omics data integration* (pp. 1513–1521). PMLR.

Lee, D., & Sebastian Seung, H. (2000). Algorithms for non-negative matrix factorization. *Advances in Neural Information Processing Systems, 13.*

Lee, K., Hyoung-oh Jeong, S. L., & Jeong, W.-K. (2019). CPEM: Accurate cancer type classification based on somatic alterations using an ensemble of a random forest and a deep neural network. *Scientific Reports, 9*(1), 16927.

Lee, Y., Park, J. H., Oh, S., Shin, K., Sun, J., Jung, M., Lee, C., et al. (2022). Derivation of prognostic contextual histopathological features from whole-slide images of tumours via graph deep learning. *Nature Biomedical Engineering.* https://doi.org/10.1038/s41551-022-00923-0.

Levene, M. J., Korlach, J., Turner, S. W., Foquet, M., Craighead, H. G., & Webb, W. W. (2003). Zero-mode waveguides for single-molecule analysis at high concentrations. *Science (New York, N. Y.), 299*(5607), 682–686.

Li, B., Kugeratski, F. G., & Kalluri, R. (2023). A novel machine learning algorithm selects proteome signature to specifically identify cancer exosomes. *BioRxiv,* 2023–07.

Lin, D., Wang, Y., Wang, T., Zhu, Y., Lin, X., Lin, Y., & Feng, S. (2020). Metabolite profiling of human blood by surface-enhanced raman spectroscopy for surgery assessment and tumor screening in breast cancer. *Analytical and Bioanalytical Chemistry, 412,* 1611–1618.

Ling, C., & Rönn, T. (2019). Epigenetics in human obesity and type 2 diabetes. *Cell Metabolism, 29*(5), 1028–1044.

Lu, M., & Zhan, X. (2018). The crucial role of multiomic approach in cancer research and clinically relevant outcomes. *EPMA Journal, 9*(1), 77–102. https://doi.org/10.1007/s13167-018-0128-8.

Lu, M. Y., Chen, T. Y., Williamson, D. F. K., Zhao, M., Shady, M., Lipkova, J., & Mahmood, F. (2021). AI-based pathology predicts origins for cancers of unknown primary. *Nature, 594*(7861), 106–110. https://doi.org/10.1038/s41586-021-03512-4.

Lu, W., Fu, D., Kong, X., Huang, Z., Hwang, M., Zhu, Y., Chen, L., et al. (2020). FOLFOX treatment response prediction in metastatic or recurrent colorectal cancer patients via machine learning algorithms. *Cancer Medicine, 9*(4), 1419–1429. https://doi.org/10.1002/cam4.2786.

Lundberg, S. M., & Lee, S.-I. (2017). A unified approach to interpreting model predictions. *Advances in Neural Information Processing Systems, 30.*

Luo, Ping, Li, Y., Tian, L.-P., & Wu, F.-X. (2019). Enhancing the prediction of disease–gene associations with multimodal deep learning. *Bioinformatics (Oxford, England), 35*(19), 3735–3742.

Ma, T., & Zhang, A. (2019). Integrate multi-omics data with biological interaction networks using multi-view factorization autoencoder (MAE). *BMC Genomics, 20,* 1–11.

Ma, T., Liu, Q., Li, H., Zhou, M., Jiang, R., & Zhang, X. (2022). DualGCN: A dual graph convolutional network model to predict cancer drug response. *BMC Bioinformatics, 23*(4), 1–13.

Mann, M., & Jensen, O. N. (2003). Proteomic analysis of post-translational modifications. *Nature Biotechnology, 21*(3), 255–261.

Martin, A. R., Kanai, M., Kamatani, Y., Okada, Y., Neale, B. M., & Daly, M. J. (2019). Clinical use of current polygenic risk scores may exacerbate health disparities. *Nature Genetics, 51*(4), 584–591.

Marusyk, A., Janiszewska, M., & Polyak, K. (2020). Intratumor heterogeneity: The Rosetta stone of therapy resistance. *Cancer Cell, 37*(4), 471–484. https://doi.org/10.1016/j.ccell.2020.03.007.

Maugis, C., Celeux, G., & Magniette, M.-L. M. (2009). Variable selection for clustering with Gaussian mixture models. *Biometrics, 65*(3), 701–709.

Maulud, D., & Abdulazeez, A. M. (2020). A review on linear regression comprehensive in machine learning. *Journal of Applied Science and Technology Trends, 1*(4), 140–147.

Maung, R. (2016). Pathologists' workload and patient safety. *Diagnostic Histopathology, 22*(8), 283–287. https://doi.org/10.1016/j.mpdhp.2016.07.004.

Mbemi, A., Khanna, S., Njiki, S., Yedjou, C. G., & Tchounwou, P. B. (2020). Impact of gene–environment interactions on cancer development. *International Journal of Environmental Research and Public Health, 17*(21), 8089.

McDonald, G. C. (2009). Ridge regression. *Wiley Interdisciplinary Reviews: Computational Statistics, 1*(1), 93–100.

McInnes, L., Healy, J., & Melville, J. (2018). Umap: Uniform manifold approximation and projection for dimension reduction. arXiv Preprint arXiv:1802.03426.

McMahan, B., Moore, E., Ramage, D., Hampson, S., & Aguera y Arcas, B. (2017). *Communication-efficient learning of deep networks from decentralized data* (pp. 1273–1282). PMLR.

Meacham, C. E., & Morrison, S. J. (2013). Tumour heterogeneity and cancer cell plasticity. *Nature, 501*(7467), 328–337.

Menyhárt, O., & Győrffy, B. (2021). Multi-omics approaches in cancer research with applications in tumor subtyping, prognosis, and diagnosis. *Computational and Structural Biotechnology Journal, 19*, 949–960.

Moon, I., LoPiccolo, J., Baca, S. C., Sholl, L. M., Kehl, K. L., Hassett, M. J., ... Gusev, A. (2023). Machine learning for genetics-based classification and treatment response prediction in cancer of unknown primary. *Nature Medicine, 29*(8), 2057–2067.

Muralidhar, N., Islam, M. R., Marwah, M., Karpatne, A., & Ramakrishnan, N. (2018). *Incorporating prior domain knowledge into deep neural networks* (pp. 36–45). IEEE.

Murdoch, B. (2021). Privacy and Artificial Intelligence: Challenges for protecting health information in a new era. *BMC Medical Ethics, 22*(1), 1–5.

Murtagh, F., & Contreras, P. (2012). Algorithms for hierarchical clustering: An overview. *Wiley Interdisciplinary Reviews: Data Mining and Knowledge Discovery, 2*(1), 86–97.

Newgard, C. B. (2017). Metabolomics and metabolic diseases: Where do we stand? *Cell Metabolism, 25*(1), 43–56.

Ng, A., Jordan, M., & Weiss, Y. (2001). On spectral clustering: Analysis and an algorithm. *Advances in Neural Information Processing Systems, 14*.

Park, C., Ahn, J., Kim, H., & Park, S. (2014). Integrative gene network construction to analyze cancer recurrence using semi-supervised learning. *PLoS One, 9*(1), e86309.

Peterson, R. E., Kuchenbaecker, K., Walters, R. K., Chen, C.-Y., Popejoy, A. B., Periyasamy, S., ... Brick, L. (2019). Genome-wide association studies in ancestrally diverse populations: Opportunities, methods, pitfalls, and recommendations. *Cell, 179*(3), 589–603.

Qiao, S., Shen, W., Zhang, Z., Wang, B., & Yuille, A. (2018). *Deep co-training for semi-supervised image recognition*, In Proceedings of the european conference on computer vision (eccv), (pp. 135–152).

Qiu, B., Shen, Z., Wu, S., Qin, X., Yang, D., & Wang, Q. (2023). A machine learning-based model for predicting distant metastasis in patients with rectal cancer. *Frontiers in Oncology, 13*.

Queirós, A. C., Villamor, N., Clot, G., Martinez-Trillos, A., Kulis, M., Navarro, A., Penas, E. M. M., et al. (2015). A B-cell epigenetic signature defines three biologic subgroups of chronic lymphocytic leukemia with clinical impact. *Leukemia: Official Journal of the Leukemia Society of America, Leukemia Research Fund, U. K, 29*(3), 598–605. https://doi.org/10.1038/leu.2014.252.

Ramón y Cajal, S., Sesé, M., Capdevila, C., Aasen, T., De Mattos-Arruda, L., Diaz-Cano, S. J., ... Castellví, J. (2020). Clinical implications of intratumor heterogeneity: Challenges and opportunities. *Journal of Molecular Medicine, 98*(2), 161–177. https://doi.org/10.1007/s00109-020-01874-2.

Rappoport, N., & Shamir, R. (2018). Multi-omic and multi-view clustering algorithms: Review and cancer benchmark. *Nucleic Acids Research, 46*(20), 10546–10562.

Reddy, Y. C. A. P., Viswanath, P., & Eswara Reddy, B. (2018). Semi-supervised learning: A brief review. *International Journal of Engineering & Technology, 7*(1.8), 81.

Rish, I. (2001). An empirical study of the Naive Bayes classifier. *IJCAI, 3*, 41–46.

Rule, G. S., & Hitchens, T. K. (2006). *NMR spectroscopy*. Springer.

Rumelhart, D. E., & McClelland, J. L. (1987). *Learning Internal Representations by Error Propagation. Parallel Distributed Processing: Explorations in the Microstructure of Cognition: Foundations*. MIT Press, 318–362.

Sammut, S.-J., Crispin-Ortuzar, M., Chin, S.-F., Provenzano, E., Bardwell, H. A., Ma, W., ... Abraham, J. E. (2022). Multi-omic machine learning predictor of breast cancer therapy response. *Nature, 601*(7894), 623–629.

Sanger, F. (1981). Determination of nucleotide sequences in DNA. *Science (New York, N. Y.), 214*(4526), 1205–1210.

Sarker, I. H. (2021). Machine learning: Algorithms, real-world applications and research directions. *SN Computer Science, 2*(3), 160.

Sarker, I. H. (2022). Ai-based modeling: Techniques, applications and research issues towards automation, intelligent and smart systems. *SN Computer Science, 3* (2), 158.

Schadt, E. E., Turner, S., & Kasarskis, A. (2010). A window into third-generation sequencing. *Human Molecular Genetics, 19*(R2), R227–R240.

Schmitz, S. U., Grote, P., & Herrmann, B. G. (2016). Mechanisms of long noncoding RNA function in development and disease. *Cellular and Molecular Life Sciences, 73*, 2491–2509.

Schulte-Sasse, R., Budach, S., Hnisz, D., & Marsico, A. (2021). Integration of multiomics data with graph convolutional networks to identify new cancer genes and their associated molecular mechanisms. *Nature Machine Intelligence, 3*(6), 513–526.

Schuster, M., & Paliwal, K. K. (1997). Bidirectional recurrent neural networks. *IEEE Transactions on Signal Processing, 45*(11), 2673–2681.

Seyfried, T. N., & Shelton, L. M. (2010). Cancer as a metabolic disease. *Nutrition & Metabolism, 7*, 1–22.

Seyhan, A. A., & Carini, C. (2019). Are innovation and new technologies in precision medicine paving a new era in patients centric care? *Journal of Translational Medicine, 17*, 1–28.

Shao, Z., Bian, H., Chen, Y., Wang, Y., Zhang, J., & Ji, X. (2021). Transmil: Transformer based correlated multiple instance learning for whole slide image classification. *Advances in Neural Information Processing Systems, 34*, 2136–2147.

Sharifi-Noghabi, H., Zolotareva, O., Collins, C. C., & Ester, M. (2019). MOLI: Multi-omics late integration with deep neural networks for drug response prediction. *Bioinformatics (Oxford, England), 35*(14), i501–i509.

Shi, L., Zhang, Y., & Wang, H. (2023). Prognostic prediction based on histopathologic features of tumor microenvironment in colorectal cancer. *Frontiers in Medicine, 10*. https://doi.org/10.3389/fmed.2023.1154077.

Shmatko, A., Ghaffari Laleh, N., Gerstung, M., & Kather, J. N. (2022). Artificial Intelligence in histopathology: Enhancing cancer research and clinical oncology. *Nature Cancer, 3*(9), 1026–1038. https://doi.org/10.1038/s43018-022-00436-4.

Sinaga, K. P., & Yang, M.-S. (2020). Unsupervised K-means clustering algorithm. *IEEE Access, 8*, 80716–80727.

Stark, G. F., Hart, G. R., Nartowt, B. J., & Deng, J. (2019). Predicting breast cancer risk using personal health data and machine learning models. *PLoS One, 14*(12), e0226765.

Stricker, S. H., Köferle, A., & Beck, S. (2017). From profiles to function in epigenomics. *Nature Reviews. Genetics, 18*(1), 51–66.

Sun, D., Wang, M., & Li, A. (2018). A multimodal deep neural network for human breast cancer prognosis prediction by integrating multi-dimensional data. *IEEE/ACM Transactions on Computational Biology and Bioinformatics, 16*(3), 841–850.

Sun, Y., Selvarajan, S., Zang, Z., Liu, W., Zhu, Y., Zhang, H., Chen, W., et al. (2022). Artificial Intelligence defines protein-based classification of thyroid nodules. *Cell Discovery, 8*(1), 85. https://doi.org/10.1038/s41421-022-00442-x.

Sun, Y., Zhu, S., Ma, K., Liu, W., Yue, Y., Hu, G., ... Chen, W. (2019). Identification of 12 cancer types through genome deep learning. *Scientific Reports, 9*(1), 17256.

Swanson, K., Wu, E., Zhang, A., Alizadeh, A. A., & Zou, J. (2023). From patterns to patients: Advances in clinical machine learning for cancer diagnosis, prognosis, and treatment. *Cell, 186*(8), 1772–1791. https://doi.org/10.1016/j.cell.2023.01.035.

Tam, V., Patel, N., Turcotte, M., Bossé, Y., Paré, G., & Meyre, D. (2019). Benefits and limitations of genome-wide association studies. *Nature Reviews. Genetics, 20*(8), 467–484.

Tang, Y.-C., & Gottlieb, A. (2021). Explainable drug sensitivity prediction through cancer pathway enrichment. *Scientific Reports, 11*(1), 3128.

Taudt, A., Colomé-Tatché, M., & Johannes, F. (2016). Genetic sources of population epigenomic variation. *Nature Reviews. Genetics, 17*(6), 319–332.

Tibshirani, R. (1996). Regression shrinkage and selection via the lasso. *Journal of the Royal Statistical Society Series B: Statistical Methodology, 58*(1), 267–288.

Tranchevent, L.-C., Azuaje, F., & Rajapakse, J. C. (2019). A deep neural network approach to predicting clinical outcomes of neuroblastoma patients. *BMC Medical Genomics, 12*, 1–11.

Tsai, P.-C., Tsung-Hua Lee, K.-C. K., Fang-Yi Su, T.-L. M. L., Eliana Marostica, T. U., et al. (2023). Histopathology images predict multi-omics aberrations and prognoses in colorectal cancer patients. *Nature Communications, 14*(1), 2102. https://doi.org/10.1038/s41467-023-37179-4.

Tseng, Y.-J., Hsin-Yao Wang, T.-W. L., Jang-Jih Lu, C.-H. H., & Liao, C.-T. (2020). Development of a machine learning model for survival risk stratification of patients with advanced oral cancer. *JAMA Network Open, 3*(8), e2011768.

van Engelen, J. E., & Hoos, H. H. (2020). A survey on semi-supervised learning. *Machine Learning, 109*(2), 373–440. https://doi.org/10.1007/s10994-019-05855-6.

Vale-Silva, L. A., & Rohr, K. (2021). Long-term cancer survival prediction using multimodal deep learning. *Scientific Reports, 11*(1), 13505.

Van Der Maaten, L., & Hinton, G. (2008). Visualizing data using T-SNE. *Journal of Machine Learning Research, 9*(11).

Van Dijk, E. L., Auger, H., Jaszczyszyn, Y., & Thermes, C. (2014). Ten years of next-generation sequencing technology. *Trends in Genetics, 30*(9), 418–426.

Vanguri, R. S., Luo, J., Aukerman, A. T., Egger, J. V., Fong, C. J., Horvat, N., ... Rizvi, H. (2022). Multimodal integration of radiology, pathology and genomics for prediction of response to PD-(L) 1 blockade in patients with non-small cell lung cancer. *Nature Cancer, 3*(10), 1151–1164.

Vaswani, A., Shazeer, N., Parmar, N., Uszkoreit, J., Jones, L., Gomez, A. N., ... Polosukhin, I. (2017). Attention is all you need. *Advances in Neural Information Processing Systems, 30*.

Vizza, P., Aracri, F., Guzzi, P. H., Gaspari, M., Veltri, P., & Tradigo, G. (2024). Machine learning pipeline to analyze clinical and proteomics data: Experiences on a prostate cancer case. *BMC Medical Informatics and Decision Making, 24*(1), 93. https://doi.org/10.1186/s12911-024-02491-6.

Vogeser, M., & Parhofer, K. G. (2007). Liquid chromatography tandem-mass spectrometry (LC-MS/MS)-technique and applications in endocrinology. *Experimental and Clinical Endocrinology & Diabetes, 115*(09), 559–570.

Wan, N., Weinberg, D., Liu, T.-Y., Niehaus, K., Ariazi, E. A., Delubac, D., ... Bertin, M. (2019). Machine learning enables detection of early-stage colorectal cancer by whole-genome sequencing of plasma cell-free DNA. *BMC Cancer, 19*, 1–10.

Wan, S., Mak, M.-W., & Kung, S.-Y. (2016). Ensemble linear neighborhood propagation for predicting subchloroplast localization of multi-location proteins. *Journal of Proteome Research, 15*(12), 4755–4762. https://doi.org/10.1021/acs.jproteome.6b00686.

Wan, S., Mak, M.-W., & Kung, S.-Y. (2017). Transductive learning for multi-label protein subchloroplast localization prediction. *IEEE/ACM Transactions on Computational Biology and Bioinformatics, 14* (1), 212–224. https://doi.org/10.1109/TCBB.2016.2527657.

Wang, C., Lue, W., Kaalia, R., Kumar, P., & Rajapakse, J. C. (2022). Network-based integration of multi-omics data for clinical outcome prediction in neuroblastoma. *Scientific Reports, 12*(1), 15425. https://doi.org/10.1038/s41598-022-19019-5.

Wang, F., & Zhang, C. (2008). Label propagation through linear neighborhoods. *IEEE Transactions on Knowledge and Data Engineering, 20*(1), 55–67. https://doi.org/10.1109/TKDE.2007.190672.

Wang, G., Qiu, M., Xing, X., Zhou, J., Yao, H., Li, M., ... Pan, S. (2022). Lung cancer scRNA-seq and lipidomics reveal aberrant lipid metabolism for early-stage diagnosis. *Science Translational Medicine, 14*(630), eabk2756.

Wang, H., Subramanian, V., & Syeda-Mahmood, T. (2021). *Modeling uncertainty in multi-modal fusion for lung cancer survival analysis* (pp. 1169–1172). IEEE.

Wang, R., Li, B., Lam, S. M., & Shui, G. (2020). Integration of lipidomics and metabolomics for in-depth understanding of cellular mechanism and disease progression. *Journal of Genetics and Genomics, 47*(2), 69–83.

Wang, T., Shao, W., Huang, Z., Tang, H., Zhang, J., Ding, Z., & Huang, K. (2021). MOGONET integrates multi-omics data using graph convolutional networks allowing patient classification and biomarker identification. *Nature Communications, 12*(1), 3445.

Wang, W., Zhao, Y., Teng, L., Yan, J., Guo, Y., Qiu, Y., Ji, Y., et al. (2023). Neuropathologist-level integrated classification of adult-type diffuse gliomas using deep learning from whole-slide pathological images. *Nature Communications, 14*(1), 6359. https://doi.org/10.1038/s41467-023-41195-9.

Wang, Z., & Wang, Y. (2019). Extracting a biologically latent space of lung cancer epigenetics with variational autoencoders. *BMC Bioinformatics, 20*(18), 1–7.

Want, E. J., Cravatt, B. F., & Siuzdak, G. (2005). The expanding role of mass spectrometry in metabolite profiling and characterization. *Chembiochem: A European Journal of Chemical Biology, 6*(11), 1941–1951.

Warnat-Herresthal, S., Schultze, H., Shastry, K. L., Manamohan, S., Mukherjee, S., Garg, V., ... Aziz, N. A. (2021). Swarm learning for decentralized and confidential clinical machine learning. *Nature, 594*(7862), 265–270.

Wilhelm, M., Schlegl, J., Hahne, H., Gholami, A. M., Lieberenz, M., Savitski, M. M., ... Marx, H. (2014). Mass-spectrometry-based draft of the human proteome. *Nature, 509*(7502), 582–587.

Wu, R., Haas, W., Dephoure, N., Huttlin, E. L., Zhai, B., Sowa, M. E., & Gygi, S. P. (2011). A large-scale method to measure absolute protein phosphorylation stoichiometries. *Nature Methods, 8*(8), 677–683.

Wu, S., Zhu, W., Thompson, P., & Hannun, Y. A. (2018). Evaluating intrinsic and non-intrinsic cancer risk factors. *Nature Communications, 9*(1), 3490. https://doi.org/10.1038/s41467-018-05467-z.

Wu, X., Li, W., & Tu, H. (2023). Big data and Artificial Intelligence in cancer research. *Trends in Cancer.* https://doi.org/10.1016/j.trecan.2023.10.006.

Xie, Y., Wei-Yu Meng, R.-Z. L., Yu-Wei Wang, X. Q., Chang Chan, Z.-F. Y., Xing-Xing Fan, H.-D. P., & Xie, C. (2021). Early lung cancer diagnostic biomarker discovery by machine learning methods. *Translational Oncology, 14*(1), 100907.

Xu, G., Zhang, M., Zhu, H., & Xu, J. (2017). A 15-gene signature for prediction of colon cancer recurrence and prognosis based on SVM. *Gene, 604*, 33–40. https://doi.org/10.1016/j.gene.2016.12.016.

Xu, Y., & Chen, H. (2023). *Multimodal optimal transport-based co-attention transformer with global structure consistency for survival prediction*, Proceedings of the IEEE/CVF International Conference on Computer Vision (ICCV), (pp. 21241–21251).

Xu, Y., Liu, X., Cao, X., Huang, C., Liu, E., Qian, S., ... Qiu, C.-W. (2021). Artificial Intelligence: A powerful paradigm for scientific research. *The Innovation, 2*(4).

Yang, L., Zhang, Z., Song, Y., Hong, S., Xu, R., Zhao, Y., ... Yang, M.-H. (2023). Diffusion models: A comprehensive survey of methods and applications. *ACM Computing Surveys, 56*(4), 1–39.

Yang, X.-L., Yi Shi, D.-D. Z., Rui Xin, J. D., Ting-Miao Wu, H.-M. W., Pei-Yao Wang, J.-B. L., & Li, W. (2021). Quantitative proteomics characterization of cancer biomarkers and treatment. *Molecular Therapy-Oncolytics, 21*, 255–263.

Yu, J., Wu, X., Lv, M., Zhang, Y., Zhang, X., Li, J., ... Zhang, Q. (2020). A model for predicting prognosis in patients with esophageal squamous cell carcinoma based on joint representation learning. *Oncology Letters, 20*(6) 1–1.

Zhu, Z., Yu, L., Wu, W., Yu, R., Zhang, D., & Wang, L. (2023). MuRCL: Multi-instance reinforcement contrastive learning for whole slide image classification. *IEEE Transactions on Medical Imaging, 42*(5), 1337–1348. https://doi.org/10.1109/TMI.2022.3227066.

Zadeh Shirazi, A., McDonnell, M. D., Fornaciari, E., Bagherian, N. S., Scheer, K. G., Samuel, M. S., Yaghoobi, M., et al. (2021). A deep convolutional neural network for segmentation of whole-slide pathology images identifies novel tumour cell-perivascular niche interactions that are associated with poor survival in glioblastoma. *British Journal of Cancer, 125*(3), 337–350. https://doi.org/10.1038/s41416-021-01394-x.

Zhang, X., Zhu, X., Wang, C., Zhang, H., & Cai, Z. (2016). Non-targeted and targeted metabolomics approaches to diagnosing lung cancer and predicting patient prognosis. *Oncotarget, 7*(39), 63437.

Zheng, W., Pu, M., Li, X., Du, Z., Jin, S., Li, X., ... Zhang, Y. (2023). Deep learning model accurately classifies metastatic tumors from primary tumors based on mutational signatures. *Scientific Reports, 13*(1), 8752.

Zheng, Z., Xie, S., Dai, H., Chen, X., & Wang, H. (2017). *An overview of blockchain technology: Architecture, consensus, and future trends* (pp. 557–564). IEEE.

Zhou, Z.-H., & Li, M. (2005). Semi-supervised regression with co-training. *IJCAI, 5*, 908–913.

Zhu, J., Adli, M., Zou, J. Y., Verstappen, G., Coyne, M., Zhang, X., ... De Jager, P. L. (2013). Genome-wide chromatin state transitions associated with developmental and environmental cues. *Cell, 152*(3), 642–654.

Printed and bound by CPI Group (UK) Ltd, Croydon, CR0 4YY
02/12/2024
01798497-0005